Advances in biodiesel production

Related titles:

Handbook of biofuels production
(ISBN 978-1-84569-679-5)
The global increase in the use of biofuels as substitute transportation fuels has prompted research and development into advanced chemical, biochemical and thermochemical biofuels production. This research and development is aimed at improving the quality and environmental impact of biofuels production, as well as the overall efficiency and output of biofuels production plants. The range of biofuels has also increased, with market developments leading to the increased production and utilisation of advanced biofuels such as biosyngas, biohydrogen and biobutanol. Handbook of biofuels production provides a thorough reference on the range of biomass conversion processes and technologies.

Biodiesel science and technology
(ISBN 978-1-84569-591-0)
Biodiesel fuel is increasingly being used in diesel supplies worldwide, as it provides both an environmentally preferable option and improves diesel engine performance through greater lubricity. The development and use of appropriate feedstocks to avoid food/fuel concerns, and technological developments directed towards improved quality and capacity, are crucial to the environmental impact of biodiesel production and utilisation and to the future of the industry. This book provides a thorough timely reference on the entire biodiesel production chain, from soil to oil, detailing the range and development of biodiesel feedstocks as well as state-of-the-art biodiesel production technology.

Bioalcohol production
(ISBN 978-1-84569-510-1)
Bioethanol is one of the main biofuels used as a petroleum substitute in transportation. Second generation bioalcohol production technology, based on highly efficient biochemical conversion, offers advantages over existing, energy-intensive bioethanol production processes. This book covers the process engineering, technology, modelling and integration of the entire production chain, from feedstock pretreatment on to hydrolysis, fermentation, and purification. Primarily reviewing bioethanol production, coverage extends to the production of longer-chain bioalcohols which will be elemental in the future of the industry.

Woodhead Publishing Series in Energy: Number 39

Advances in biodiesel production

Processes and technologies

Edited by

Rafael Luque and Juan A. Melero

WP
WOODHEAD PUBLISHING

Oxford Cambridge Philadelphia New Delhi

Published by Woodhead Publishing Limited,
80 High Street, Sawston, Cambridge CB22 3HJ, UK
www.woodheadpublishing.com
www.woodheadpublishingonline.com

Woodhead Publishing, 1518 Walnut Street, Suite 1100, Philadelphia,
PA 19102-3406, USA

Woodhead Publishing India Private Limited, G-2, Vardaan House,
7/28 Ansari Road, Daryaganj, New Delhi – 110002, India
www.woodheadpublishingindia.com

First published 2012, Woodhead Publishing Limited

British Library Cataloguing in Publication Data
A catalogue record for this book is available from the British Library.

Library of Congress Control Number: 2011944656

ISBN 978-0-08-101664-0 (print)
ISBN 978-0-85709-586-2 (online)
ISSN 2044-9364 Woodhead Publishing Series in Energy (print)
ISSN 2044-9372 Woodhead Publishing Series in Energy (online)

The publisher's policy is to use permanent paper from mills that operate a sustainable forestry policy, and which has been manufactured from pulp which is processed using acid-free and elemental chlorine-free practices. Furthermore, the publisher ensures that the text paper and cover board used have met acceptable environmental accreditation standards.

Typeset by RefineCatch Limited, Bungay, Suffolk, UK

Contents

Contributor contact details

(* = main contact)

Editors and Chapter 1

Dr Rafael Luque
Department of Organic Chemistry
Universidad de Córdoba
Campus de Rabanales
E-14014
Córdoba
Spain
E-mail: q62alsor@uco.es

Professor Juan Antonio Melero
Department of Chemical and Environmental Technology
Universidad Rey Juan Carlos
C/ Tulipán s/n. E–28933
Móstoles
Madrid
Spain
E-mail: juan.melero@urjc.es

Chapter 2

Dr Harish K. Jeswani and Professor Adisa Azapagic*
School of Chemical Engineering and Analytical Science
Room C16, The Mill
Sackville Street
The University of Manchester
Manchester M13 9PL
UK
E-mail: Harish.jeswani@manchester.ac.uk; Adisa.Azapagic@manchester.ac.uk

Chapter 3

Dr Caterina De Lucia*
Nicholas School of the Environment
Duke University
450 Research Drive
Campus Box 90328
Durham, NC
USA
E-mail: caterina.delucia@gmail.com; caterina.delucia@duke.edu

Dr Bipasa Datta
Department of Economics
University of York
Heslington
York
YO10 5DD
UK
E-mail: bipasa.datta@york.ac.uk

Chapter 4

Dr Sara Pinzi and Professor Maria Pilar Dorado*
Dept. Physical Chemistry and Applied Thermodynamics
Edificio Leonardo da Vinci
Campus de Rabanales
Universidad de Córdoba
14071 Córdoba
Spain
E-mail: pilar.dorado@uco.es; qf1pinps@uco.es

Chapter 5

Dr Alicia Carrero*
Department of Energy and Chemical Technology
Universidad Rey Juan Carlos
C/Tulipán s/n. E–28933
Móstoles
Madrid
Spain
E-mail: alicia.carrero@urjc.es

Dr Ángel Pérez
Departamento de Ingeniería Química
Universidad de Castilla la Mancha
Facultad de Ciencias Químicas
Avenida Camilo José Cela, No 12
13005 Ciudad Real
Spain
E-mail: angel.perez@uclm.es

Chapter 6

Professor Martin Mittelbach
Institute of Chemistry (IFC)
Karl-Franzens-University Graz
Heinrichstrasse 28
A-8010 Graz
Austria
E-mail: martin.mittelbach@uni-graz.at

Chapter 7

Dr Jose Iglesias* and Dr Gabriel Morales
Department of Chemical and Environmental/Energy Technology
Universidad Rey Juan Carlos
C/Tulipán s/n. E–28933
Móstoles
Madrid
Spain
E-mail: jose.iglesias@urjc.es; gabriel.morales@urjc.es

Chapter 8

Dr Luis Fernando Bautista and Dr Gemma Vicente*
Department of Chemical and Environmental Technology
Universidad Rey Juan Carlos
C/ Tulipán s/n. E–28933
Móstoles
Madrid
Spain
E-mail: fernando.bautista@urjc.es; gemma.vicente@urjc.es

Dr Victoriano Garre
Departamento de Genética y Microbiología
Facultad de Biología
Universidad de Murcia
E30071 Murcia
Spain
E-mail: vgarre@um.es

Chapter 9

Professor Teresa M. Mata*
LEPAE – Laboratory for Process, Environmental and Energy Engineering
Faculty of Engineering
University of Porto (FEUP)
R. Dr. Roberto Frias S/N
4200-465 Porto
Portugal
E-mail: tmata@fe.up.pt

Antonio A. Martins
CEFT – Center for Transport Phenomena Studies
Faculty of Engineering
University of Porto (FEUP)
R. Dr. Roberto Frias S/N
4200-465 Porto
Portugal
E-mail: amartins@fe.up.pt

Nídia S. Caetano
LEPAE – Laboratory for Process, Environmental and Energy Engineering
School of Engineering (ISEP)
Polytechnic Institute of Porto (IPP)
4200-072 Porto
Portugal

Chapter 10

Professor Claudio J. A. Mota
Universidade Federal do Rio de Janeiro
Instituto de Química
Av Athos da Silveira Ramos 149
CT Bloco A
Cidade Universitária
Rio de Janeiro, 21941-909
Brazil
E-mail: cmota@iq.ufrj.br

Appendix

Devrim Murat Yazan*, Joy Clancy and Jon C. Lovett
Twente Centre for Studies in Technology and Sustainable Development
University of Twente
School of Management and Governance
PO Box 217
7500 AE Enschede
The Netherlands
E-mail: d.m.yazan@utwente.nl; j.s.clancy@utwente.nl; j.lovett@utwente.nl

Woodhead Publishing Series in Energy

14 **Materials for energy efficiency and thermal comfort in buildings**
Edited by Matthew R. Hall
15 **Handbook of biofuels production: Processes and technologies**
Edited by Rafael Luque, Juan Campelo and James Clark
16 **Developments and innovation in carbon dioxide (CO_2) capture and storage technology Volume 2: Carbon dioxide (CO_2) storage and utilisation**
Edited by M. Mercedes Maroto-Valer
17 **Oxy-fuel combustion for power generation and carbon dioxide (CO_2) capture**
Edited by Ligang Zheng
18 **Small and micro combined heat and power (CHP) systems: Advanced design, performance, materials and applications**
Edited by Robert Beith
19 **Advances in clean hydrocarbon fuel processing: Science and technology**
Edited by M. Rashid Khan
20 **Modern gas turbine systems: High efficiency, low emission, fuel flexible power generation**
Edited by Peter Jansohn
21 **Concentrating solar power (CSP) technology: Developments and applications**
Edited by Keith Lovegrove and Wes Stein
22 **Nuclear corrosion science and engineering**
Edited by Damien Féron
23 **Power plant life management and performance improvement**
Edited by John E. Oakey
24 **Direct-drive renewable energy systems**
Edited by Markus Mueller and Henk Polinder
25 **Advanced membrane science and technology for sustainable energy and environmental applications**
Edited by Angelo Basile and Suzana Pereira Nunes
26 **Irradiation embrittlement of reactor pressure vessels (RPVs) in nuclear power plants**
Edited by Naoki Soneda
27 **High temperature superconductors (HTS) for energy applications**
Edited by Ziad Melhem
28 **Infrastructure and methodologies for the justification of nuclear power programmes**
Edited by Agustín Alonso
29 **Waste to energy (WtE) conversion technology**
Edited by Marco Castaldi
30 **Polymer electrolyte membrane and direct methanol fuel cell technology Volume 1: Fundamentals and performance of low temperature fuel cells**
Edited by Christoph Hartnig and Christina Roth
31 **Polymer electrolyte membrane and direct methanol fuel cell technology Volume 2: *In situ* characterization techniques for low temperature fuel cells**
Edited by Christoph Hartnig and Christina Roth
32 **Combined cycle systems for near-zero emission power generation**
Edited by Ashok Rao
33 **Modern earth buildings: Materials, engineering, construction and applications**
Edited by Matthew R. Hall, Rick Lindsay and Meror Krayenhoff

54 **Geoscience of carbon dioxide (CO_2) storage**
Edited by Jon Gluyas and Simon Mathias

55 **Handbook of membrane reactors Volume 1: Fundamental materials science, design and optimisation**
Edited by Angelo Basile

56 **Handbook of membrane reactors Volume 2: Industrial applications and economics**
Edited by Angelo Basile

57 **Alternative fuels and advanced vehicle technologies: Towards zero carbon transportation**
Edited by Richard Folkson

58 **Handbook of microalgal bioprocess engineering**
Christopher Lan and Bei Wang

59 **Fluidized-bed technologies for near-zero emission combustion and gasification**
Edited by Fabrizio Scala

60 **Managing nuclear projects: A comprehensive management resource**
Edited by Jas Devgun

61 **Handbook of process integration: Energy, water, waste and emissions management in processing and power industries**
Edited by Jiří Klemeš

62 **Membranes for clean and renewable power applications**
Edited by Annarosa Gugliuzza and Angelo Basile

1
Introduction to advanced biodiesel production

R. LUQUE, Universidad de Córdoba, Spain and
J. A. MELERO, Universidad Rey Juan Carlos, Spain

Abstract: This introduction chapter highlights the environmental and economic benefits of second generation biodiesel production based on advanced processing strategies using non-edible oleaginous feedstock. An overview of the different technologies, including the development of innovative catalysts and new reactor concepts will be provided. Moreover, a list of potential feedstocks and new advances in glycerol valorisation will be also briefly reviewed.

Key words: advanced processes, glycerol, non-edible oleaginous feedstock, second generation biodiesel.

1.1 Introduction

The transport sector currently accounts for over 30% of the total primary energy consumption in the European Union (EU) and is 98% dependent on fossil fuels. The EU is a net crude oil importer, with an increasing oil dependency ratio (over 80% in 2007, Eurostat). One of the most important energy targets for the EU is thus the reduction of oil use and dependency. A recent EU Directive (2009/28/EC) establishes a 20% target share of renewable energy in primary energy consumption by 2020, with a 10% share of energy from renewable sources for transport. In turn, another EU Directive (2003/96/EC) allows the Member States to have exemptions from or reductions in excise duties so as to promote biofuels and, consequently, there is a growing interest in biofuels in Europe. In this context, biodiesel constitutes a renewable fuel that is almost compatible with commercial diesel engines and has clear environmental benefits relative to diesel fuel.

Biodiesel consists of a mixture of fatty acid methyl esters (FAMEs) obtained from renewable resources, such as vegetable oils and animal fats, by transesterification with methanol in the presence of an acid or basic catalyst. It can be synthesised from a variety of feedstocks, but refined vegetable oils (such as soybean, rapeseed, palm and others) are currently the primary industrial feedstock (first generation biodiesel). The relatively high and unstable prices of these food-grade raw materials, which are linked to the transfer of land from food to energy crops and also to doubts as to whether biofuels are as environmentally friendly as originally thought, constitute a great obstacle for biodiesel production and commercialisation. Likewise, EU Directive (2009/30/CE) establishes, with effect from 1 January 2017, that the greenhouse emission savings from the use of biofuels taken into account for the purpose of 10% shall be at least of 50% (being this value increased up to 60% from 1 January 2018). Annex IV of this Directive indicates that these greenhouse emission savings are in general lower than 50% for the production of biodiesel using conventional edible oils as feedstock, but in contrast, this value is increased to over 80% using waste vegetable or animal oils. Therefore, it is of major importance to study alternative non-edible waste feedstocks for the sustainable production of second generation biodiesel.

Non-edible oils like those derived from jatropha, pongamia, jojoba, and so on are promising feedstocks especially in developing countries where edible oils are in short supply.[1,2] Likewise, waste cooking oil (WCO) is considered an economic and increasingly available resource for biodiesel production.[3,4] The estimated total waste cooking oil capacity in the EU is 700 000–1 000 000 t/year. Although part of the WCO produced is collected for adequate disposal or treatment, most of the domestic WCO is usually discharged into the environment and is the origin of several problems in wastewater depuration systems. The valorisation of this used oil through the production of biodiesel contributes to increasing its recovery rate from the waste cycle.

Low-grade or waste animal fats can also be used as raw materials for the production of biodiesel.[5–7] The production of biodiesel from microalgae, as well as from oleaginous microorganisms, is already considered as interesting feedstock for biodiesel production. Other potential low-cost environmentally friendly sources of biodiesel have been reported. For instance, the solvent extraction of waste coffee grounds yields 10–15% oil depending on the coffee species, which makes this waste a potential candidate for raw material for biodiesel production.[8] Municipal sewage sludge (MSS) is also gaining attention as a lipid feedstock for biodiesel production owing to its significant concentration of lipids.[9] MSS is plentiful all around the world and its management already poses formidable environmental problems. However, there are still numerous challenges faced in biodiesel production from waste sludge.

Nevertheless, most technologies for the production of biodiesel using low-quality feedstock are still in their infancy and research efforts are needed in the future. But the development of these technologies in the future will lead to important benefits for the biodiesel industry: less environmental impact, safer processes, higher glycerol purity, reduction of land use and non interference in the food chain (Fig. 1.1).

In this book, we aim to provide an overview of the different processes and technologies suitable for the production of second generation biodiesel with special emphasis on the development of innovative catalysts as well as new reactor concepts. In the first introductory section, details of the environmental (Chapter 2), political and socio-economic (Chapter 3) implications of the implementation of second generation biodiesel as well as of the different available non-edible feedstocks (Chapter 4) and new trends in quality control and pre-normative standards in biodiesel characterisation (Chapter 5) will be presented.

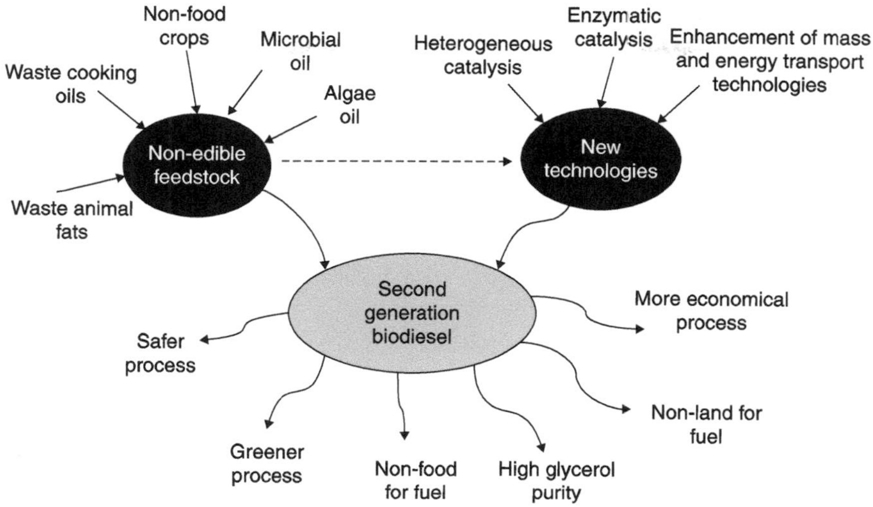

1.1 Advances in biodiesel production.

The second part of the book will be focused on an overview of the processes and technologies used for the production of second generation biodiesel (Chapter 6) using low cost non-edible feedstock (waste oils and animal fats, Chapter 7; microbial oil, Chapter 8; microalgae, Chapter 9). The last section of the book will deal with the valorisation of glycerol as by-product of biodiesel production (Chapter 10).

1.2 Processes and technologies for advanced biodiesel production

The current technology for first generation biodiesel production has serious drawbacks. It uses homogenous alkaline catalysts (NaOH, KOH, NaOMe and KOMe) which make it compulsory to separate the spent catalyst from FAME and the glycerol phase by additional washing steps. Hence, although biodiesel is a green product, it is not currently manufactured by a green process.

Likewise, the other difficulty of using alkaline homogenous catalysts is their high sensitivity to free fatty acids (FFAs) and water in the oil source. FFAs react with alkaline catalysts to form soaps, complicating the glycerol separation and reducing FAME yield whereas water in the feedstock results in the hydrolysis of FAME to yield FFAs. The FFA and water content is limited in the feedstock to 0.5 wt% and 0.06 wt%, respectively. Thus, low-quality non-edible feedstock which generally contains a higher content of FFA and water cannot be directly utilised in the existing biodiesel facilities with homogeneous catalysts. Moreover, we must have in mind that the cost of the oil accounts for about 80% of biodiesel production costs.

An important technological challenge in biodiesel production is the design of heterogeneous catalysts.[10,11] The use of heterogeneous catalysts would result in simpler, cheaper separation processes, a reduced water effluent load and reduced capital and energy costs. There would be fewer inputs and less waste, as no soap would be formed and the catalyst would not have to be continuously added, would be easier to reuse and could be arranged in fixed-bed reactors. Furthermore, there would be no neutralisation products, so a higher grade of glycerol might be produced without additional purification steps.

The development of basic heterogeneous catalysts for biodiesel production has been widely described in the literature including carbonates and hydro-carbonates of alkaline metals, alkaline metal oxides, alkaline metal hydroxides, anionic resins, basic zeolites, and so on. Some of these catalysts have shown a good catalytic performance even under reaction conditions similar to those of the homogenous catalysts. However, these kinds of catalysts are still far from being used in an industrial application since their evaluation has been carried out in stirred batch reactors but there are few studies in continuous process using packed bed flow reactors. Currently, a unique heterogeneous commercial process is based on the Esterfip-H technology developed by the Institute Français du Petrole (IFP) with a pioneering plant that started up in 2006 with 200 000 metric tonnes per year (France). Several more plants are currently in design and under construction worldwide.

Currently, biodiesel research is focused on exploring new and sustainable solid acid catalysts. Acid catalysts can simultaneously carry out esterification of FFAs and transesterification of triglycerides. Several catalysts have been described so far in the literature: different metal oxides (zirconium, titanium and tin), acid zeolites, sulphonic ion-exchange resin, sulphonated carbon-base catalyst and heteropolyacids. Nevertheless, the commercial introduction of these catalysts still needs important advances to have a positive impact on biodiesel synthesis technologies. These improvements include: increasing the stability of acid sites to avoid their leaching, increasing thermal stability, enhancement of mass transfer avoiding diffusional limitations, milder operation conditions and increasing the resistance to water.

Biodiesel can be also obtained by enzymatic transesterification especially using lipase as catalyst. Enzyme catalysis proceeds without the generation of by-products, with easy recovery of biodiesel product, mild reaction conditions, and is insensitive to FFAs. The catalyst can be reused, thus contributing to reducing energy consumption and waste generation compared to the conventional base-catalysed chemical process. However, this enzymatic technology suffers from a crucial drawback arising from the high cost of the enzymes.

In addition, non-catalytic processes have also been described in literature with supercritical methanol.[12] In this case, the reactions take place in a single homogeneous phase since the supercritical methanol is fully mixable with the vegetable oils; moreover, the reaction rates are very fast and the subsequent purification is much simpler than that of conventional processes. The supercritical route is also characterised by a high yield because of simultaneous transesterification of triglycerides and methyl esterification of fatty acids. The main drawback is that the non-catalytic supercritical process requires very high temperatures (350–400°C) and pressures (200–400 bar).

As presented earlier, heterogeneous and enzymatic catalysts have the potential to overcome the problems of homogeneous catalysts. However, these processes need to improve mass and heat transfer limitations to be scaled up to the industrial scale. Several technologies have been described in the literature to overcome these limitations. Microwave technology for the transesterification reaction has gained great interest from many research groups. This technology has been demonstrated to be more energy efficient than conventional heating in continuous biodiesel production leading to high FAME yields in a short time. However, the main drawback of microwave technology is the scaling-up of the process from the laboratory scale to the industrial production scale as well as process safety.

In addition, other technological improvements have been addressed to enhance the mass transfer between immiscible liquid–liquid phases within a heterogeneous/enzymatic system. Ultrasonic technology in transesterification has proven to be an efficient mixing tool leading to shorter reaction times as well as to a molar ratio of alcohol to oil and a reduction in energy consumption compared to conventional mechanical stirring. But, up to now, there are very few studies applying ultrasonic technology to oils transesterification. Other researchers have studied the introduction of co-solvents in the reaction mixture with the aim of increasing the solubility and subsequently improving the mass transfer rate between the oil and alcohol phase as well as enhancing diffusion within the pores of the solid catalyst. Different researchers have also been focusing on novel reactors to produce biodiesel through some improvement in mixing intensity between reactants and especially when heterogeneous catalysts are used. These technologies include oscillatory flow reactors, microchannel reactors, membrane reactors, static mixers and rotating reactors.

For more specific details of these technologies, the reader is referred to Chapter 6, in which more detailed information about catalysts and processes will be given.

1.3 New feedstocks for second generation biodiesel production

A major economic challenge for the biodiesel industry is the high cost of refined vegetable oils which constitutes between 70 and 85% of the overall production cost. Likewise, the competition of these edible oils with the food market is also an important concern. Thus, in order to reduce the cost of biodiesel and the social impact, alternative feedstocks that are readily available in large quantities and at low cost must be considered. Chapter 4 will give an overview of these non-conventional feedstocks for the production of second generation biodiesel, whereas Chapters 7, 8 and 9 will focus specifically on the technologies and future perspectives for the production of biodiesel from waste oils and fats, oleaginous microorganisms and microalgae.

1.3.1 Non-edible vegetable oils

One of the ways to reduce dependence on edible oil to make biodiesel is to use non-edible vegetable oils. Among these, jatropha is being considered as one of the most promising potential oil sources for biodiesel production in Asia, Europe and Africa. Jatropha can grow under a wide variety of climatic conditions and in marginal and waste lands with a high oil yield per hectare. Thus, it is not necessary to use land that would normally be dedicated to food crops.

1.3.2 Waste oils and fats

Low cost animal fats and waste oils are attracting attention as possible feedstocks. This feedstock represents an attractive alternative for the production of biodiesel owing to its low cost, its supposedly environmentally friendly use and the fact that it does not compete with the food industry, thus avoiding social concerns. But unfortunately it usually contains a large amount of FFA and a pre-treatment is necessary before it can be processed through the conventional homogeneous alkaline process. Chapter 7 will deal with the exploitation of this alternative feedstock for the production of biodiesel with special focus on its availability and catalytic technologies for its transformation into biodiesel.

1.3.3 Algae-based biodiesel

There is a growing interest in algae-based biodiesel production. Microalgae use CO_2 and sunlight as carbon and energy sources (autotrophic organisms) and present higher yields per hectare compared with conventional oils crops. However, the industrial commercialisation of biodiesel from algae has currently serious obstacles that will be reviewed in Chapter 9, which will also include life cycle assessment (LCA) studies on the viability of algal oils for biodiesel production.

1.3.4 Biodiesel from other microbial oils

Oleaginous microorganisms (e.g. yeast, fungi, bacteria) are another interesting feedstock for biodiesel production that will be discussed in detail in Chapter 8. These microorganisms accumulate lipids contents up to 20 wt% and, in contrast to microalgae, use organic compounds as a source of carbon and energy (heterotrophic organisms).

1.4 Glycerol as by-product of biodiesel production

The main side product of biodiesel production has also attracted recent interest from research scientists and companies in the sense that it can be transformed in a series of relevant chemicals and fuels through a range of chemical transformations. The rapid growth of the biodiesel industry (world production reached 16 billion L in 2009 and it is expected to increase to 45 billion L by 2020) generated a surplus of crude glycerol in the market which resulted in a significant drop in the price of this feedstock (from US$0.55/kg in 2004 to US$0.055/kg in 2006).[13] Glycerol is thus becoming a cheap and abundant renewable feedstock and novel technologies for large-scale processing of this resource into valuable products have been increasingly investigated in recent years and are considered to be of primary importance. Figure 1.2 summarises some of the main chemical transformations for glycerol, which will be discussed to some extent in Chapter 10.

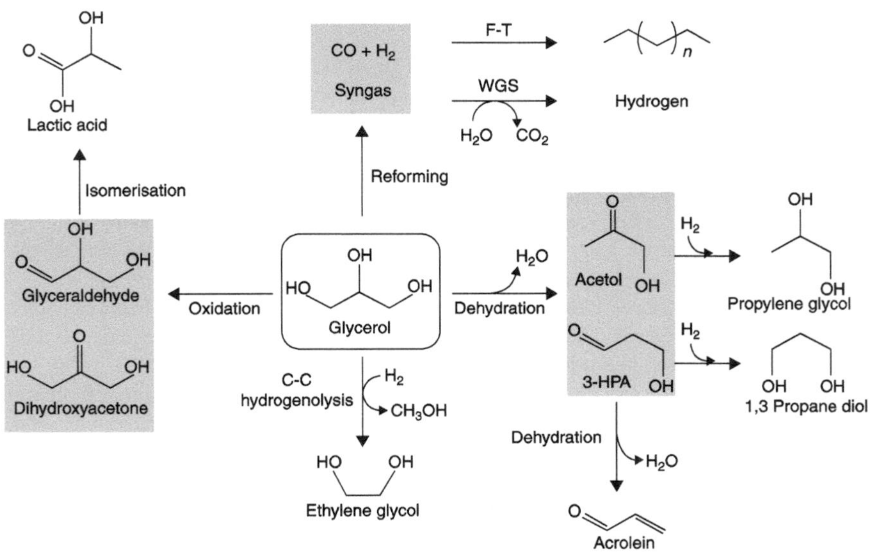

1.2 Main routes for the aqueous-phase transformation of glycerol into fuels and chemicals.

1.5 Conclusions and outlook

Biodiesel production has experienced several advances over the past few years, evolving from the conventional base-catalysed transesterification process with virgin vegetable oils as feedstocks to advanced processing strategies using non-edible (e.g. microalgae, oleaginous seeds and microorganisms) as well as waste raw materials (e.g. waste oils and fats). The aim of this book is therefore to provide an overview and starting point for research scientists and companies interested in recent advances in the utilisation of alternative feedstocks and novel processing technologies for cleaner and more efficient biodiesel production. The interest in glycerol valorisation in recent years is also reflected in the content of this work as well as regulatory and quality control issues. As editors, we sincerely hope both the topic and the content of the book will be attractive to readers in the bioenergy arena and we look forward to further developments in this area in future years.

1.6 Acknowledgements

J. A. Melero thanks the Ministerio de Ciencia e Innovacion for financial support through project CTQ2008-01396 and the Regional Government of Madrid through project S2009-ENE1743. R. Luque is grateful to Ministerio de Ciencia e Innovacion, Gobierno de España for the concession of a Ramon y Cajal contract (ref. RYC-2009-04199) and Consejeria de Ciencia e Innovacion, Junta de Andalucia for funding under Project P10-FQM-6711.

1.7 References

1. Karmakar A., Karmakar S. and Mujherjee S. (2010). Properties of various plants and animals feedstocks for biodiesel production. *Bioresour. Technol.*, **101**, 7201–10.
2. Jain S. and Sharma M.P. (2010). Biodiesel production from Jatropha curcas oil. *Renewable Sustainable Energy Rev.*, **14**, 3140–57.
3. Kee Lam M., Teong Lee K. and Rahman Mohamed A. (2010). Homogeneous, heterogeneous and enzymatic catalysis for transesterification of high free fatty acid oil (waste cooking oil) to biodiesel: a review. *Biotechnol. Adv.*, **28**, 500–18.
4. Han M., Yi W., Wu Q., Liu Y., Hong Y. and Wang D. (2009). Preparation of biodiesel from waste oils catalyzed by a Bronsted acidic ionic liquid. *Bioresour. Technol.*, **100**, 2308–10.
5. Canoira L., Rodríguez-Gamero M., Querol E., Alcántara R., Lapuerta M. and Oliva F. (2008). Biodiesel from low-grade animal fat: production process assessment and biodiesel properties characterization. *Ind. Eng. Chem. Res.*, **47**, 7997–8004.
6. Gürü M., Artukoglu B.D., Keskin A. and Koca A. (2009). Biodiesel production from waste animal fat and improvement of its characteristics by synthesized nickel and magnesium additive. *Energy Convers Manage.*, **50**, 498–502.
7. Lapuerta M., Rodríguez-Fernández J., Oliva F., Canoira L. (2009). Biodiesel from low-grade animal fats: diesel engine performance and emissions. *Energy Fuels*, **23**, 121–9.
8. Kondamudi N., Mohapatra S.N. and Misra M. (2008). Spent coffee grounds as a versatile source of green energy. *J. Agric. Food. Chem.*, **56**, 11757–60.
9. Kargbo D.M. (2010). Biodiesel production from municipal sewage sludges. *Energy Fuels*, **24**, 2791–4.
10. Yan S., DiMaggio C., Mohan S., Kim M., Salley S.O. and Simon Ng K. Y. (2010). Advancements in heterogeneous catalysis for biodiesel synthesis. *Top. Catal.*, **53**, 721–36.

11. Semwal S., Arora A. K., Badoni R. P. and Tuli D. K. (2011). Biodiesel production using heterogeneous catalysts. *Bioresour. Technol.*, **102**, 2151–61.
12. Lee J. S. and Saka S. (2010). Biodiesel production by heterogeneous catalysts and supercritical technologies. *Bioresour. Technol.*, **101**, 7191–200.
13. Lugue R., Lovett J. C., Datta B., Clancy J., Campelo J. M. and Romero A. A. (2010). *Energy Environ. Sci.*, **3**, 1706–21.

Part I
Fundamental issues and assessment in advanced biodiesel production

2

Life cycle sustainability assessment of second generation biodiesel

H. K. JESWANI and A. AZAPAGIC,
The University of Manchester, UK

Abstract: Second generation biodiesel could contribute to significant reductions in carbon dioxide emissions from transport because the biofeedstock used for their production is considered to be carbon neutral. This chapter examines the life cycle sustainability of second generation biodiesel derived from different feedstocks and produced in different production systems, including integrated biorefineries. The environmental sustainability aspects considered include water use, global warming, acidification, eutrophication and loss of biodiversity. The socio-economic impacts are also reviewed, including feedstock and capital costs, value added through production of by-products as well as the social acceptability of biofuels. The future viability of biodiesel is also discussed.

Key words: biodiesel, economic costs, environmental impacts, life cycle assessment, sustainability assessment.

2.1 Introduction

Biofuels are seen as one of the most feasible options for reducing carbon emissions in the transport sector because the biomass used in their production is considered to be carbon neutral. This is based on the assumption that the amount of carbon released during combustion of biofuels in the use phase is equivalent to the amount of carbon sequestered during the growth of biomass from which the fuels were derived. However, there are several concerns related to first generation biodiesel (derived from food crops), including unsustainable production owing to deforestation, water use and land management (FAO, 2008; IEA, 2010). Other issues associated with the large-scale production of first generation biofuels include competition with food production, leading to increased costs of food and in some cases, food poverty (Bird *et al.*, 2008; Escobar *et al.*, 2009; Fargione *et al.*, 2008; Searchinger *et al.*, 2008). As a result, there is a growing interest in second generation biofuels (derived from non-food sources) which have the potential to avoid the above concerns.

Global production of biofuels has been growing rapidly in recent years. In 2008, it reached about 83 billion litres, a more than four-fold increase compared to production in 2000 (IEA, 2009). The production is dominated by bioethanol, which accounted for approximately 84% of total biofuel production in 2008. Biofuels currently contribute about 1.5% of global transport fuel consumption, with demand projected to rise steadily over the coming decades as countries start to implement policies for promoting liquid biofuel development (IEA, 2009). For example, various IEA scenarios predict a dramatic increase in the production of second generation biodiesel by 2050 (IEA, 2008). According to the IEA's Blue Map Scenario (which assumes 50% reduction in global energy-related CO_2 emissions by 2050 on 2005 levels by deploying least-cost low-carbon technologies) (IEA, 2008), the global biofuel demand is projected to increase to about 610 Mtoe (one million tonnes of oil equivalent) in 2050, a share of about 26% of the total road transportation fuel. About two-thirds of this demand is expected to be met by second generation biodiesel.

Second generation biodiesel is produced from non-food sources such as dedicated energy crops (e.g. jatropha, perennial grasses, short rotation coppice (SRC) and other lignocellulosic plants), waste biomass (e.g. agricultural and forestry residue) and waste materials (e.g. waste cooking oil). Two main processing routes can be used to produce second generation biodiesel: physicochemical and thermochemical. Physicochemical processing involves transesterification and is used for seed oil feedstock such as jatropha, castor and waste cooking oil, while thermochemical conversion by gasification followed by fuel synthesis (e.g. Fischer-Tropsch) is mainly used for lignocellulosic feedstock. Physicochemical processing is well developed and is being used for first generation biodiesel, whereas the thermochemical route is still under development.

Depending on the feedstock choice and the cultivation technique, second generation biodiesel production has a considerable potential to provide benefits such as reduction of waste residues and use of abandoned land, in addition to the reduction of greenhouse gas (GHG) emissions by replacing fossil fuels (IEA, 2010). This in turn could create jobs, promote rural development and improve economic conditions in emerging and developing regions. However, their production could become unsustainable if they compete with food crops for available land and water resources. Moreover, currently second generation biofuels, including biodiesel, have high initial investment costs as well as higher costs for the end-product compared to fossil fuels or many first generation biofuels (Eisentraut, 2010). Thus sustainability of biodiesel should be assessed carefully, considering all relevant environmental, economic and social aspects.

This chapter discusses how the main sustainability issues associated with second generation biodiesel can be assessed on a life cycle basis, considering different biofeedstocks and production routes.

2.2 Environmental sustainability of biodiesel

Life cycle assessment (LCA) is used as the main tool for evaluating the environmental sustainability of second generation biodiesel on a life cycle basis. The life cycle of biodiesel is outlined in Fig. 2.1; for comparison, the equivalent life cycle of conventional fossil diesel is also shown. The life cycle of biodiesel

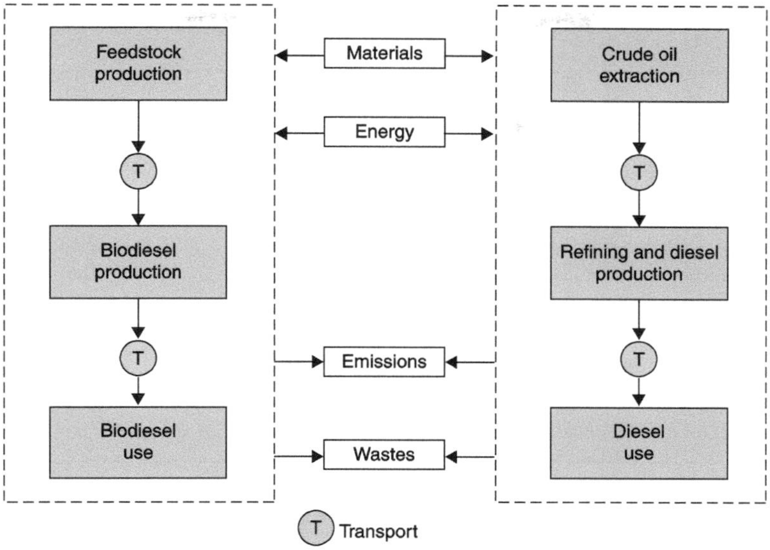

2.1 Life cycles of biodiesel and fossil diesel from 'cradle to grave'.

encompasses planting, growing and harvesting of biomass (if applicable), conversion to biodiesel and its use, also including all transport in the system. Each stage in the life cycle of biodiesel is associated with several environmental issues, depending on the type of the feedstock and production route. The following sections discuss the global warming potential (caused by GHG emissions) and other environmental impacts of second generation biodiesel produced from different feedstocks.

2.2.1 Global warming potential

Approaches to estimating the GHG emissions from biofuels vary from one country to another. Until recently, the UK, Germany and other EU countries were using different approaches to estimate GHG emissions from biofuels (Azapagic and Stichnothe, 2010). However, these approaches have recently been synchronised with the EU Renewable Energy Directive (RED). Table 2.1 provides an overview of the RED and US EPA GHG calculation methodologies. The key methodological differences include allocation of environmental impact and indirect land-use change. These approaches also use different reference values for conventional diesel to evaluate GHG savings over the fossil-based fuels.

While studies of GHG emissions from biofuel systems abound, only a limited number of studies have attempted to estimate GHG emissions from second

Table 2.1 Overview of the EU RED and US EPA GHG calculation methodologies for biodiesel

	EU RED (EC, 2009)	US EPA (US EPA, 2010)
Base year	2008	2005
Reference value for diesel (g CO_2 eq MJ^{-1})	Most recent value or 83.8	91.4
System boundary	Cradle to grave	Cradle to grave
Capital goods/infrastructure	Excluded	Excluded
Pesticides	Included	Included
Fertilisers	Included	Included
Allocation method	Net calorific value	System expansion/ economic value[a]
Direct land-use change	Included (IPCC, 2007)	Included (IPCC, 2007)
Indirect land-use change	Excluded	Included
GHG saving threshold (%)	35–60[b]	50–60[c]

[a] The allocation procedure used depends on the system.
[b] The current limit is 35%, which will increase to 50% in 2017 and 60% for thereafter.
[c] 50% for biodiesel from waste oil and 60% cellulosic biodiesel.

generation biodiesel and net savings in comparison to the fossil diesel (see Table 2.2 and Fig. 2.2). Since the commercial production of second generation biodiesel has yet to start, these studies are based on either the theoretical production concepts or experience from pilot/demonstration projects. The unit of analysis for comparison is based on the equivalent energy content of the fuels and is usually defined (arbitrarily) as '1 MJ of fuel'. The global warming potential associated with GHG emissions is

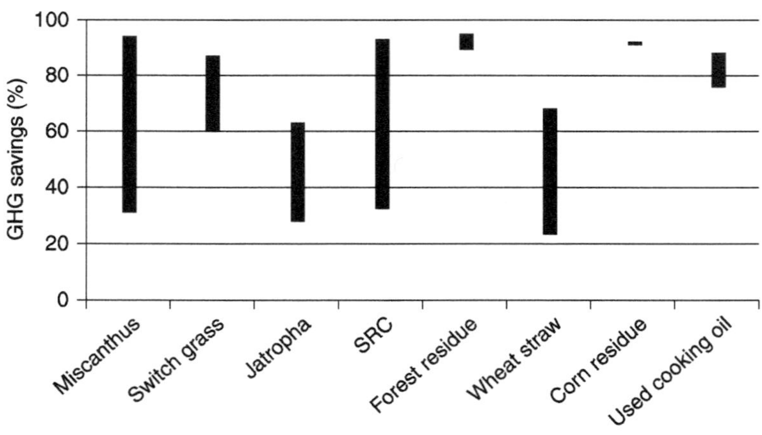

2.2 GHG savings for biodiesel from different feedstocks and country of origin, compared to fossil diesel (data sources as in Table 2.2).

expressed either in g or kg CO_2 eq MJ^{-1} and the GHG savings from biodiesel compared to fossil diesel are calculated by (EC, 2009):

$$GHG_{saving} = \frac{GHG_{fossil\ diesel} - GHG_{biodiesel}}{GHG_{fossil\ diesel}} \times 100\ (\%) \qquad [1]$$

The results for GHG emissions and savings from biodiesel shown in Table 2.2 and Fig. 2.2 reveal that, depending on the feedstock and the conversion route, second generation biodiesel has a GHG mitigation potential of 23–95% over fossil diesel. The significant variations in the results are also partly due to the assumptions made in the studies and the methodology used for the consideration of co-products (allocation or substitution, see Section entitled GHG emissions from co-products, for details). For example, the difference in Table 2.2 between the results for biodiesel from SRC obtained by Jungbluth *et al.* (2007) and Edwards *et al.* (2008) can be explained, amongst other reasons, by a different assumption for the amount of nitrogen fertiliser used for SRC cultivation: in the former study, this is more than

Table 2.2 GHG emissions from second generation biodiesel from different feedstocks

Feedstock	Origin	GHG emissions (g CO_2 eq MJ^{-1})	Key assumptions	Reference
Miscanthus	Europe	57.9	Allothermal internal circulating fluidised bed (ICFB) gasification	Jungbluth *et al.*, 2007
	EU15	5.4	System expansion credit	Hoefnagles *et al.*, 2010
Switch grass	EU15	11.3	System expansion credit	Hoefnagles *et al.*, 2010
	Various	22–37	System expansion credit; future (2020) technologies	US EPA, 2010
Jatropha	Tanzania	33.4	System expansion credit	Hoefnagles *et al.*, 2010
	India	60	System expansion credit	Reinhardt *et al.*, 2007
	Various	31	Default value[a]; energy allocation	RFA, 2010
SRC/farmed wood	Europe	29.6–57.2	Different process route; centralised entrained flow gasification (cEF); centralised autothermal circulating fluidised bed gasification (CFB) and ICFB	Jungbluth *et al.*, 2007

(Continued)

Table 2.2 Continued

Feedstock	Origin	GHG emissions (g CO_2 eq MJ^{-1})	Key assumptions	Reference
Waste wood/forest residue	Europe	7.3–13.8	System expansion credit	Edwards *et al.*, 2008
	UK	6	Default value[a]; energy allocation	RFA, 2010
Waste wood/ forest residue	Europe	4–9	System expansion credit	Edwards *et al.*, 2008
	UK	4	Default value[a]; energy allocation	RFA, 2010
Wheat straw	Europe	26.5–64.6	Economic allocation; different process routes: decentralised entrained flow gasification (dEF), cEF and CFB	Jungbluth *et al.*, 2007
Corn residue	US	8.53	System expansion credit; future (2020) technologies	US EPA, 2010
Waste cooking oil	UK	14	Default value[a]; energy allocation	RFA, 2010
	Spain	10.05	System expansion credit	Lechón *et al.*, 2009
	Australia	10.9		Beer *et al.*, 2007
	US	13.3	System expansion credit; future (2020) technologies	US EPA, 2010
	Ireland	20.22	System expansion credit	Thamsiriroj and Murphy, 2011

[a] Renewable Fuels Agency defines a default value to be used in the absence of specific information (RFA, 2010).

double that reported in the latter. The amount of electricity, heat and other products produced as co-products and the way the impacts are allocated among them also influences overall GHG emissions in these studies.

Owing to a limited number of available LCA studies, it is difficult to draw general conclusions regarding the biodiesel GHG emissions from different feedstocks. Nevertheless, the results indicate that biodiesel produced from waste, particularly forest residue and waste cooking oil, offers the highest GHG savings (around 90%), with average GHG emissions of 8 and 13 g CO_2 eq MJ^{-1}, respectively (Table 2.2 and Fig. 2.2). This is due to the use of waste which avoids emissions from cultivation, application of fertilisers and land-use change otherwise incurred by other feedstocks. The only GHG emissions in the life cycle of biodiesel from waste are thus from the energy required to collect, pre-process and convert the feedstock into biodiesel.

By comparison, biodiesel from wheat straw offers a lower GHG saving (20–65%). Moreover, removal of agricultural residues from fields for bioenergy production could have other implications, including lower yields of future crops owing to the removal of nutrients in the straw, increased N_2O emissions owing to a greater need for fertilisers and decline in soil carbon pools as removal of residues reduces the soil carbon storage capacity (Cherubini *et al.*, 2009). Also by comparison, first generation biodiesel from rapeseed and palm oil offers a GHG saving in the range of 20–80% (IEA, 2010).

The results show that biodiesel from all energy crops (miscanthus, switch grass, jatropha and SRC) have the GHG mitigation potential in the range of 28–90% over conventional diesel. The wide range of the estimated GHG savings from energy crops reflects the diversity of potential feedstocks and production routes and the uncertainty regarding the technology and the methodological differences (such as allocation of emissions between co-products and land-use change) in various studies. These methodological issues are discussed further below.

GHG emissions from co-products

Biodiesel production systems generate various co-products and, depending on the approach used to allocate the emissions and impacts between the biodiesel and co-products, the results of LCA may vary significantly (Menichetti and Otto, 2009). In LCA studies, two approaches are commonly applied: system expansion and allocation on the basis of either mass, energy content or market value (price) of the products.

As shown in Table 2.1, the US EPA prefers system expansion. In this method, the system is credited with producing additional output and displacing the need for another system that produces the same or equivalent product. However, the methodological difficulty is in identifying the 'correct' way to credit the system. For example, if electricity is co-produced with biodiesel in an EU country, the question is what electricity mix should be used to credit the system: best available technology or the average national or EU energy mix? The choice of the allocation method and the 'credit' are of the utmost importance as often very different results are obtained using different approaches. In any case, this should be examined as part of sensitivity analysis.

The EU RED favours allocation based on the energy content of biofuels, although other allocation procedures, such as system expansion or economic value might be more appropriate in particular cases. For example, energy allocation cannot be applied in systems where co-products do not have an energy value but have an economic value, for example ash and fertilisers. In these cases, allocation based on the economic value may be more appropriate. However, this produces volatile results in line with the economic values of commodities and should be used only where other allocation methods cannot be applied (ISO, 2006).

Land-use change

Land-use change (LUC) is probably the most controversial and uncertain issue associated with biofuels (Fargione *et al.*, 2008). The main concern is related to possible additional GHG emissions when carbon stored in the soil is disturbed and released as CO_2 owing to LUC. Two types of LUC are considered: direct and indirect. Direct LUC involves conversion of existing land from a current use to cultivation, in this instance, of biomass feedstocks for biodiesel production. As shown in Table 2.1, direct LUC is considered in both the EU RED and US EPA approaches and the IPCC factors are used for these purposes (IPCC, 2007).

Indirect LUC is associated with the displacement of an existing agricultural activity (Searchinger *et al.*, 2008). This is often difficult to assess owing to the uncertainties involved, particularly at the international level. Currently, only the US approach considers indirect LUC (US EPA, 2010). Both direct and indirect LUC can have a significant impact on the GHG balance of biofuels, owing to GHG emissions from changes in soil organic carbon, litter and dead wood (Gallagher, 2008).

It is important to note that agricultural and forestry residues and waste cooking oil do not necessitate LUC beyond what occurs for food production and existing forestry activities. Similarly, use of degraded or marginal lands for energy crop production could avoid impacts related to LUC. In addition to LUC, different crop management practices can also influence emissions of carbon from soil. However, there is still considerable uncertainty and lack of knowledge regarding the loss from or sequestration of carbon in soils.

2.2.2 Other environmental impacts

The following sections provide a brief discussion of other environmental impacts, in addition to global warming, that are associated with the life cycle of second generation biodiesel.

Biodiversity

Over the past 50 years, there has been a rapid loss of biodiversity globally, mainly due to land conversion for agriculture and forestry (MEA, 2005). Biofuel crops can also alter local habitats and resources in a way that could affect native species. These effects will depend on the crop, its density, duration and distribution in the landscape and any regular inputs, including water and chemicals (The Royal Society, 2008).

The biodiversity impacts of second generation biodiesel feedstocks can be both negative and positive, depending on the type of feedstock and the existing land use (CBD, 2008; FAO, 2008). Biodiversity losses are higher when forests and grasslands are converted to biofuel crops. On the other hand, if degraded lands are restored for biofuel feedstock production, biodiversity can increase.

Second generation energy crops, under minimum tillage systems and reduced fertiliser input, are in general resistant to pests and diseases and would thus maintain biodiversity. A recent study noted that SRC and miscanthus plantations could be generally regarded as beneficial for some bird species, small mammals, butterflies, insects and flowering plants in agriculturally dominated landscapes (Rowe *et al.*, 2009). However, second generation energy crops present another set of problems: replacement of native forest with mono-crop plants could result in loss of wildlife habitat owing to a significant reduction in plant biodiversity (Timilsina and Shrestha, 2011). Similarly, introducing non-native, particularly more invasive, species into an area could lead to the displacement of local biodiversity (IEA, 2010). Eucalyptus, some miscanthus species and switchgrass all exhibit some features of invasiveness (The Royal Society, 2008).

The use of forest and agricultural residues as biofuel feedstock is expected to have lower negative impacts on biodiversity than dedicated energy crops (IEA, 2010). Some of the impact on biodiversity associated with the use of forestry residues includes reduction in the amount of decaying wood – a niche habitat and disturbance of wildlife caused by increased forest access. Removal of agricultural residue from fields may increase weed growth, which could lead to the increased use of herbicides and thus affect local biodiversity.

Biodiversity loss can also occur from the direct effects of LUC. For example, if set-aside land in Europe is used to grow biofuel crops, the impacts on biodiversity will need to be evaluated because some of these areas are more biodiverse than farmlands (Critchley and Fowbert, 2000). Intensified cultivation of biofuel crops could also lead to new pests and diseases which could in turn lead to increased use of pesticides/herbicides, causing further environmental damage.

Therefore, it is important that the overall risks and benefits of biodiversity be evaluated appropriately for bioenergy feedstocks. The EU Renewable Energy Directive (RED) requires that biofuel feedstock must not be grown on environmentally sensitive land, including protected areas and land with high biodiversity value. The Royal Society (2008) recommends using a risk assessment framework that covers the following:

- the full life cycle of biofuel production;
- the invasiveness potential of the crop;
- potential interactive effects of the biofuel crop with other pressures in the area (for example, drought stress);
- the impacts on ecosystems;
- changes in these risks under a future climate.

However, the lack of data represents a significant barrier in addressing biodiversity on a life cycle basis as biofuel crops have not yet been assessed for their impacts on biodiversity. Furthermore, there is currently no agreed methodology for estimating the impacts on biodiversity in LCA. However, some methods, such as Impact 2002+ and Eco-indicator 99, consider the impacts on biodiversity by taking into account damage to ecosystem quality (Jolliet *et al.*, 2003; Goedkoop and Spriensma, 2000). The loss of biodiversity is estimated in terms of potential disappeared fraction (PDF) of species, which represents the rate of extinction of species caused by LUC, toxicity, global warming, or eutrophication.

Water use

There have been significant concerns related to the impacts of water use in the production of biofeedstocks, particularly for first generation biofuels (Dominguez-Faus *et al.*, 2009; FAO, 2008). These concerns are primarily related to the requirements for irrigation water for certain feedstocks which might compete with water used for other purposes. However, in the case of second generation biodiesel, water use could vary widely depending on the type of feedstock. For example, agricultural and forest residue, as well as waste cooking oil, have little or no negative impact on water use. On the other hand, energy crops such as jatropha, eucalyptus, SRC, switch grass and miscanthus, generally have higher water demands than arable crops owing to a combination of higher growth and transpiration rates, longer seasonal growth and increased rooting depth (Rowe *et al.*, 2009). Although in most places these crops are not irrigated currently, perennial plants such as jatropha grown in semi-arid areas on marginal land may require some irrigation during critical growth stages (Payne, 2010). The energy crops could have implications for groundwater in areas where the rainfall is not sufficient to meet their evapotranspiration requirements.

Biodiesel refineries also require water, although use is relatively modest, ranging from 1–3 litres per litre of biodiesel (Pate *et al.*, 2007). Nevertheless, this can have a localised impact on water availability because water use is concentrated in a smaller area.

The impacts of water use are usually not included in LCA or other evaluations of environmental sustainability of biofuels. The main reasons are the lack of data and an agreed methodology for estimating the water footprint. Although there are some data available on water use for common crops, fewer data are available for some of the cellulosic feedstocks. Currently there are little or no reliable data on water usage in life cycle databases for almost any productive system. For further information on the methodologies for assessing the impacts of water use of biofuels in LCA, see Jeswani and Azapagic (2011).

Other impacts

Most LCA studies of second generation biodiesel, mentioned in Table 2.2, focus on GHG emissions and energy balances. However, as already discussed, biofuels have wider environmental impacts, including resource depletion, acidification, eutrophication and toxicity. Figure 2.3 compares other selected environmental impacts of biodiesel (from SRC) and conventional diesel.

According to the study considered here, biodiesel from SRC has 1.8 times higher acidification and 2.4 times higher eutrophication than fossil diesel. (Note that the use stage of fuels is not considered as these data are not available in literature. In general, the use of biodiesel results in a reduction of carbon monoxide, hydrocarbons, sulphur oxides and particulate emissions in comparison to the use of diesel.) The agricultural stage is the major contributor to these impacts owing to emissions of ammonia, nitrogen leaching and emissions from the use of fuel in agricultural machinery.

These results illustrate that while second generation biodiesel can provide GHG savings, their wider impacts can be greater than that of conventional fossil fuels, especially for biodiesel from energy crops (Fig. 2.3). The use of agricultural as well as forestry residues in addition to waste cooking oil is expected to have lower acidification and eutrophication than energy crops (IEA, 2010). Nevertheless, from this and the earlier discussion it is clear that evaluation of environmental sustainability of biodiesel should involve consideration of all relevant environmental impacts along the whole life cycle to avoid shifting the burden and making unsustainable choices.

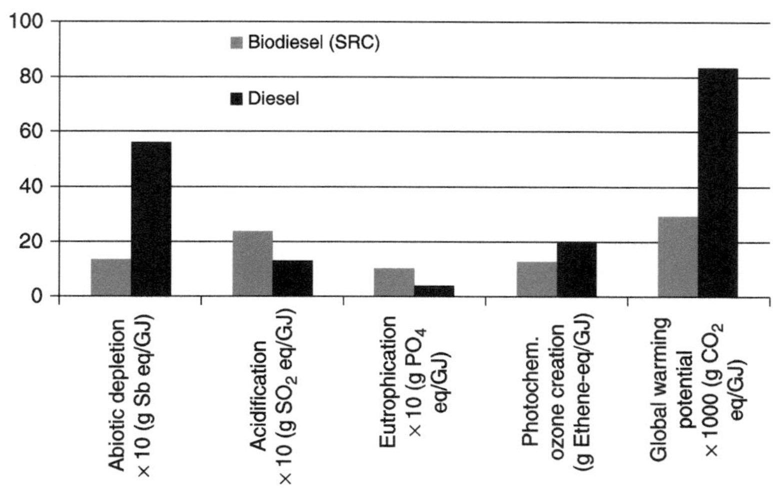

2.3 Environmental impacts of biodiesel (from SRC) and fossil diesel. Note that biodiesel (from SRC) is produced through the centralised entrained flow gasification (cEF) process; SRC is cultivated in Europe; the system boundary is from 'cradle to grave' (the use stage is not included except for GWP owing to lack of data) (data source for biodiesel (from SRC): Jungbluth *et al.*, 2007 and fossil diesel: Ecoinvent, 2010).

2.3 Socio-economic sustainability of biodiesel

2.3.1 Economic sustainability

To date, only a few studies of economic sustainability of second generation biodiesel are available in the public domain. This is mainly because economic data are confidential as the conversion technologies are still under development. Moreover, because of different assumptions for feedstocks and conversion technologies, comparison of different biofuels on economic criteria is difficult (Bridgwater, 2009).

On a life cycle basis, the costs of second generation biodiesel are mainly contributed by:

- the costs of feedstock cultivation (if applicable), preparation and delivery;
- the capital costs of manufacturing plants for conversion into biodiesel; and
- other costs such as labour, utilities, maintenance, insurance, etc.

The following sections give an overview of the feedstock and capital costs and discuss how they influence final biodiesel prices.

Feedstock costs

As shown in Table 2.3, the current costs of lignocellulosic feedstock in Europe vary greatly depending on the biomass and range from €1.32 to €8.4 per GJ (€25 to €155 per tonne of dry matter) (Müller-Langer *et al.*, 2007). This variation is due to the different energy and moisture content as well as the origin of the feedstocks. Since agricultural and forestry residues do not incur cultivation (biomass production) costs, they are at the lower end of price range, while miscanthus and SRC are at the upper end. Miscanthus has higher handling and transport costs than

Table 2.3 Production costs for different lignocellulosic feedstocks in Europe

Feedstock	Cost (€ GJ^{-1})	Cost (€ t^{-1}) (dry matter)	Assumed transport distance to the storage facility (km)
SRC	5.2–5.8	98–109	30
Miscanthus	5.9–8.4	109–155	30
Switch grass	4.7	86	30
Agricultural residue (straw)	2.15–3.8	37–67	15
Forest residue	1.32–2.74	25–52	1

Source: Müller-Langer *et al.*, 2007.

The costs include biomass production (only for energy crops), harvesting, handling, transport to the storage site, storage, brokerage and land and risk compensation costs.

woody crops owing to its lower bulk density. Straw also has higher handling costs than forest residue. Transport costs only include feedstock transport to the storage site close to the field or forest. The assumed transport distances for different feedstocks are shown in Table 2.3.

There are also relatively large differences between the costs of different feedstocks within different regions in Europe. The eastern region has a much lower cost compared with all other regions owing to much lower labour costs (Müller-Langer *et al.*, 2007).

Table 2.4 shows a breakdown of the total costs of the feedstock at the biodiesel plant for the example of SRC in Northern Europe. As shown, just over half of the costs (57%) are related to the cultivation of the feedstock and the rest are for storage and delivery to the biodiesel plant. The production costs are expected to decrease significantly in future with large scale cultivations as well as improved yields, technology and knowledge (Rosenqvist and Nilsson, 2007). On the other hand, the biomass delivery costs would increase with scaled-up biofuel plant sizes owing to longer transport distances and often more complex logistics requirements (RENEW, 2008).

Capital costs

Currently, second generation biodiesel plants are up to ten times more capital-intensive than first generation plants of the same capacity (IEA, 2010). However, estimates of capital costs for biofuel plants (or any other developing technology), are uncertain owing to the many influencing factors. This is clear from the significant variation in cost estimates in different studies. A comprehensive study carried out under the European project RENEW (2008) puts the cost of

Table 2.4 Production and delivery costs for SRC in Northern Europe

	Amount (€ GJ^{-1})	Contribution (%)
Establishment	0.73	10.81
Fertilisation	0.74	10.96
Harvest	0.89	13.19
Field transport	0.15	2.22
Costs of land	0.5	7.41
Costs of risk compensation for growing new crops	0.9	13.33
Total farm costs	*3.91*	*57.93*
Transport to storage	0.69	10.22
Storage and other costs	0.84	12.44
Transport to biodiesel plant by road	1.31	19.41
Total field to plant	*2.84*	*42.07*
Total	**6.75**	**100.00**

Source: Müller-Langer *et al.*, 2007; Rosenqvist and Nilsson, 2007.

Table 2.5 Capital investment costs for Fischer-Tropsch (FT) biodiesel produced using different technologies, in million Euros

Production route	Centralised entrained flow gasification (cEF)	Decentralised entrained flow gasification (dEF)[a]	Centralised circulating fluidised bed gasification (cCFB)
Storage and pre-treatment	69.21	68.57	94.24
Pyrolysis	0	45.14	0
Gasification	51	44.06	33.64
Gas cleaning	31.44	45.77	120.14
Gas conditioning	42.86	48	17.6
Synthesis	94.37	68.1	72.54
Upgrading	84.9	65.88	63.56
Utilities	59.34	23.84	100.8
Total capital investment	*433.12*	*409.38*	*502.52*
Biomass input energy (MWth)	500	500	500
Product output (diesel and naphtha) (t yr^{-1})	178 000 (straw as a feedstock) 180 000 (SRC as a feedstock)	140 000 (straw as a feedstock)	120 000 (straw as a feedstock) 133 000 (SRC as a feedstock)

Source: RENEW, 2008.

[a]Pre-treatment and pyrolysis are carried out at a decentralised facility for the dEF technology.

thermochemical Fischer-Tropsch plants from €410–500 million for plants treating 500 MWth (thermal megawatts) of biomass input. The corresponding liquid fuel production capacity ranges from 120 000–180 000 tonnes of biodiesel per year, depending on the type of feedstock, efficiency of the process and the amount of electricity produced as a co-product. Table 2.5 provides the breakdown of capital costs. As shown in the table, the centralised entrained flow gasification (cEF) process appears to be the most cost-effective processing route.

The above-mentioned estimates are lower than those estimated by DENA (2006), which puts the cost of thermochemical FT plants from €525–650 million for a capacity of 106 000–118 000 tonnes of biodiesel per year. The study found out that integration of biofuel production into an existing refinery can lower investment costs by around 25% (DENA, 2006). In addition to the economic benefits, this provides operational and organisational synergies and lowers significantly the plant availability risk.

Biofuel prices

Table 2.6 provides estimates of second generation biodiesel prices in Europe for different production processes and feedstocks (RENEW, 2008). As shown, the production costs range from €0.69–2.59 per litre. The main reason for the variation between cost estimates for the same processing route is the difference in feedstock

Table 2.6 Estimated prices of second generation biodiesel in Europe compared with prices of oil and oil products (biofuels exclusive of taxes) (RENEW, 2008)

Biodiesel	€/litre
Price of oil[a]	0.25–0.40
Corresponding pre-tax price of petroleum products[b]	0.28–0.48
Corresponding price of petroleum products in Europe with taxes include, (retail price)	1.20–1.59
FT diesel cEF straw	0.69–1.08
FT diesel cEF SRC	0.95–1.72
FT diesel dEF straw	1.11–2.21
FT diesel cCFB straw	1.46–2.59
FT diesel cCFB SRC	1.1–1.85

[a] based on 2006 price of 50–80 US$/barrel and exchange rate of 1.255 US$/€.
[b] based on 2006 price of 0.35–0.60 US$/litre (The Royal Society, 2008).

costs: the lowest production costs are for biodiesel from agricultural residues and the highest from energy crops.

According to the RENEW (2008) estimates, the provision costs of biomass are the main influencing factor followed by the capital cost. Therefore, the economic prospects of second generation biodiesel will depend on the efficiency of the conversion processes and the reduction in feedstock costs, for example through improvements in the yields of energy crops. Widespread deployment in the future is expected to lead to reduced costs for plant construction and operation experience and other 'learning-by-doing' effects (IEA, 2010). Furthermore, higher oil prices as well as tax credits or other incentives for biofuels could make second generation biofuels commercially more attractive (The Royal Society, 2008).

2.3.2 Social sustainability

Positive impacts associated with second generation biodiesel include job creation and regional development (IEA, 2010). The creation of jobs along the value chain depends on the type of feedstock. The use of dedicated energy crops would create jobs in the cultivation of the feedstock, whereas the use of agricultural residues has a limited potential to create jobs since existing farm labour could be used. On the other hand, the use of agricultural residues for production of biodiesel (and other biofuels) could be beneficial to farmers as it would add value to these by-products. Similarly, the use of forest residues could provide an additional source of income with positive impacts on local economies and rural development. However, this could affect farmers or a rural population that is dependent on residues for animal fodder or domestic fuel.

Although second generation biodiesel is produced from feedstocks that are not used for food, the main concern about the social impact of energy crops production is the occupation of arable land for cultivation, especially in developing countries and thus competition with current agricultural production (IEA, 2010). This in turn could affect food prices. A recent study indicated that to meet the biofuel targets in the OECD and some developing countries, accelerated introduction of second generation biofuels (cellulosic) could increase cereal crops prices by around 15% in 2020 (OFID/IIASA, 2009). However, the price impact on cereal crops would be around 30% if biofuel targets are realised with first generation biofuels.

2.4 Conclusions and future trends

There is a considerable potential for the production of second generation biofuels. On one hand, the feedstocks are widespread, relatively cheap and easily available; on the other hand, their use could allow co-production of chemicals, electricity and heat, leading to the development of biorefineries (Cherubini *et al.*, 2009). A successful deployment of second generation biodiesel technologies will require intensive research and development effort over the next few years. The sustainability of large-scale industrial developments will depend on a range of environmental, economic and social factors. Some of these include:

- availability of land for dedicated energy crops;
- impact on soil nutrients owing to the use of agricultural and forest residues as biofeedstock;
- GHG emissions savings over fossil fuels, particularly when a LUC is involved;
- impacts on biodiversity and water resources;
- biofeedstock and investment costs and the resulting biofuel prices;
- human health caused by air pollution and human toxicity in the biodiesel life cycle; and
- the impact on food security.

There is a need to ensure that only marginal/abandoned land or waste and residues are used to supply the feedstocks to avoid potential negative impacts (Gallagher, 2008). In an attempt to ensure future sustainability of biofuels, an internationally accepted certification system has been proposed via the International Sustainability and Carbon Certification project (ISCC, 2009). The project aims to develop an international certification concept together with representatives from industry, trade, agriculture, policy makers and NGOs. Initiatives like this, in addition to the growing international awareness and legislation related to biofuels, may contribute towards more sustainable biodiesel in the future.

2.5 Acknowledgements

Part of the work presented in this chapter has been funded within the project Carbon Calculations over the Life Cycle of Industrial Activities (CCaLC) by the EPSRC, NERC and Carbon Trust (grant no. EP/F003501/1). This funding is gratefully acknowledged.

2.6 Sources of further information and advice

Biofuels and sustainability in Europe: http://www.biofuelstp.eu/sustainability.html
Biofuels sustainability scorecard: http://idbdocs.iadb.org/wsdocs/getdocument.aspx?docnum=2152669
CCaLC, Carbon and water footprinting tool, including biofuels and bio-feedstocks: www.ccalc.org.uk
IEA Bioenergy: http://www.ieabioenergy-task38.org
Integrated project RENEW: Renewable fuels for advanced power trains: http://www.renew-fuel.com/home.php
Tool for estimating GHG emissions from the production of transport fuels, electricity and heat from biomass: http://www.agentschapnl.nl/en/programmas-regelingen/determine-biofuel-greenhouse-gas-emissions-production-transport-fuels-made-bio
Well-to-wheels evaluation of biofuels: http://ies.jrc.ec.europa.eu/jec-research-collaboration/activities-jec/jec-well-to-wheels-analyses-wtw.html

2.7 References

Azapagic, A. and Stichnothe H. (2010) Life cycle sustainability assessment of biofuels. In: *Handbook of biofuels production: Processes and technologies*, Luque, R., Campelo J. and Clark J. (eds), Woodhead Publishing Series in Energy No. 15, Cambridge, UK.

Beer, T., Grant T. and Campbell P. (2007) *The Greenhouse and Air Quality Emissions of Biodiesel Blends in Australia*. CSIRO http://www.csiro.au/files/files/phim.pdf. Accessed on 11 January 2012.

Bird, D. N., Cherubini F., Cowie A., Downing M., Kojakovic A. *et al.* (2008) *Ten Years of Analysing the Greenhouse Balances of Bioenergy Systems*. IEA Task 38, International Energy Agency. http://www.ieabioenergy-task38.org/publications/T38_OC6_2_17_EU_Conf_2009.pdf. Accessed on 11 January 2012.

Bridgwater, A. V. (2009) *Technical and Economic Assessment of Thermal Processes for Biofuels*. NNFCC Report, York. Accessed on 11 January 2012.

CBD (2008) The potential impact of biofuels on biodiversity. Note by the Executive Secretary for the *Conference of the Parties to the Convention on Biological Diversity*, 19–30 May 2008. Convention on Biological Diversity, Bonn, Germany.

Cherubini, F., Bird N. D., Cowie A., Jungmeier G., Schlamadinger B. and Woess-Gallasch S. (2009) Energy- and greenhouse gas-based LCA of biofuel and bioenergy systems: Key issues, ranges and recommendations. *Resources, Conservation and Recycling*, **53**(8), 434–47.

Critchley, C. and Fowbert J. (2000) Development of vegetation on set-aside for up to nine years from a national perspective. *Agriculture, Ecosystems and Environment*, **79**(2), 159–74.

DENA (2006) *Biomass to Liquid – BTL Implementation Report (Executive Summary)*. Deutsche Erneuerbare Energieagentur (DENA), Berlin.

Dominguez-Faus, R., Powers S. E., Burken J. G. and Alvarez P. J. (2009) The water footprint of biofuels: A drink or drive issue? *Environmental Science and Technology*, **43**(9), 3005–10.

EC (2009) *Directive 2009/28/EC of the European Parliament and of the Council of 23 April 2009 on the Promotion of the Use of Energy from Renewable Sources and Amending and Subsequently Repelling Directives 2001/77/EC and 2003/30/EC.*

European Commission, Brussels. http://eur-lex.europa.eu/LexUriServ/LexUriServ.do?uri=OJ:L:2009:140:0016:0062:EN:PDF. Accessed on 11 January 2012.

Ecoinvent (2010) *Ecoinvent v2.0 database*. Swiss Centre for Life Cycle Inventories, Dübendorf, Switzerland.

Edwards, R., Larive J.-F., Mahieu V. and Rouveirolles P. (2008) Well-to-wheels analysis of future automotive fuels and power trains in the European context. *JRC, Eucar and Concawe*, V3. November 2008. http://ies.jrc.ec.europa.eu/jec-research-collaboration/activities-jec/jec-well-to-wheels-analyses-wtw.html. Accessed on 11 January 2012.

Eisentraut, A. (2010) *Sustainable Production of Second-Generation Biofuels: Potential and perspectives in major economies and developing countries*. International Energy Agency, IEA/OECD, Paris, www.iea.org.

Escobar, J. C., Lora E. S., Venturini O. J., Yáñez E. E., Castillo E. F. and Almazan O. (2009) Biofuels: Environment, technology and food security. *Renewable and Sustainable Energy Reviews*, **13**(6–7), 1275–87.

FAO (2008) *The State of Food and Agriculture – Biofuel: Prospects, Risk and Opportunities*. Food and Agriculture Organization of The United Nations, Rome.

Fargione, J., Hill J., Tilman D., Polasky S. and Hawthorne P. (2008) Land clearing and the biofuel carbon debt. *Science*, **319**(5867), 1235–8.

Gallagher (2008) *The Gallagher Review of the Indirect Effects of Biofuels*. Renewable Fuels Agency, London, UK.

Goedkoop M. and R. Spriensma R. (2000) *The Eco-Indicator 99: A Damage Oriented Method for Life Cycle Impact Assessment*. PRé Consultants, Amersfoort, The Netherlands. www.pre.nl/eco-indicator99/.

Hoefnagels, R., Smeets E. and Faaij A. (2010) Greenhouse gas footprints of different biofuel production systems. *Renewable and Sustainable Energy Reviews*, **14**(7) 1661–94.

IEA (2008) *Energy Technology Perspectives*. International Energy Agency, IEA/OECD, Paris, www.iea.org.

IEA (2009) *World Energy Outlook 2009*. International Energy Agency, OECD/IEA, Paris.

IEA (2010) *Sustainable Production of Second-Generation Biofuels – Potential and perspectives in major economies and developing countries*. International Energy Agency, Paris, www.iea.org/papers/2010/second_generation_biofuels.pdf. Accessed on 11 January 2012.

IPCC (2007) *Climate Change 2007: Synthesis Report. Contribution of Working Groups I, II and III to the Fourth Assessment Report of the Intergovernmental Panel on Climate Change* (Core Writing Team, Pachauri, R.K and Reisinger, A. eds). IPCC, Geneva, Switzerland.

ISCC (2009) *International Sustainability and Carbon Certification Project*. http://www.iscc-system.org.

ISO (2006) *ISO 14044: Environmental Management – Life Cycle Assessment – Requirements and Guidelines*. International Standard Organization (ISO), Geneva.

Jeswani, H. K. and Azapagic A. (2011). Water footprint: methodologies and a case study for assessing the impacts of water use. *Journal of Cleaner Production*, **19**(12) 1288–99.

Jolliet O., Margni M., Charles R., Humbert S., Payet J., *et al.* (2003) IMPACT 2002+: A new life cycle impact assessment methodology. *International Journal of LCA*, **8**(6), 324–30.

Jungbluth, N., Frischknecht R., Emmenegger M., Steiner R. and Tuchschmid M. (2007) *Life Cycle Assessment of BTL-Fuel Production: Life Cycle Impact Assessment and Interpretation*. ESU Services GmbH, Uster, Germany.

Lechón, Y., Cabal H., de la Rúa C., Caldés N., Santamaría M. and Sáez R. (2009) Energy and greenhouse gas emission savings of biofuels in Spain's transport fuel. The adoption of the EU policy on biofuels. *Biomass and Bioenergy*, **33**(6–7), 920–32.

MEA (2005) *Ecosystems and Human Well-being: Our Human Planet. Summary for Decision Makers*. Millennium Ecosystem Assessment, Island Press, Washington DC, USA, http://www.maweb.org/documents/document.356.aspx.pdf.

Menichetti, E. and Otto M. (2009) Energy balance and greenhouse gas emissions of biofuels from a life cycle perspective. In: *Biofuels: Environmental Consequences and Interactions with Changing Land Use. Proceedings of the Scientific Committee on Problems of the Environment (SCOPE) International Biofuels Project Rapid Assessment*, 22–25 September 2008, Howarth, R. and Bringezu S. (eds), Gummersbach, Germany. Cornell University, NY, USA, 81–109.

Müller-Langer, F., Thrän D., Gańko E. and Jarowski L. (2007) *RENEW Deliverable 5.3.6, Biomass supply costs – Final report*. EC Baltic Renewable Energy Centre, Poland; Institut für Energetik und Umwelt, Germany.

OFID/IIASA (2009) *Biofuels and Food Security*. OPEC Fund for International Development/International Institute for Applied Systems Analysis, Vienna, Austria.

Pate, R., Hightower, M., Cameron, C. and Einfeld W. (2007) *Overview of Energy–Water Interdependencies and the Emerging Energy Demands on Water Resources*. Sandia National Laboratories, Los Alamos, NM, USA.

Payne, W. A. (2010) Are biofuels antithetic to long-term sustainability of soil and water resources? In: *Advances in Agronomy*, Donald, L. S. (ed). Academic Press, San Diego, USA, 1–46.

Reinhardt, G. A., Gärtner S., Rettenmaier N., Münch J. and von Falkenstein E. (2007) *Screening Life Cycle Assessment of Jatropha Biodiesel*, Final Report. IFEU, Heidelberg, Germany.

RENEW (2008) *Renewable Fuels for Advanced Powertrains*, Final report. www.renew-fuel.com.

RFA (2010) *Carbon and Sustainability reporting within the Renewable Transport Fuel Obligation, Technical Guidance Part One Version 3.3 April 2010*. Renewable Fuels Agency. http://webarchive.nationalarchives.gov.uk/20110407094507/http://renewablefuelsagency.gov.uk/sites/rfa/files/RFA_C_and_S_TG_%20Part_One_v3%203.pdf. Accessed on 11 January 2012.

Rosenqvist, H. and Nilsson L. (2007) *RENEW Deliverable 5.3.4, Energy crop production cost in the EU*. Lund University, Sweden.

Rowe, R. L., Street N. R. and Taylor G. (2009) Identifying potential environmental impacts of large-scale deployment of dedicated bioenergy crops in the UK. *Renewable and Sustainable Energy Reviews*, **13**(1), 271–90.

Searchinger, T., Heimlich R., Houghton R. A., Dong F., Elobeid A., *et al.* (2008) Use of US croplands for biofuels increases greenhouse gases through emissions from land-use change. *Science*, **319**(5867), 1238–40.

Thamsiriroj, T. and Murphy J. D. (2011) The impact of the life cycle analysis methodology on whether biodiesel produced from residues can meet the EU sustainability criteria for biofuel facilities constructed after 2017. *Renewable Energy*, **36**(1), 50–63.

The Royal Society (2008) *Sustainable Biofuels: Prospects and Challenges*. Royal Society, London, UK.

Timilsina, G. R. and Shrestha A. (2011) How much hope should we have for biofuels? *Energy*, **36**(4), 2055–69.

US EPA (2010) *Regulation of Fuels and Fuel Additives: Changes to Renewable Fuel Standard Program; Final Rule, 40 CFR Part 80, Part II, March 2010*. http://www.epa.gov/otaq/fuels/renewablefuels/regulations.htm. Accessed on 11 January 2012.

3

Socio-economic, environmental, and policy perspectives of advanced biodiesel production

C. DE LUCIA, Duke University, USA and
B. DATTA, University of York, UK

Abstract: This chapter illustrates and discusses the main potentials and the limitations of second and third generation biodiesel and biodiesel policies viewed with respect to the multi-directional effects of the economy and the environment. The analysis hinges upon the multiple effects of next generation biodiesel such as the need to guarantee energy and food security, environmental protection (particularly carbon emission reductions) and price stability in international trade and institutional support for biodiesel policies including the contribution of these to sustainable development.

Key words: biodiesel, feedstock, food safety, international trade, sustainable development.

3.1 Introduction

Over the last 40 years an increasing awareness of resource scarcity and energy security had led to the adoption of alternative energy practices in current production and use of energy commodities. Biodiesel is one of the main alternatives to fossil fuel for production and energy goods. It has environmental attractiveness because it is renewable and sustainable (owing to its lower toxicity levels compared to fossil fuels: see http://www.green-trust.org/biodiesel1.htm) over time as well as contributing to reducing carbon emissions. It also contributes to enhancing agricultural practices and sustaining the rural economy particularly in developing countries. In the latter case, the impact of job creation in rural areas is considerably higher owing to the multiplier effect of additional agricultural demand for feedstock conversion (Hazell and Pachauri, 2006). This increased liquidity can boost development paths and narrow income gaps with richer countries.

This picture is nonetheless not without drawbacks. The positive synergies that occur among a multitude of first generation biodiesels are counterbalanced by several negative aspects. First, most conversion feedstocks are obtained from edible oils and this reduces the availability of food resources to satisfy primary needs; second, owing to substitutability with food crops, first generation biodiesel is also responsible for food price increases and inflationary distortions, particularly in developing countries. As a result, developing countries face higher input costs which affect the agriculture sector. This causes competition with edible oil markets which, in turn, increases the costs of edible oils and biodiesel over subsequent rounds of agricultural commodity transactions. Third, the imbalances of land availability occurring for biodiesel feedstock may increase deforestation, carbon emissions and the impoverishment of land productivity over time.

The combined effects of first generation biodiesel, as described above, have stimulated a greater interest and even some sense of urgency, in the development of biodiesel produced from non-food biomass, commonly referred to as second generation biodiesel. The second generation biodiesels are generally less land and water intensive and are usually manufactured from agricultural and forest residues and from non-food crop feedstocks. It is expected that these next generation biodiesels, though yet to be technologically and economically fully developed, should have good potential for cost reduction and increased production efficiency levels as more experience is gained. Depending partly upon future oil prices, these biodiesels are therefore likely to become a part of the solution to the challenge of shifting the transport sector towards more sustainable energy sources in the medium run and more generally contributing towards combating the ever pressing problem facing the world today, that is, of controlling greenhouse gas emissions.

Despite increased interest in expanding second and third generation biodiesels (and biofuels more generally) and the progress made with respect to massive investments in research and development (R&D) observed in recent years, significant hurdles still need to be resolved before these next generation biodiesels can be produced on a commercial scale (OECD/IEA *et al.*, 2008). Furthermore, where the lignocellulosic feedstock is to be produced from specialist energy crops grown on arable land, several concerns still remain over competing land use, although energy yields are likely to be higher than if crops grown for first generation biodiesel (and its co-products) were produced on the same land (OECD/IEA *et al.*, 2008).

Whilst significant investments are currently being made to gain improvements in the technology, it is suspected that, at least in the medium to long run, the next generation biodiesel industry will grow only at a steady rate. Therefore both first generation and second generation biodiesels are expected to coexist in the foreseeable future in order to meet agreed environmental, sustainability, and economic policy goals. Eventually though, future generation biodiesels and more generally biofuels are expected to replace the first generation ones. Therefore, although the potential benefits of producing and consuming next generation biodiesel in the future can be massive, there are also limitations that will be encountered during this transition period. The extent and rate at which the limitations of future generation biodiesels and biofuels can be overcome however depend to a great extent on the adoption and implementation of suitable domestic as well as international policies regarding production, consumption and trade in these next generation green fuels.

The objective of this chapter is therefore to provide a comprehensive analysis of both the potentials and limitations of second and third generation biodiesels, to evaluate them in the light of existing first generation ones, and to provide explicit policy recommendations both in the domestic and international context.

The chapter is organised as follows. Section 3.2 analyses socio-economic, environmental and policy limitations arising from the use of first generation biodiesels. Section 3.3 discusses the economic potential and limitations of second and third generation biodiesels by paying particular attention to the cases of algae biodiesel, wood biodiesel and biomethanol. Sections 3.4 and 3.5, respectively, analyse the potential and limitations of second and third generation biodiesels in proving energy security, land and food safety. Section 3.6 discusses the potential of second generation biodiesels in the context of international trade and how they can serve as an important driver in promoting economic growth. Section 3.7 discusses various policy issues emerging from the analysis provided in the previous sections. Finally, section 3.8 provides some concluding remarks.

3.2 Socio-economic, environmental and policy limitations of first generation biodiesels

Whilst there has been substantial growth in production and consumption of biodiesels in the last few years (see Table 3.1 below), it has been becoming increasing clear that, despite the numerous environmental and socio-economic benefits of the first generation biodiesels produced primarily from food crops, there are also several direct and indirect costs involved in making first generation biodiesel production and consumption a truly viable option worldwide. As such, first generation biodiesels have received considerable criticisms on the grounds of their limited greenhouse gas (GHG) reduction potential and their high marginal carbon abatement costs; on their being a catalyst in creating an upward surge in the world food prices; on being an expensive option for energy security and for their continuing need for suitable government policies ensuring their economic viability; and their direct and indirect impact on land use change and other environmental factors.

Table 3.1 World biodiesel production by selected regions 2005–2009 (thousand barrels per day)

Region	2005	2006	2007	2008	2009
North America	6.1	17.1	33.7	45.9	35.2
Central and South America	0.5	2.2	15.2	38.6	–
Europe	68.1	113.2	137.5	155.0	172.6
Eurasia	0.3	0.3	0.7	2.5	3.8
Asia and Oceania	2.2	9.1	15.8	28.8	38.5
World	**77.2**	**142.0**	**202.9**	**270.9**	**308.2**

Source: *Biomass Energy Data Book* (2010) http://cta.ornl.gov/bedb/biofuels/biodiesel/World_Biodiesel_Production_by_Region_Selected_Countries.pdf

While each of these drawbacks listed is a problem in its own right, they are however all directly and/or indirectly related. For example, the value of biodiesels as a viable GHG reduction option depends crucially upon their production costs, as well as their full GHG impact. With high production costs and relatively less net GHG reduction potential, the marginal carbon abatement costs of using some biofuels have been quoted to be as high as US$ 200–300 per tonne CO_2 avoided (OECD/IEA *et al.*, 2008). We discuss these factors in detail below.

3.2.1 High costs of production of first generation biodiesels

Despite continued improvements in biodiesel production efficiencies and yields, high production costs of biodiesels still remain a major barrier to their commercial development. Biodiesel production costs are highly dependent upon feedstock prices as well as the scale of operations and, hence, high feedstock costs (including its opportunity costs) naturally hike up biodiesel production costs. This has affected particularly the OECD countries who are currently the major producers of first generation biodiesels. In the EU, the current cost of producing biodiesels from rapeseed oil, the major source of production of biodiesels in the EU, varies from US$0.35 to US$0.80 per diesel-equivalent litre depending upon the scale of operation. In the USA, there are fewer large-scale production facilities and hence costs are even higher, ranging from US$0.48 to US$0.73 per-diesel equivalent litre using soy-oil as the major source of biodiesel production (Oregon Department of Energy, 2003).

Although the costs of biodiesels produced from waste grease and oils are usually lower in these countries, quantities of biodiesel produced from these sources are also quite limited and hence the amount of biodiesel produced at a very low cost is quite small relative to the diesel fuel use. While the average production costs can be expected to decline slowly in the future owing to (gradual) technological improvements and 'learning by doing' effects, without a substantial increase in large-scale biodiesel production brought about by a major technological breakthrough, it is unlikely that there will be significant reduction in the production costs of first generation biodiesels in the foreseeable future. In fact, any costs reduction due to mere improvement in technology could easily be offset by higher crop prices and/or decline in the value of co-products such as glycerine.

3.2.2 Costs in terms of rising food prices

According to Mitchell (2008) and reports from the International Monetary Fund (IMF), prices of internationally traded food commodities have risen sharply since 2002 and especially since late 2006, with prices of major staples like grain and oilseeds doubling in the recent past. The IMF's index of internationally traded food commodity prices increased 130% from January 2002 to June 2006 and 56% from January 2007 to June 2008 as shown in Fig. 3.1, where the circled shaded area highlights the particularly sharp increase in prices taking place since January 2007.

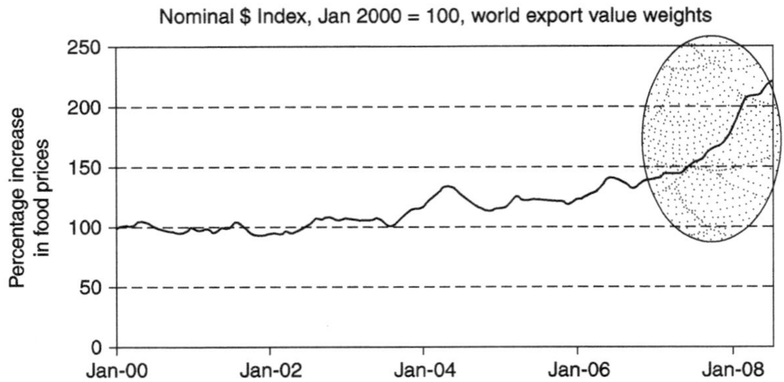

3.1 IMF index of internationally traded food commodities (source: Mitchell, 2008).

This sharp increase in food prices beginning after 2004 also coincides with the time of rapid expansion of biodiesel production worldwide. The IMF estimated that the increased demand for biodiesels and biofuels accounted for 70% of the increase in maize prices and 40% of the increase in soybean prices (Mitchell, 2008). According to the October 2007 *World Economic Outlook* (IMF, 2007), 'higher biofuel demand in the US and EU has not only led to higher corn and soybean prices, it has also resulted in price increases on substitution crops and increased the cost of livestock feed by providing incentives to switch away from other crops'. While there has been a debate about various factors other than biofuel (in which biodiesels constitutes a major part) contributing to this trend of rising prices, very few quantitative estimates are however available to assess their impact (Mitchell, 2008).

Undoubtedly, such increases in global food prices affect the poorest households the most as food accounts for a major part of their expenditure. In addition, higher food prices put upward pressure on inflation directly and, through their impact on non-food prices, affect the poorest households even harder. According to the IMF survey (IMF, 2007) the direct first round contribution of food to inflation for the world as a whole has risen from about one-fourth in 2000–2006 to more than one-third in the first four months of 2007, with the impact being harder in the poorer countries (Fig. 3.2).

3.2.3 Greenhouse gas reduction potential and limitations of first generation biodiesels

Several studies analysing the potential of first generation biodiesels for reducing greenhouse gases have claimed that replacing gasoline with biodiesels reduces

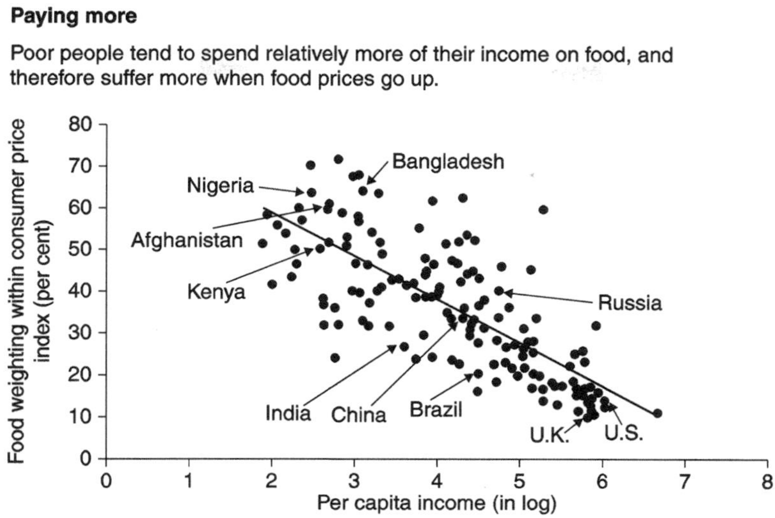

3.2 Food weighting within the consumer price index (source: http://www.imf.org/external/pubs/ft/survey/so/2007/RES1017A.htm).

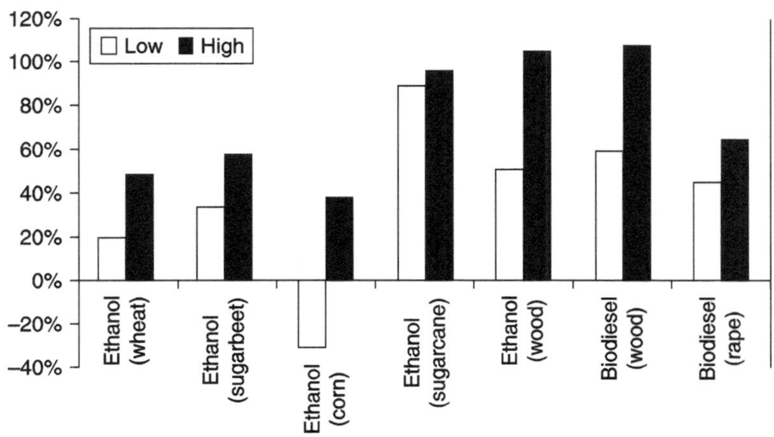

3.3 GHG reduction for different biofuels (source: Dufey, 2006).

greenhouse gases. For example, Hill *et al.* (2006) show that the life cycle GHG emission of soybean biodiesels are 59% those of diesel fuel. Figure 3.3 shows the extent of GHG reductions attainable from biodiesels and other biofuels.

However, many of these studies obtain such results by excluding emissions from changes in land use and hence they only provide a partial analysis in the sense that they only count the carbon benefits of using land for biodiesels or biofuels crops and thus do not include other factors such as carbon costs, carbon storage and sequestration sacrificed by diverting land from its existing use (Searchinger *et al.*, 2008). In fact, Searchinger *et al.* (2008), using a worldwide agricultural model to estimate emissions from land use changes, find that biofuels (primarily corn-based ethanol), instead of producing 20% savings, nearly double GHG emissions over 30 years! They also argue that such costs could largely be avoided by using second and third generation sources such as grass harvests from reserve lands and algae.

Furthermore, estimates in the literature for GHG mitigation from first generation biodiesels (as well as corn ethanol) vary from country to country depending upon the underlying technology and are in the range of US$200–300/t CO_2 avoided and can even be up to US$1700 (OECD/IEA *et al.*, 2008). One can therefore infer that, given the limited scope for cost reduction and growing global demand for food, little improvement in mitigation costs can be expected in the short run. In fact, according to a study by OFID (2009), anticipated GHG savings from first generation biodiesels and biofuels can only be expected to accrue after 30 to 50 years which is about the time when increased agricultural vulnerability will be at its peak particularly in a number of developing countries.

3.2.4 Land use and the 'food versus fuel' issue

First generation biodiesel is produced mainly from edible vegetable oils all over the world. Currently, more than 95% of the world biodiesel is produced from edible oils such as rapeseed, soybean, sunflower and palm, which are easily available on the large scale from the agricultural industry. However, continuous and large-scale production of biodiesels from edible oils has recently been of great concern because they compete with food materials as well as land usage giving rise to the 'food versus fuel' dispute. The global use of edible oils, which increased faster than its production between years 2004 and 2007, has caused the world demand for edible oils to outpace its supply. The estimated increase in edible oil use for biodiesel production was 6.6 million tonnes between 2004 and 2007, attributing 34% of the increase in global consumption to biodiesel. As a result this has created a chain reaction: excess demand for edible oil has caused the price of such oils to increase which has caused the production cost of biodiesels to increase. This in turn has not only caused the price of biodiesels to rise but has also created rises in food prices (see Figure 3.2 and the discussion in the previous subsection) which, on the other hand, has created inflation. Between 2005 and 2017, biodiesel use of edible oils is projected to account for more than one-third of the expected growth in edible oil use (FAO (Food and Agriculture Organisation of the United Nations), 2009). If this occurs, the problems of inflation and rising food prices are going to be even worse in the future. It has been suggested that about (an estimate of) 7.8 million hectares were used to provide

Table 3.2 Biodiesel production and land use by major producing countries 2006/07

Country	Biodiesel feedstocks	Implied feedstock area (Mha)[a]	Country total (Mha)[a]	Arable land	
				Area (Mha)[a]	Biodiesel share (%)
Argentina	Soybean (100%)	0.73	0.73	28.3	2.6
Brazil	Soybean (66%)	0.45	0.45	59.1	0.8
EU–27	Rapeseed (64%)	2.75	4.33	113.8	3.8
	Soybean (16%)	1.58			
USA	Soybean (74%)	2.31	2.31	174.5	1.3
Total		7.82	7.82	357.7	2.1

Source: Trostle (2008).

biodiesel feedstock in the four major biodiesel producing countries in 2007, see Table 3.2.

It is alleged that industrialised countries with biofuels targets, such as the United States and the EU countries, are unlikely to have the agricultural land base needed to meet their growing demand for current production of biodiesels. Currently, biodiesel production uses around 4.4 million hectares of arable land in the EU. Replacing 10% of EU diesel with biodiesel would account for around 19% of world edible oil production in 2020 which means more land will be needed for planting such crops which then implies that more land in other parts of the world will need to be converted into farmland, which may further aggravate the GHG emissions problem. Added to this is the problem of population growth globally. An increasing population growth rate simply implies increasing demand for both food and biodiesels (and other biofuels) making the problem of food and fuel shortages even worse and leading to other significant problems such as starvation in developing countries. With nearly 60% of humans in the world now currently malnourished, the need for grains and other basic food crops continues to be critical. Growing crops for fuel squanders land, water and energy resources vital for the production of food for people (Balat, 2011).

3.2.5 Other environmental issues: biodiversity and deforestation

In addition to the not-so-favourable GHG balance of first generation biodiesels there are other environmental issues that need to be taken into account. Water usage (for irrigating crops and evapotranspiration), eutrophication (run off of fertilisers into natural waters) and soil erosion are some of them (Luque *et al.*, 2010). It has been suggested that prolonged dependence on first generation crops for biodiesels (and biofuels more generally) will increase the risk of deforestation. The recent UNEP (2009) report emphasises this risk and points out that two-thirds of the current expansion of palm oil cultivation in Indonesia is based on the conversion of rainforests. If that trend continued, it says, the total rainforest area of Indonesia would be reduced by 29% in the future, compared to 2005 levels.

Recently, Koh (2007) investigated the potential habitat and biodiversity losses that may result from an increase in global biodiesel production capacity to meet future biodiesel demands (an estimated 277 million tonnes per year by 2050). Koh estimated substantial increases in cultivated area for all major biodiesel feedstocks, including soybean in the USA (33.3–45.3 million ha), sunflower seed in Russia (25.7–28.1 million ha), rapeseed in China (10.6–14.3 million ha) and oil palm in Malaysia (0.1–1.8 million ha). Based on land cover data compiled by the FAO, Koh and Wilcove (2008) estimated that between 1990 and 2005, 55–59% of oil palm expansion in Malaysia and at least 56% of that in Indonesia occurred at the expense of forests. Furthermore, the authors reported that the conversion of either primary or secondary (logged) forests to oil palm would result in significant biodiversity losses in the future (Koh and Wilcove, 2008).

As a result, environmentalists have become increasingly concerned about the impact of rapidly expanding feedstock agriculture in the tropics. For example, several non-governmental organisations (NGO) have accused oil palm growers in southeast Asia of destroying large tracts of tropical forests and threatening the survival of many native species, including the orangutan (Koh and Wilcove, 2007). Hence increased biodiesel and other biofuel production could have a negative impact on biodiversity through habitat loss following land conversion, agrochemical pollution and the dispersion of invasive species. The degree of impact depends on the extent of associated land use changes and conversions, as well as the type of biofuel stocks (FAO, 2009). Because palm oil is widely used both as food and fuel, the spread of oil palm agriculture is a particularly worrying threat to tropical biodiversity.

3.2.6 Policy limitations of first generation biodiesels

Biodiesels, and more generally biofuels, have been and are being produced in many countries because together with other policies they offer the potential to tackle one of the most pressing problems of the world today, the issue of climate change. However the ability to tackle this issue through increased production and consumption of biodiesels depends crucially on at least two aspects: whether they are effective – at an acceptable cost – in achieving what they are supposed to achieve and, more importantly, whether the policies are designed and applied suitably to make them effective in achieving such goals. In the previous subsections we have discussed several 'downside' aspects of the production and consumption of biodiesels. Now the question is how far governments' domestic as well as international policies have been successful in combating the above problems? The answer to this question is unfortunately mixed. While governments in many countries have taken several measures to promote the expansion of biodiesel use, the policies have not always worked out well. Furthermore, often the government of an individual country was more concerned with promoting the welfare of its own country and as a result did not take into account the full implication of implementing certain domestic and foreign policies in the context of the whole world.

To illustrate the above points, consider the policies undertaken by the US government to promote biodiesel production and consumption. They have been cited as a 'means to improve farm incomes, reduce tax costs and enhance rural development' (Rajagopal and Zilberman, 2007) with some notable means being biodiesel tax credit, small agri-biodiesel producer credit and renewable diesel tax credit (Congressional Research Report (CRS) 18 March 2008, updated 15 September 2010). However, some of these policies did not work well from the EU perspective when biodiesel trade between EU and USA was concerned. According to the European Biodiesel Board (EBB) press release of 15 July 2009, one of the main reasons why the production levels of biodiesels in 2007 and 2008 was still well behind the EU targets is the 'unfair' international trade competition faced by EU countries. EU biodiesel producers had to compete with heavily subsidised US biodiesels (USB99) in the EU/world market. USB99 has been sold in the EU at much lower prices (even at a price lower than the raw material soybean oil). The situation has seriously affected the profitability of EU biodiesels producers since 2007 and so has acted as a disincentive for further investment. Given the binding target of 10% by 2020 which requires production of at least 30–35 million tonnes of biodiesel by 2020, the European Commission in its 2009 biofuel progress report acknowledged that the current situation in the market has been a deterrent to increased production. To take another example, changes in the domestic policies in Germany, one of the leading biodiesel producers in the EU, to phase out excise tax exemptions for biodiesels owing to the total cost has resulted in several plants closing down, causing a decline in the EU production.

According to a study by OFID (2009), first generation biofuel development as promoted by national policies conflicts with the goals of achieving food security, results in only modest increases in agricultural value added in developing countries, achieves net greenhouse gas savings only after 2030 and creates risks of deforestation and threats to biodiversity.

Prompted by the objective to reduce GHG emissions, global biodiesel trade between several countries has increased significantly since 2005. Yet, the current trade policies have so far been far from 'ideal' in the sense that they have also created significant barriers to trade in terms of (i) being protectionist; (ii) lacking clear classification; as well as (iii) lacking in sustainability criteria (Luque *et al.*, 2010) (also see Section 3.6.4):

(i) Protectionist tendencies: For example, in the EU, with no internationally agreed or even EU-wide agreed criteria for a biodiesel (and biofuel) support programme, each country has set up their own schemes, creating a non-uniform market both internationally and within the EU region. The schemes are geared towards protecting domestic agricultural feedstock and interests. For example, in France, tax exemptions are available only for biofuels (and biodiesels as well) that are produced and sold domestically.

(ii) Lack of clear classification: The lack of clear classification of biofuels within the multilateral trading system is another key factor hindering biodiesel trade. At the moment there is no clear cut agreement whether biofuels belongs to industrial, agricultural, or environmental goods. The lack of such classification has implications for which tariff rate they would then be subjected to.

(iii) Sustainability issues: The lack of coordination between developed (industrialised) and developing countries is being detrimental to the welfare of developing nations as the industrialised countries are not being fully able to grasp the nature of the economic, social, and environmental problems faced by the developing nations.

Given the limitations of first generation oils discussed above, we now turn to examine the potential of the next generation biodiesels and discuss their effectiveness in alleviating some of the problems associated with the first generation biodiesels described above.

3.3 Economic potential and limitations of second and third generation biodiesels

To overcome the dependence of biodiesel production on land conversion of food commodities, second generation biodiesels have been developed to reduce international food price increases and deforestation problems. Energy crops such as jojoba, jatroba oil, waste cooking oil or animal fats provide diverse economic and environmental advantages. The main economic impact is the reduction of the competition between 'land for food' and 'land for energy'. Most second generation biodiesel feedstock is grown on less productive or marginal land which is not suitable for growing food crops. This reduces management costs and increases actual net returns which translate into a sustainable agricultural income to promote development paths particularly in developing countries. Nonetheless, second generation biodiesel suffers for being not abundant enough to satisfy the world biodiesel demand entirely. Third generation biodiesels have the advantage of performing with higher efficiency rates for biomass production compared to conventional biodiesel practices. This contributes to higher yields for the agriculture sector and reduces the controversy of 'land for food' versus 'land for energy'. For these reasons, continuous research and development technologies are moving fast to implement advanced biodiesel production. High costs remain the major issue for the commercialisation of second and third generation biodiesels despite the high potential in terms of carbon and energy savings. This section discusses the main advantages and limitations of current experimental biodiesels such as biodiesels from algae, wood biodiesels and biodiesels from methanol.

3.3.1 The case of algae biodiesels

Micro-algae production as a feedstock for biodiesels has grown fast in the current debate about alternative renewable energy techniques. The main argument in favour of adopting micro-algae production is its relatively high reproduction rates in wastewater and its use as a feedstock for hydrogen, methane and bioethanols other than biodiesels (Ahmad *et al.*, 2011).

Biodiesel produced from micro-algae has the potential to increase yields by 50–100 times more than those of current biodiesels obtained from soybeans or related feedstocks. This high level of production not only has a small impact on food prices, but it is also sustainable in terms of recycling CO_2 emissions from other plants in the production process. Nonetheless, when micro-algae transform into energy for oil use they reduce their reproduction rates, affecting the performance of overall biodiesel production. For this reason, international attention is currently being devoted to improving the trade-off between micro-algae as feedstock and biodiesel productivity. One of the main steps of current research developments (Brennan and Owende, 2010) is to choose proper micro-algae species for the lipid content of the output. Table 3.3 shows five different algae species including biomass productivity, lipid content and lipid productivity. In terms of lipid productivity, the highest value of 61.0 mg l^{-1} day^{-1} is that of *Nannochloropsis* sp. F&M-M26 (Rodolfi *et al.*, 2009).

Table 3.3 Lipid content and productivity of 30 micro-algae strains cultivated in 250 ml flacks

Algal group	Microalgae species	Habitat	Biomass productivity (g l^{-1} day^{-1})	Lipid content (% biomass)	Lipid productivity (mg l^{-1} day^{-1})
Diatoms	*Chaetoceros muelleri* F&M-M43	Marine	0.07	33.6	21.8
	Chaetoceros calcitrans CS 178	Marine	0.04	39.8	17.6
	P. tricornutum F&M-M40	Marine	0.24	18.7	44.8
	Skeletonema costatum CS 181	Marine	0.08	21.0	17.4
	Skeletonema sp. CS 252	Marine	0.09	31.8	27.3
	Thalassiosira pseudonana CS 173	Marine	0.08	20.6	17.4
	Chlorella sp. F&M-M48	Freshwater	0.23	18.7	42.1

(Continued)

Table 3.3 Continued

Algal group	Microalgae species	Habitat	Biomass productivity (g l^{-1} day^{-1})	Lipid content (% biomass)	Lipid productivity (mg l^{-1} day^{-1})
	Chlorella sorokiniana IAM–212	Freshwater	0.23	19.3	44.7
	Chlorella vulgaris CCAP 211/11b	Freshwater	0.17	19.2	32.6
	C. vulgaris F&M-M49	Freshwater	0.20	18.4	36.9
Green algae	*Chlorococcum* sp. UMACC 112	Freshwater	0.28	19.3	53.7
	Scenedesmus quadricauda	Freshwater	0.19	18.4	35.1
	Scenedesmus F&M-M19	Freshwater	0.21	19.6	40.8
	Scenedesmus sp. DM	Freshwater	0.26	21.1	53.9
	Tetraselmis suecica F&M-M33	Marine	0.32	8.5	27.0
	Tetraselmis sp. F&M-M34	Marine	0.30	14.7	43.4
	T. suecica F&M-M35	Marine	0.28	12.9	36.4
	Ellipsoidion sp. F&M-M31	Marine	0.17	27.4	47.3
	Monodus subterraneus UTEX 151	Freshwater	0.19	16.1	30.4
	Nannochloropsis sp. CS 246	Marine	0.17	29.2	49.7
Eustigmatophytes	*Nannochloropsis* sp. F&M-M26	Marine	0.21	29.6	61.0
	Nannochloropsis sp. F&M-M27	Marine	0.20	24.4	48.2
	Nannochloropsis sp. F&M-M24	Marine	0.18	30.9	54.8
	Nannochloropsis sp. F&M-M29	Marine	0.17	21.6	37.6
	Nannochloropsis sp. F&M-M28	Marine	0.17	35.7	60.9
	Isochrysis sp. (T-ISO) CS 177	Marine	0.17	22.4	37.7
	Isochrysis sp. F&M-M37	Marine	0.14	27.4	37.8
Prymnesiophytes	*Pavlova salina* CS 49	Marine	0.16	30.9	49.4
	Pavlova lutheri CS 182	Marine	0.14	35.5	50.2
Red algae	*Porphyridium cruentum*	Marine	0.37	9.5	34.8

Source: Rodolfi *et al.* (2009).

Second, researchers are interested in processing biodiesels from micro-algae, in essence choosing the best option for oil extraction. Four methods are currently used in the processing phase: press, extraction through chemical solvents, transesterification, and supercritical fluid extraction. Details of processing techniques are not expressed here because it is beyond the scope of the present chapter. However, the optimal combination in terms of costs/biodiesel output is to process biodiesels with the use of solvents (e.g. hexane) with a relatively small extraction time. Overall, the average extraction costs in 2009 dollars based on a study conducted by Trostle (2008) gives a productivity of 100 mt ha^{-1} yr^{-1}, a lipid concentration of 35% by weight, a biodiesels yield of 10421 per gallon per ha; capital costs of US$112400 per ha and operating costs of US$39000 per ha.

Biodiesel from micro-algae is also attractive owing to a series of other applications and products deriving from the micro-algae production chain. When combined with wastewater and biological gas treatments (CO_2 recycling), these can be sustainable for the environment, cost-effective, and largely profitable. Biological (flue gas) treatments (compared to conventional chemical reaction CO_2 treatments) have the potential to reduce CO_2 emissions and serve, at the same time, as a source of micro-algae cultivation (up to 15% CO_2 mitigation) (Brennan and Owende, 2010). Zeiler *et al.* (1995), argue that when a biological treatment takes place to produce biomass, micro-algae absorb CO_2 as well as other pollutants (e.g. SO_2 and NO_x) in their reproduction process and contribute to considerably lower CO_2 emissions in the atmosphere. Wastewater treatments seem to be promising for micro-algae reproduction owing to the presence of organic compounds in some industrial wastewater which would enhance the growth of micro-algae. This effect, combined with the absorption of nitrogenous waste in wastewater, helps to reduce eutrophication and protect the biodiversity of fish populations in aquacultures. Finally, micro-algae production is also a good source for human health owing to its use in probiotic supplements in various dietary aids which help protect against a wide range of diseases.

Despite some positive impacts of biodiesels from micro-algae, a number of uncertainties arise for the economic analysis of estimated costs and economic viability. These uncertainties can be broadly classified into uncertainties arising from the harvesting process, market prices and estimation of biomass yield. The cost of harvesting relates to the complexity of selecting the optimal mix of micro-algae nutritional and physiological components which serve to decide the harvesting procedure adopted at the second stage. Concern about market prices mainly arises from accepting a worldwide price for biodiesels from micro-algae and possibly increasing, through legislation, the minimum blending content in actual fuels, see Section 3.7. This also depends on how fast R&D moves from experimentation to commercialisation of biodiesels products. Finally, the third uncertainty resides in estimating the biomass yield. If biomass yields vary too much, this undermines predictions for adopting optimal and robust biodiesels economic models.

3.3.2 The case of wood diesels

Lignocellulosic ethanol (bioethanol) is one of the most important sources of advanced biofuels owing to its promising feedstock availability and low production costs. Global bioethanol production accounts for 94% of global biofuel supply. Brazil and the USA, as depicted in Table 3.4, are current world leaders owing to the availability of corn crops or sugar cane, which serve currently as a major feedstock in both countries.

According to the Renewable Fuels Association (2007), bioethanol use has significantly increased in the USA from 3.4 to about 4.8 billion gallons in the period 2004–2006. In Brazil, bioethanol production is mainly aimed at satisfying external demand, mostly from the USA or European countries. Current production levels are approximately to 4.5 billion gallons with the potential to replace almost one-third of global gasoline (Balat *et al.*, 2008). Current bioethanol world production is in the region of 0.1% (OECD/IEA *et al.*, 2008) and this value is still low compared to first generation bioethanol as shown in Fig. 3.4.

The assessment of bioethanol feedstock is an essential factor in improving large-scale commercial production. Substantial problems exist for land availability since the current feedstock used to produce bioethanol is essentially energy crops (sugar cane and sugar beet). Several conflicts arise in the production of bioethanol owing to increasing competition of land resources for food crops. More than 60% of world bioethanol production is obtained from energy crops. Bioethanol production from sugar cane in Brazil is relatively cheap owing to the governmental support in terms of blending mandates with gasoline which lowered sugar cane costs and created additional demand for bioethanol. In Europe, bioethanol is obtained from sugar beet crops which are grown by all Member States owing to their high adaptability to various climatic conditions, low water requirements (35–45% less compared to sugar cane) and high yield per ha (100 l bioethanol/tonne sugar beet) (Eguídanos *et al.*, 2002).

Table 3.4 Top ten bioethanol producers (billion gallons)

Country	2004	2005	2006
USA	3.54	4.26	4.85
Brazil	3.99	4.23	4.49
China	0.96	1.00	1.02
India	0.46	0.45	0.50
France	0.22	0.24	0.25
Germany	0.07	0.11	0.20
Russia	0.20	0.20	0.17
Canada	0.06	0.06	0.15
South Africa	0.11	0.10	0.10
Thailand	0.07	0.08	0.09

Source: Balat *et al.* (2008).

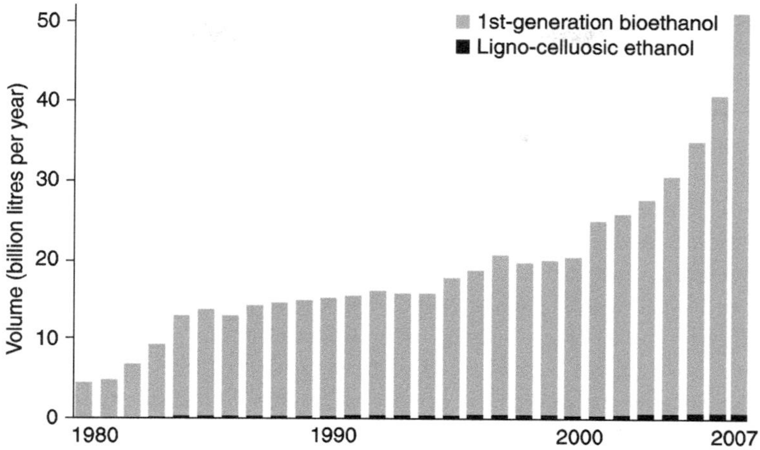

3.4 Comparison between first and second generation world bioethanol production (source: OECD/IEA *et al.*, 2008).

Lignocellulosic biomass such as wheat straw and wood is a viable alternative to energy crops feedstock. This biomass could produce up to 400 billion litres per year of bioethanol (Bohlmann, 2006) with a 16-fold growth compared to bioethanol obtained from conventional feedstock. Wood perennial crops are a highly promising feedstock for enhancing second generation biodiesel production owing to the high potential yield, adaptability to marginal lands and low environmental impact. The costs of biodiesel from lignocellulosic feedstock are still relatively too high and development of new technology is needed in view of the large-scale commercialisation of biodiesels from wood residues. Generally, 60% of total bioethanol costs derive from feedstock. This actually varies between US$22 and US$61 per tonne dry matter (Balat *et al.*, 2008) and contributes largely to total production costs.

3.3.3 The case of biodiesel from methanol

As a result of the recent oil crisis and economic recession, interest in biodiesel from methanol has notably increased owing to its carbon neutral characteristics. Biodiesel from methanol is made by energy crops and renewable resources and can be produced in existing gasoline infrastructures (reducing operative costs). A further advantage is that in addition to being an exceptional fuel for fire engines it can also be used to run cell fuel vehicles (CFV) and its production does not need to use extensive areas of land (Dekker and Lanting, 2009). The production process uses glycerine, a by-product of industrial processes, which is purified and used in fermentation to produce biogas to generate electricity and reduce carbon emissions. In Europe, biodiesel from methanol production was early experimented in Sweden and France (Faaji, 2006) and later adopted in Germany. Vogt *et al.* (2008) analysed the viability of biodiesels from methanol production in the USA (particularly in the western states) from landfill, agricultural waste and forest biomass. These authors estimated the share of energy demand that could be met by using biomethanol production (and the resulting carbon emission reduction) analysing various renewables and non renewables options in current consumption of gasoline and electricity. Results indicated the existence of high potential of biomethanol production from biomass from all sources. The amount of biomethanol that can be replaced in gasoline consumption is notable: it ranges from 36–102% in the worst case scenario and from 72–204% in the best case scenario. The electricity savings are, however, not as high as gasoline savings. They range from 12–25% in the best case scenario. To manage biodiesels from methanol at best it would be necessary to consider all feedstock sources. However, the US policy favours agricultural energy crops rather than forests or waste resources as feedstock in second generation biodiesel production. To broaden the possibility of biodiesel production, the USA should strongly consider incentives for biomass conversion rather than using energy crops.

3.4 New impacts on energy security

Recent world scenarios delineated in the latest World Energy Outlook (OECD/IEA, 2010a) outline how energy supply and demand vary considerably depending on the way energy strategies are adopted by current governments.

Figure 3.5 illustrates the above mentioned scenarios. These can be summarised into current policies, New Policy Scenarios which account for both actual and immediate future policies such as the new EU and US energy policies and the 450 scenario which means keeping world carbon emissions to 450 ppm and global temperature to 2°C. As it can be noted, if there is no change in the current *status quo* energy demand would rise to 18 000 Mtoe (million tonnes of oil equivalent)

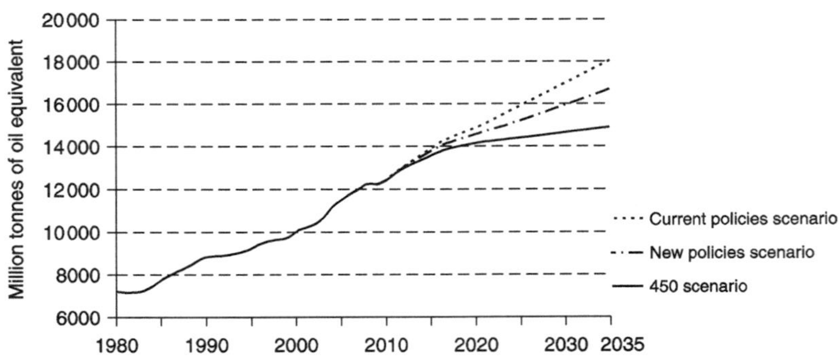

3.5 World energy demand by scenario (source: OECD/IEA, 2010a).

with dramatic consequences for the world energy supply. The implementation of the New Policy Scenario would decrease world energy demand to just above 16 000 Mtoe while the correct establishment and effectiveness of climate change policies would bring global energy demand to just above 14 000 Mtoe.

How can new future generation biodiesels contribute to change the *status quo* and be a potential source for global energy supply? On the global scale, the main reason to engage in next generation biodiesel production is to ensure energy security and reduce dependence on oil rich fossil fuels. There is no doubt that over the next years, fossil fuel-based energy inputs will still dominate world energy supplies. Given the embodied resource scarcity of fossil fuels, production costs will tend to rise and be subject to market price volatility, particularly in the short-term. This will pose serious threats to the security of energy supplies.

Great steps forward have been currently made in alternative energy sources such as biodiesel production. This is therefore expected to play a great role in future energy supply. To analyse the potential role of next generation biodiesels it is crucial to identify the contribution and determine the factors for development of current and potential biodiesels generation in future energy production. The main factors can be summarised in terms of land availability, biodiversity issues and harvesting rotations including marginal lands, carbon balances, and cost-effective opportunities between current and future biodiesels technologies (OECD/IEA, 2010a). The issue of land availability, which can also be expressed in terms of 'bioenergy potential', includes different land availability for growing feedstock such as energy crops, forest residues and agricultural waste. Under an optimistic scenario, Smeets *et al.* (2007) estimate the possibility for current agricultural production to increase the share of land available for second generation energy crops. The potential availability of next generation biodiesels (and biomass) could meet between 6 and 300% of energy demand. This wide range is owing to uncertainties in the time needed for experimentation and commercialisation of next generation biodiesel products given current and future technologies. A cost-effective analysis could also determine the demand for available biodiesels (and other biofuels) resources in future scenarios. The end-user price from marginal lands or forest residues can vary between US\$0.67–0.94 per l_{ge} and decrease to US\$0.46–0.71 per l_{ge} if oil prices double to US\$120/bbl (l_{ge} refers to 'litre of gasoline equivalent' (OECD/IEA, 2010b)) (Smeets *et al.*, 2007).

Various policies across the globe aim to promote biodiesel as a secure energy source. In the USA, the Energy Policy Act of 2005 and the Energy Independence and Security Act of 2007 clearly address the question that energy security is the major reason for the promotion of biodiesels. Likewise, the EU Renewable Energy Directive 2009/28/EC highlights the importance of energy diversification to secure energy supply in the EU.

Table 3.5 shows past, current, and future scenarios of world energy requirements. The majority of non-OECD countries would more than double global energy demand by 2035 compared to OECD regions. Nonetheless this trend should be

Table 3.5 World Energy demand by fuels and scenario in million tonnes of oil equivalent (Mtoe)

	1980 (Mtoe)	2000 (Mtoe)	2008 (Mtoe)	2015 (Mtoe)	2020 (Mtoe)	2030 (Mtoe)	2035 (Mtoe)	2008–2035 (%)
OECD	**4050**	**5233**	**5421**	**5468**	**5516**	**5578**	**5594**	**0.1**
North America	2092	2670	2731	2759	2789	2836	2846	0.2
United States	*1802*	*2270*	*2281*	*2280*	*2290*	*2288*	*2272*	*−0.0*
Europe	1493	1734	1820	1802	1813	1826	1843	0.0
Pacific	464	829	870	908	914	916	905	0.1
Japan	*345*	*519*	*496*	*495*	*491*	*482*	*470*	*−0.2*
NON-OECD	**3003**	**4531**	**6516**	**7952**	**8660**	**10002**	**10690**	**1.9**
E. Europe/Eurasia	1242	1019	1151	1207	1254	1344	1386	0.7
Caspian	*n.a*	*128*	*169*	*205*	*220*	*241*	*247*	*1.4*
Russia	*n.a*	*620*	*688*	*710*	*735*	*781*	*805*	*0.6*
Asia	1067	2172	3545	4609	5104	6038	6540	2.3
China	*603*	*1107*	*2131*	*2887*	*3159*	*3568*	*3737*	*2.1*
India	*208*	*459*	*620*	*778*	*904*	*1204*	*1405*	*3.1*
Middle East	128	381	596	735	798	940	1006	2.0
Africa	274	502	655	735	781	868	904	1.2
Latin America	292	456	569	667	723	812	855	1.5
Brazil	*114*	*185*	*245*	*301*	*336*	*386*	*411*	*1.9*
World	**7229**	**1 0031**	**12 271**	**13 776**	**14 556**	**16 014**	**16 748**	**1.2**
European Union	*n.a*	*1682*	*1749*	*1722*	*1723*	*1719*	*1732*	*−0.0*

Source: OECD/IEA (2010a).

considered with caution given the uncertainties of world population growth, industrial production rates particularly in developing countries and technology advances. In the case that current EU and USA energy policies successfully reach their targets, these will contribute to a small annual growth of 0–0.02% in global energy demand. On the other hand, emerging economies such as China, Latin America, India, Brazil and sub-Saharan countries will suffer bigger annual increases ranging from 1.5–3.1% as shown in Table 3.5.

Therefore it is not an easy task to evaluate how much biodiesel will be necessary to meet future world energy requirements. It would be a wise recommendation to continue to invest in R&D for the next generation biodiesels such that the developments made on the production side 'must be based on differences in feedstock resources and energy consumption mix in different countries' (Lin *et al.*, 2011, page 1026).

3.5 New impacts on land and food safety

Implementation of second and third generation biodiesels will be sustainable when feedstock is cultivated in an environmentally friendly way to avoid subtracting land for food crops. Most of new generation biodiesels crops can be grown in unfertile land and guarantee a final biodiesel yield higher than that obtained from first generation biodiesels. Given the as yet not fully established commercialisation of second generation biodiesels, food crops for biodiesels will still remain a problem in the immediate future. Some crops such as sugar cane or palms are being used as feedstock for biodiesel production. This is causing cereal prices to increase and become a strong signal of land scarcity which mostly affects developing countries. Consequently, it would be advisable to ensure a clear designation of world land surface for food in relation to biodiesel crops (Escobar *et al.*, 2009). Alternatively, most wood biodiesel feedstock (jatroba and/or switchgrass) can grow on wasteland using low pollution impact fertilisers (Groom *et al.*, 2008) as well as micro-algae biodiesels that can grow in wastewater treatments. Jatroba or switchgrass, for example, are perpetual crops that can guarantee a biological harvest for more than 30 years; they can adapt to growing in various environmental conditions and produce biodiesel yields for a long time span.

Advances in new technologies have been providing the possibility to produce biodiesels from inedible biomass feedstock. This could ensure substitutability with conventional fossil fuels and contribute, at the same time, to a larger energy supply which would also have positive effects on job creation and the economy of less developed areas. By 2030, ethanol yields are estimated to produce 400 litres per day compared to the actual 270 litres per day with current technology (Larson, 2008). The International Energy Agency (OECD/IEA, 2010b) argues for a substantial role for biomass production over the next few years and the consequent potential for land availability. Notable advances are possible for grazing lands, where the implementation of second generation biodiesels would leave these lands free from competition with first generation biodiesels feedstock. To ensure success, substantial planning (in particular in developing countries) should be carried out for designating infertile lands for the cultivation of biodiesel feedstock. Figure 3.6 illustrates the state of world arable with respect to infertile land.

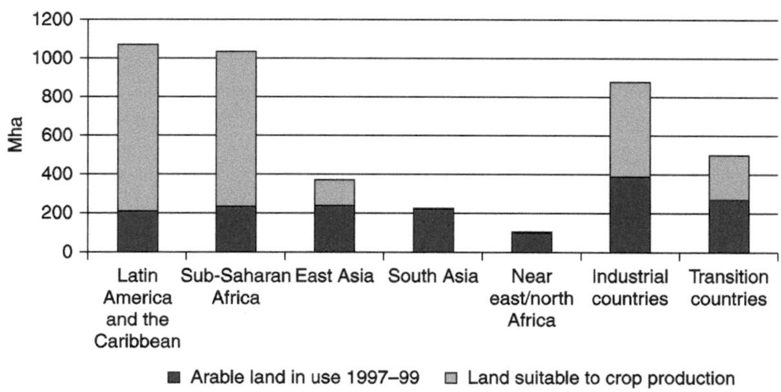

3.6 State of global arable versus infertile land (source: OECD/IEA, 2010b and FAO, 2003).

In Fig. 3.6, only 250–800 Mha are estimated to be available worldwide, the remaining being covered by forests and/or wetlands. Estimates show that Latin American and sub-Saharan countries could play a fundamental role in the production of feedstock for second generation biodiesels crops. Various economic models are used to simulate future scenarios of land availability. A study conducted by the International Energy Agency (OECD/IEA, 2006) assumes world availability of infertile land of 1.7 gigahectares (Gha) and agricultural land of 4 Gha. This is a rather promising scenario compared to that designed by Hoogwijk *et al.* (2003) where no agricultural land would be available by 2050. There is no doubt that uncertainties about commercialisation of advanced biodiesels largely affect the results obtained in these macro-economic models.

A common factor in the development of next generation biodiesels is the use of wasteland or wastewater to grow potential feedstock. Forest residues could contribute to expected increases in wood biodiesel demand through simulated bio-economic models. Simulation analysis varies from considering an optimistic scenario of full availability of forest biomass and residues to a less optimistic situation where forests and residues are not fully available but forests are managed in a sustainable way. Results for forest biomass can be in the range of 10–16 Exajoules (EJ) or 43 EJ in the more optimistic scenario (VTT, 2007).

In the European Union, Smeets *et al.* (2007) estimated that under an optimistic scenario a surplus of 129–592 Mha of marginal land would have an impact on potential biodiesels production of between 100 EJ and 303 EJ. In particular, new Member States and other former Soviet Union countries could have a great potential to serve as European suppliers of next generation biodiesels. This result also finds validation in a macro-economic computable general equilibrium (CGE) model for biorefineries in the EU carried out for the recent completed Seventh Research Framework Programme research project SUSTOIL (see http://www.york.ac.uk/res/sustoil/) (De Lucia and Bartlett, 2011).

For the USA and Canada, Smeets *et al.* (2007) estimate a surplus of agricultural land availability between 54 Mha (pessimistic scenario) and 348 Mha (optimistic scenario) depending on the scenario of land availability hypothesised. The potential for advanced biodiesel production ranges from 39–204 EJ in 2050 where 11 EJ comes from wasteland and 6 EJ from forest residues.

Sub-Saharan countries could provide a vast potential for biodiesels feedstock cultivation over the next 40 years if large scale commercialisation of advanced biodiesels takes place and logistics solutions are available. In particular, Smeets *et al.* (2007) suggest that there may be a land surplus of 104–717 Mha and the production of biodiesels from forest residues or wasteland of 16–21 EJ. Sub-Saharan countries could therefore provide a considerable share in world second generation biodiesels production if the above conditions are met.

In Asian countries, China leads potential future production of biodiesels from marginal agricultural and forest lands in the region of 10 EJ. This would imply expansion of the agricultural sector without compromising food scarcity. Less optimistic is the situation regarding expected expansion of advanced biodiesels cultivation in developing countries. Unless foreign investments or carbon offset projects for biofuels production are implemented, it is unlikely that developing countries will set up feedstock crops for second generation biodiesels production. Instead, considering the growing population and scarce agricultural land availability, land use would satisfy food demand rather than energy purposes. In Smeet *et al.* (2007), future development in the 'land for food' versus the 'land for energy' debate also considers that the potential increase in agricultural yield would result in shifts from land use for food to land use for biodiesel crops. Assuming the occurrence of improved economic development and foreign capital aid in developing countries, the establishment of sustainable farming practices could better allocate crop areas and enable sufficient land to be assigned to the production of advanced biodiesels concurrently ensuring food safety.

The beneficial effects of potential wood feedstock cultivations can also provide higher yields using lower energy input costs compared to first generation biodiesels crops. Opportunities for improving feedstocks for second and third generation biodiesels lie in the advantage that perennial crops can grow in a multitude of climatic conditions and the propensity to recycle nutrients increases annual feedstock growth and oil yield. This has a positive impact on estimates of land availability in current bio-economic models (Murphy *et al.*, 2011) providing a new direction to narrow the gap between food and biodiesel crop conflicts and has a potentially positive impact on GHG savings as well as re-establishing traditional agricultural land use for food production.

The issue of using marginal lands for second generation biodiesel production is not without limitations. In theory, the remaining available land used for food crops could be used for energy crop plantations for advanced biofuels feedstock. Biomass production requires a large amount of lignocellulosic material and with present technology the integration between food and energy systems appears not yet to be an economically viable solution (OECD/IEA, 2010b). Furthermore, because of the inaccuracy of land use data in developing countries, further research is needed to indentify marginal land for sustainable energy crops. Estimates and expectations about available marginal land should be taken with caution to avoid biased results in bio-economic models and raising conflicts with food safety. The issue of food security in developing countries should be considered separately for each country because of the complexity caused by different resource allocation, poverty status, availability of infrastructures, economic growth, income distribution, market prices of agricultural products and food consumption patterns.

3.6 Effects on international trade and sustainable development

3.6.1 Economic theory of international trade and economic growth

The positive effects of international trade and economic growth were first pointed out by Adam Smith in 1776 (Smith, 1776), subsequently leading other economists like Ricardo (1817) and Schumpeter (1912, 1942) to develop the idea further by providing a more concrete theory based on ideas of comparative advantage and innovation-led growth, respectively. The basic idea is that international trade is essential for the prosperity and economic growth of trading nations as every country lacks some vital resources and, given a country's specific geographic and economic conditions, a nation can be efficient in producing some goods while inefficient in others. Specialisation can then permit production of larger outputs and offer economies of large scale production. According to the Law of Comparative Advantage, gain from trade is always possible as long as countries specialise in production of the goods in which they have a *relative* advantage. Free international trade then leads to economic growth by expanding the possible production frontiers of all countries concerned. Hence economists argue that there is a link between the liberalisation of trade and economic growth (Krueger, 1997; Frankel and Romer, 1999). Lucas (1988) in his model 'learning by doing and comparative advantage' pioneered the idea that in international trade, each country should specialise in the good for which the autarky donation of human capital presents a comparative advantage. This learning and specialisation then gives rise to different rates of endogenous growth in different countries.

A second generation of models (Romer, 1990, 1993; Grossman and Helpman, 1990, 1991a, 1991b; and Aghion and Howitt, 1992) considered innovations to be the key factor for endogenous growth: innovations are the result of explicit R&D activities and hence the result of R&D becomes the main driver of endogenous growth. These authors stress that because international trade is necessary for the diffusion of knowledge, open economies have better access to a wide base of technological knowledge which would lower their costs of product development and that stronger competition would foster creativity, innovation, and the exploration of economies of scale. In particular, Romer (1993) stresses that technology is the driving force for economic growth and advises that in particular the less developed countries need to be able to access openly such fruits of innovations through foreign investment and technology transfer.

In the light of the above introduction, let us now examine how successful international trade in biofuels, and more specifically in biodiesels, has been in promoting economic growth for all trading nations.

3.6.2 Current biodiesel trade scenario

Global biodiesel trade has increased strongly since 2005 with the main exporters being USA, Argentina, Indonesia and Malaysia, while the main importer is the EU. Around 2.8 Mt of bioethanol and 2.9 Mt of biodiesel were traded globally in 2008, in addition to approximately 4 Mt of wood pellets (Junginger *et al.*, 2010). In the USA, while imports of biodiesel increased from less than 130 000 tonnes in 2005 to more than 200 000 tonnes in 2007, exports increased even more, from less than 130 000 tonnes in 2006 to more than 1.25 million tonnes in the first half of 2008 alone, making USA a net exporter. More than 95% of global exports in 2008 were directed towards the EU (Carriquiry and Babcock, 2008). The EU had the world's most developed biodiesels industry in 2008 with its production increasing by 6%. However, its production declined by 7% in 2009 because of strong competition from abroad. At the same time, US gross exports increased from negligible levels in 2005 to about 1.4 million tonnes in 2008, compared to net export of about 1.175 million tonnes (Junginger *et al.*, 2010). It has been suggested that such strong increases in both exports and imports have been caused by the rising trend in trade of liquid biofuels as well as (moderate) increases in trade of solid biomass.

3.6.3 Current impediments to international trade and the need for trade liberalisation

Currently, impediments to biodiesel trade primarily include high import tariffs and technical barriers to trade. In the context of EU–USA trade, whilst the USA has emerged as the major biodiesels exporter to the EU (with more than 80% of market share among all exporters) supplying about 19% of the EU domestic market for biodiesels, the European biodiesels industry has suffered from biodiesel imports from the United States over the last few years. Because of the tax credit given to the US blenders and the 'splash and dash' policy (Carriquiry and Babcock, 2008), the EU initiated anti-dumping measures and countervailing duties in March 2009 which reduced US biodiesel exports to the EU to negligible quantities. Allegedly, some of these US exports have now been replaced partially by exports from Argentina (Argentine exports to the EU increased strongly from 70 000 tonnes in 2008 to an estimated 1 million metric tonnes in 2009 (European Biodiesel Board, 2009)), Indonesia, Malaysia, as well as growing trade flows from Canada (Al-Riffai *et al.*, 2010).

Additionally, one of the greatest technical barriers in the coming years could be certification of biodiesel for environmental sustainability, prompted by concerns about burning and clearing of rainforests to plant palm and soybeans (both of which are feedstock for biodiesels) in southeast Asia and Latin America. While such technical barriers may be justified on normative grounds and in some cases may even be welfare enhancing, it could potentially reduce the volume of trade and could in principle be contrary to the Renewable Energy Directive's (European Commission, 2009) objective which states that it aims to meet European biofuels targets using a combination of domestic production and imports of biofuels and to this end, will propose 'relevant measures to achieve a balanced approach between domestic production and imports, taking into account, *inter alia*, the development of multilateral and bilateral trade negotiations, environmental, social and economic considerations and the security of energy supply'.

In order to promote growth in trade while at the same time paying attention to the issues related to environmental sustainability, at least two measures of trade policies can be recommended: the trade liberalisation policy and the *joint* international trade agreement policy. The former will foster growth in trade volumes by reducing trade barriers such as high import and export tariffs, while the latter will promote more harmonised trade policies that will enable the developing countries to produce biodiesels for export and allow developed countries to meet their bio-energy (and renewable energy) targets without sacrificing sustainability issues (Junginger *et al.*, 2010). However, to promote economic growth, a proper balance between these policies is needed as trade liberalisation policy can reduce domestic prices in the previously protected market implying an increase in demand which can then push up the prices. The net effect will therefore depend upon the strength of liberalisation policy. At the same time, conflict of interest between countries may arise if some of the domestic production is replaced by imports. Therefore in order to reap true benefits from international trade, respective governments need to pay attention not just to domestic issues but also to global welfare. In other words, nations will need to adopt a more long-term perspective about how to bring about sustainable development.

3.6.4 Trade in second generation biodiesels and biofuels as an important driver for economic development

Next generation biofuels, though still far from commercially viable, can open up many new opportunities because they can be sourced from a much wider variety of non-edible feedstock, thereby limiting the direct 'food versus fuel' competition associated with most first generation biofuels. Furthermore, second generation biofuels are supposed to have much better GHG reduction potential. While the full environmental impact of second generation biofuel production depends upon the conversion route as well as the feedstock and site-specific conditions (e.g. climate, soil type, crop management etc), current values indicate a potential of a minimum of 60% reduction in CO_2 emission which is undoubtedly better than most other first generation type biofuels, according to the report *Sustainable Production of Second Generation Biofuels* (OECD/IEA, 2010b). (In fact, IEA values indicate a GHG mitigation potential of 60–120%. However these values do not include the impact of land-use change.) Given that the Energy Technology Perspective (ETP) 2008 'ACT Scenarios' show how global CO_2 emissions could be brought back to current levels by 2050, and ETP 2008 'BLUE Scenarios' targets a 50% reduction in CO_2 emissions by 2050 (OECD/IEA, 2008), second generation biodiesels and hence biofuels will no doubt have to play a major role in meeting these targets.

In order to implement the above objectives, it is of the utmost importance to be able to design appropriate trade policies that will benefit both exporting and importing countries equally. In contrast to the first generation biodiesels where the use of feedstock and the corresponding high feedstock price constitute major components of overall production and implementation costs, this is not a major issue in the context of second generation biodiesels. With second generation biodiesels (FT-diesel), the greatest cost component is processing and not feedstock. The greatest likelihood of achieving meaningful cost reductions is therefore through the learning curve on processing costs, mostly being available in a developed country. Figure 3.7 shows how first and second generation biodiesels differ from one another with respect to their feedstock component in the present and in the future.

At present, the major input suppliers of second and third generation biodiesels are the developing countries such as Southeast Asian, Latin American and African countries. Therefore within the current constraints, optimum trade policies should be classified according to the objective of promoting economic growth in all countries concerned as follows:

- Short to medium term: According to the *theory of comparative advantage* in the economics of international trade, optimal trade policies in the short to medium term should include import of biomass from developing countries to the developed countries who, having access to the advanced technological

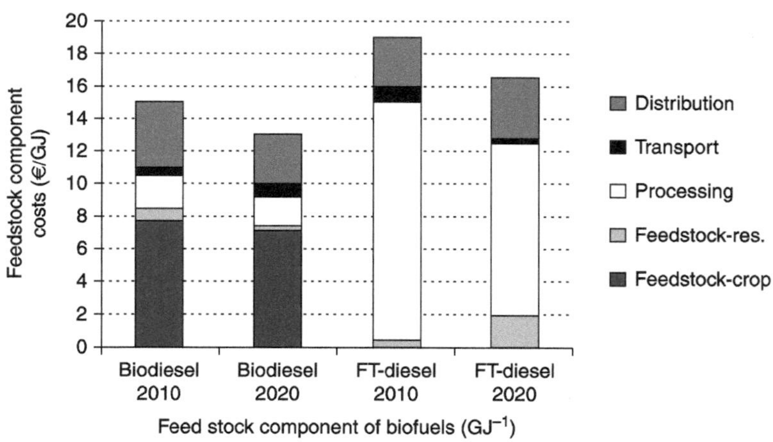

3.7 Feedstock component of biofuels (€ GJ^{-1}) (REFUEL) (source: Bradley *et al.*, 2009).

know-how, should be able to process and convert the raw materials to next generation biodiesels in the most cost effective way. This would simply imply that the price of (second and third generation) biodiesels does not become excessively (and unnecessarily) high. In order to facilitate this, trade restrictions need to be kept as minimum as possible.
- Long run: More substantial cost reduction enabling the biodiesels price to be kept down *globally* can be achieved by technology transfer from the developed countries to the developing countries who not only would have the necessary supply of raw materials but can also then learn to produce the final goods themselves. This will be feasible only through suitable trade policies that should not only encourage free market for biodiesels trades (so that opportunistic behaviour can be avoided) but also cooperative joint ventures between partnering countries.

One of the main lessons emerging from the above is that, even though there are positive impacts of international trade, it is important to recognise that international trade alone cannot bring about economic growth and prosperity in any country. There are many other factors like flexible trade policies, a favourable macroeconomic scenario and political stability that need to be in place in order to complement the gains from trade. It can therefore be said that international trade leads to economic growth provided the policy measures and economic infrastructure are accommodating enough to cope with the changes in the social and financial scenario that result from it.

3.7 What are the right policies for next generation biodiesels?

The increasing development of worldwide biodiesel production has often been the consequence of support policies (De Lucia, 2010) either in developed or in developing countries owing to the competitiveness of biodiesels compared to conventional fossil fuels. Blending mandates are undoubtedly a conventional measure ensuring acceptability of biodiesels in current production. According to the latest energy security legislation (Energy Independence and Security Act of 2007), the USA is the only country to adopt a blending policy for second generation biodiesels. This is mostly based on lignocellulosic feedstock from 2010 to 2022. The Renewable Fuels Standards requires the use of wood feedstock up to 60.6 billion l^{-1} yr^{-1} to ensure a reduction of 100 million tonnes of CO_2 per year by 2022 (OECD/IEA, 2010b).

The European Union does not have yet a blending mandate for second generation biodiesels. The Renewable Energy Directive (Directive 2009/28/EC) defines sustainability standards which require savings in GHG for biofuels of at least 35% compared to conventional fossil fuels from 2013, increasing to 60% from 2018. Directive 2009/28/EC also determines that biodiesel feedstock should be produced outside protective areas to strengthen biodiversity and reinforce soil productivity. Other countries, such as China, India or South Africa, have followed the example of the European Union setting minimum requirements for GHG savings or sustainability in support of enhancing the adoption of and boosting demand for advanced biodiesels.

Most of these support policies affect both supply and use of biodiesels. In the USA, the International Energy Agency (OECD/IEA, 2010b) forecasts that domestic supply in 2012 will be able to reach blending targets for second generation biodiesels. As a consequence, trade in biodiesels will play an important role. Imports (primarily from Brazil and other Latin American countries) should narrow the excess in domestic demand for biodiesels by 2020 to meet blending requirements. To serve as second generation biodiesel (mostly cellulosic ethanol) suppliers to the USA is an important opportunity for developing countries to promote a sustainable growth. Likewise, the European Union has only met 3% of its blending targets and it is expected that an increase of 10% in total renewable energy in the transport sector by 2020 would affect the production of second generation biodiesels.

Furthermore, second generation biodiesels and conventional fossil fuel prices will also be influential in determining the optimal amount of imported and domestic biodiesel feedstock. The dependence of both the European Union and the USA on these imported biodiesel feedstocks and products over the next few years will also affect international trade. Tariffs and quotas on biodiesels commodities to protect domestic markets will reduce competitiveness with foreign markets. These should be set such that they do not distort the export oriented policies of developing countries and prevent growth mechanisms from taking place. The European Union already favours, through tax-free policies, the access of ethanol and biodiesel imports in Member States. The USA, on the other hand, imposes a biodiesel commodity duty of US$0.14 per litre on imported ethanol or biodiesels products. All exporting countries should comply with sustainability and/or GHG savings criteria established by the Directive 2009/28/EC or the Renewable Fuels Standards. Based on this legislation, emerging exporting countries have the possibility to investigate which second generation biodiesel feedstocks or products meet the aforementioned requirements best. In so doing, the two pieces of legislation aim potentially to accelerate the production of second generation biodiesel feedstocks and provide at the same time sustainable trade and growth in developing countries.

A consequence of policy uncertainties and the current financial crisis have slowed down the rate at which technology development of second generation biodiesels is progressing from the experimentation phase to commercialisation. To contribute to reaching the targets of climate change and/or energy policies, investment is needed to sustain R&D in advanced biodiesels. Funding is considered to be one of the most important support policies for improving R&D in the next generation of biodiesel feedstocks and commodities.

Through the Seventh Research Framework Programme (The Programme), the European Union has devoted US$2.5 billion to R&D into second and third generation biodiesels and biofuels. The Programme runs from 2007 to 2013 and aims at project cooperation within Member States or between Member States and emerging economies. In the USA, the Energy Act of 2008 provides funding of US$1 billion for bioenergy projects including second and third generation biodiesel initiatives. The Canadian and Australian governments also have similar funding initiatives to promote R&D for second and third generation biodiesel products (see http://www.sdtc.ca/index.php?page=about-our-funds&hl=en_CA and http://www.ret.gov.au/resources/resources_programs/). Developing countries have limited access to governmental funding for biodiesel-related projects owing to other priorities facing their economies (i.e. infrastructure, energy supply, education, etc). Consequently, one major limitation of support policies that favour the adoption of second generation biodiesels is that funding is mostly clustered in developed countries. It is desirable that developed countries intensify direct investments in emerging economies to implement advanced biodiesel projects. A viable alternative would be the establishment of carbon offset projects under the form of Clean Development Mechanisms (CDM) to ensure both the achievement of the Kyoto Protocol and the set up of second and third generation biodiesel projects in emerging economies.

Numerous initiatives should also aim to boost the aquaculture sector to promote algae biodiesels. The experience of some developing countries such as Malaysia is an excellent example of governmental support encouraging private R&D into algae biodiesels. Several funding opportunities have been carried out by the Sabah province and the Malaysian government such as the Sabah Outline Perspective Plan (1995–2010), the Second Agriculture Policy and the Ninth Malaysia Plan (2006–2010) which ensured sustainable development aimed at protecting the marine environment and the aquaculture sector including the promotion of algae biodiesels projects (Goh and Lee, 2010).

3.7.1 Roles for government intervention: a synopsis

It is apparent that the second generation biodiesels systems require more sophisticated processing equipment, more investment per unit of production and larger scale facilities to capture capital-cost scale economies than do first generation ones. In addition, to achieve true economic benefits from the second generation, further research, development and demonstration work is needed on feedstock production and conversion. Even at high oil prices, the next generation biodiesels (and biofuels) is unlikely to become commercially viable without appropriate government intervention. In order to reduce direct production costs significantly, a technological breakthrough is needed. However this will not be achievable without substantial investment in R&D in this sector. But, as is common in any R&D exercise, such an investment will be subject to uncertainties which can act as a deterrent to such an exercise. Hence, one of the primary policy prescriptions of the government will be to provide direct incentives to producers and investors in these sectors through direct subsidies and tax credits.

Second generation biodiesel technologies are primarily being developed in developing countries and hence are naturally more capital intensive. In the light of our discussion on technology-driven economic growth (see Section 3.6), developing countries will need to be able to adapt these technologies which will then give rise to the need for technology transfer. Governments of all concerned countries need to be able to settle for a cooperative outcome that will facilitate such transfers without providing countries with any incentive to exercise any rent-seeking behaviour. This will not be possible without exercising regulatory mandates. One important consequence of this is that there should not be a patent right granted to developed countries engaged in R&D technology development of next generation biodiesels for too long a period, as this will be detrimental to the goal of achieving diffusion of knowledge. In economic theory, there is a debate about the role of patent rights: on one hand patent protection gives firms incentives to innovate; on the other hand this creates a barrier to knowledge diffusion. The idea therefore is to have the right balance between patent length, breadth and scope that will be suitable for the case being considered. In particular, policies supportive of international joint ventures would be particularly beneficial to firms in developing countries in order to have access to the intellectual property rights owned by developed nations.

Finally, given that any investment in R&D is capital intensive, there is clearly a (an even more) need for mobilising funds, especially in developing nations, to finance production plans for next generation biodiesels. In this respect, investment schemes, such as bio-equity trade funds, will be particularly helpful, whereby firms in developed nations hold shares in the ongoing investment projects. This aspect has recently been emphasised in a study undertaken by IEA task 40 Trade (Bradley *et al.*, 2010).

3.8 Conclusions

The present chapter aimed to present a discussion on the main socio-economic and policy aspects of second and third generation biodiesel production. The limited potential of first generation biodiesels (and more generally biofuels) to make a significant contribution to displacing fossil fuels and reduce GHG emissions has ushered in a sense of urgency in transiting towards second and third generation biodiesels. The premise is that these biodiesels would be less intensive in their demand for agricultural land, resulting in better energy balances, improved reductions in GHG emissions and less competition for prime land with food crops, compared to the first generation biodiesels. Evaluating potential benefits as well as limitations of the future generation of biodiesels in the light of the above issues has been the main objective of the chapter.

The main potential for future generation biodiesels comes from improvements in R&D. There is substantial research taking place into algae biodiesels, which, with their high reproductive rates obtained in wastewater, are also used as feedstock for other biofuels. Biodiesels produced from micro-algae have the potential to increase yields by 50–100 times those of first generation biodiesels. Their main limitation is represented by high production costs due to the choice of the right process (i.e. the chemical reaction) to employ to obtain the final biodiesel output. When this is combined using CO_2 recycling techniques, for example, they contribute considerably to lower CO_2 emissions in the atmosphere because micro-algae absorb CO_2 as well as other pollutants (e.g. SO_2 and NO_x). The cost of harvesting is due to the complexity of selecting the optimal mix of micro-algae nutritional components which will help to decide which harvesting procedure to adopt at a later stage. Concern about market prices also arises from the acceptance of a price for biodiesels from micro-algae and the possibility of regulating blending targets for this oil in the near future.

The case for biodiesels from wood is also considered to be an important source of advanced biodiesels which can potentially replace one-third of global gasoline owing to their promising feedstock availability and low production costs (Balat *et al.*, 2008). Substantial problems nonetheless exist for land availability since the current feedstock used to produce biodiesels from wood is essentially energy crops.

Recent developments in R&D are also promising for future commercialisation of biodiesels from wood. Biodiesels from wood are currently produced by energy crops (and therefore contribute to the actual conflicts of land availability for food versus energy crops) and renewable resources and is an exceptional fuel for combustion engines and cell fuel vehicles. Furthermore, when during the production process glycerin is purified to produce biogas, it generates green electricity. The amount of biomethanol that can be replaced in gasoline consumption is notable: it ranges from 36–102% in the worst case scenario and from 72–204% in the best case scenario. Further R&D is needed to commercialise this type of second generation biodiesel.

Next generation biodiesels will contribute to reducing energy dependence from oil fuels and GHG emissions and reaching the targets of current energy and climate change policies. To analyse the main contribution of advanced biodiesels in shaping the future global energy supply, the IEA (OECD/IEA, 2010b) argues that total biodiesel potential can only be assessed by correctly estimating the land use availability for growing feedstock such as energy crops, forest residues and agricultural waste, as well as by considering other potentials like geographical impact, economic assessment, and biological and technical biomass potentials. A consequence of this integrated vision of biodiesel potential suggests that future biodiesel production will increase and grow almost four-fold if current global energy policies reach their objectives and will grow almost eight-fold if the Kyoto Protocol succeeds in meeting its targets.

Additionally, improved R&D and commercialisation of advanced biodiesels will reduce the conflicts of land availability between food and energy crops with a consequent reduction in current primary food prices. Most new generation biodiesels crops can be grown on infertile land or forest residues and guarantee a final biodiesel yield higher than that obtained from first generation biodiesels. This would ensure substitutability with conventional fossil fuels and contribute, at the same time, to a larger energy supply with the consequent positive effects on job creation in less developed areas.

Therefore, it becomes apparent from our analysis that, given the current state of technology, there still remains uncertainty about future breakthroughs that would make the next generation of biodiesels a truly viable option. Policymakers therefore need to consider carefully which goals to pursue in providing support for different biodiesels and biofuels. The USA is currently adopting a blending mandate which requires the use of lignocellulosic feedstocks of up to 60.6 billion l^{-1} yr^{-1} to ensure a reduction of 100 million tonnes of CO_2 per year by 2022. The European Union does not have a blending mandate for second generation biodiesels yet. Directive 2009/28/EC defines sustainability standards which require savings in GHG for biofuels of at least 35% compared to conventional fossil fuels from 2013, increasing to 60% from 2018.

Funding is an essential support policy to improve R&D in the next generation of biodiesel feedstocks and commodities. Although several steps have already been taking place throughout Europe owing to the progress of the Seventh Research Framework Programme and, in the USA, to a number of projects funded by the Energy Act, still much remains to be done in order to promote funding in developing countries that usually have limited resource access to governmental budgets. It is desirable that developed countries intensify direct investment in second generation biodiesel projects to contribute further to sustainable development paths. A viable alternative is also represented by carbon offset projects (e.g. CDM) to ensure both GHG emission reductions and a better allocation in the already limited public spending of emerging economies. Funding should also aim at boosting the aquaculture sector to promote algae biodiesels. The experience of some developing countries such as Malaysia is an excellent example of governmental support in encouraging private R&D in second generation biodiesels.

In the light of the above discussion, it is clear that biodiesels that simultaneously advance multiple policy goals warrant greater support when designing incentive schemes. An integrated approach, combining economically sustainable development, climate change mitigation and alternative energy provision, seems like a good policy framework (Carriquiry *et al.*, 2010). It is also of the utmost importance to consider regional as well as international developments in policies for trade-based laws for comparative advantage and innovation-led endogenous economic growth, in order to maximise the potentials achievable through the policies implemented.

3.8.1 Future prospects

The direction and opportunities for second generation biodiesels delineated in the present chapter illustrate several potentials for transforming future available feedstock into second and third generation biodiesels with low waste and carbon efficiency. The development patterns of advanced biodiesels can be summarised into two strategies: first, the implementation of wood and algae crops offering the potential to adapt and grow in wastelands and wastewaters; second, improvement in crop yields through low-cost management approaches as well as the increased capacity of investments in new technologies leading to the expansion of conventional crops for both food and biodiesel production. These gains are likely to play an important role in developing countries such as Latin America or sub-Saharan Africa which should employ a strategic and robust approach to developing from basic to advanced agriculture encompassing multi-product strategies that satisfy, on one hand, both internal and external demand for food and biodiesels commodities and minimise land availability conflicts and carbon emissions on the other.

If technological progress proves to be effective, the implementation of lignocellulosic and algae biodiesels is likely to dominate future land use change including protection of biodiversity and effective management of carbon sequestration. Effective and balanced management is also required for sustainable harvesting of perennial lignocellulosic crops and crop yields improvements. Guaranteeing efficient solar radiation has, in fact, implications for carbon stocks and soil organic composition offering yield stability in the short as well as in the long term (30 years).

Support regulation and monitoring practices are also required to estimate future land availability accurately in order to meet the US target (and a potential future EU target) for biodiesel demand. It should be taken into account that current bio-economic models are biased towards actual assumptions on crops and land availability. A great potential exists for future development of biodiesels, as far as land usage (particularly marginal lands and forests residues) and waste utilisation (e.g. wastewater or wasteland to breed lignocellulosic or algae feedstock) are concerned. It is likely that wood and algae biodiesels that adapt well to growing on marginal land or water will also find optimal use in terms of enabling positive carbon balances over time. Furthermore, biodiesel development with agricultural interactions can be enhanced by intensification of agricultural practices on current arable land for both food and biodiesels crops (Pretty, 2008). These intensification practices will have beneficial effects on GHG savings, improved soil productivity and biodiversity protection.

Finally, in order to make second generation biodiesels a truly important driver for sustainable economic growth, governments in various countries need to be proactive. Most conversion processes for next generation biodiesels are being developed for industrialised country applications that are typically capital intensive, labour minimising, suitable for large-scale installations and are designed for temperate climate feedstock. To capitalise on the comparative advantages of better growing climatic conditions and lower labour costs, developing countries need to be able to adapt to such techniques by making them relatively more labour intensive and more suitable for operation in respective local conditions. Such practices can, however, only be ensured through technology transfer schemes and investments in joint research ventures. Governments can therefore create opportunities for firms in developed and developing nations to engage in this creativity by providing not just financial support in terms of R&D investment grants but also through implementation of appropriate supportive policies in international joint technology ventures. However, in order to do so, governments will need to have long-term visions about the entire global welfare and not just simply be driven by their own country's welfare maximisation problem.

3.9 References

Aghion, P and Howitt, P (1992), 'A model of growth through creative destruction', *Econometrica*, **60**(2), 323–51.

Ahmad, AL, Mat Yasin, NH, Derek, CJC and Lim, JK (2011), 'Microalgae as a sustainable energy source for biodiesels production: a review', *Renewable and Sustainable Energy Reviews*, **15**, 584–93.

Al-Riffai, P, Betina Dimaranan, B and Laborde, D (2010), 'Global trade and environmental impact study of biofuel mandate', available at: www.ifpri.org/sites/default/files/publications/biofuelsreportec.pdf [Accessed 29 April 2011].

Balat, M (2011), 'Potential alternatives to edible oil for biodiesels production – a review of current work', *Energy Conversion and Management*, **52**, 1479–92.

Balat, M, Balat, H and Oz, C (2008), 'Progress in bioethanol processing', *Progress in Energy and Combustion Science*, **34**, 551–73.

Bohlmann, GM (2006), 'Process economic considerations for production of ethanol from biomass feedstocks', *Industrial Biotechnology*, **2**, 14–20.

Bradley, M, Cuypers, D and Pelkmans, P (2009), 'Second generation biofuels and trade: an exploratory study', IEA task force 40.

Bradley, M, Hektor, B and Schouwenberg, P-S (2010), 'World bio-trade equity fund study', IEA task 40 Bio-trade.

Brennan, L and Owende, P (2010), 'Biofuels from micro-algae – A review of technologies for production, processing and extraction, of biofuels and co-products', *Renewable and Sustainable Energy Reviews*, **14**, 557–77.

Carriquiry, M and Babcock, B (2008). 'Splashing and Dashing Biodiesels', *Centre for Agricultural and Rural Development*, available at: http://www.card.iastate.edu/iowa_ag_review/fall_08/article3.aspx [Accessed 30 April 2011].

Carriquiry, MA, Du, X and Timilsina, GR (2010), 'Second generation biofuels: economics and policies', *Policy Research Working paper* # 5406, The World Bank.

CRS report Congress (18-Mar-2008, updated 15-Sep-2010), 'Biofuels incentives: a summary of Federal programs'. Available from: http://ncseonline.org/NLE/CRSreports/10Oct/R40110.pdf. [Accessed 22 July 2011].

Dekker, E and Lanting, K (2009), 'Biomethanol as a second-generation biofuel for transportation', *Biofuels* Issue 1. Available from: http://www.biofuels-tech.com. [Accessed 30 April 2011].

De Lucia, C (2010), 'Multiple objectives policies for biofuels production: environmental, socio-economic and regulatory issues', in Luque, R, Campelo, J M and Clarke, J (eds.), *Handbook of biofuels production: processes and technologies*, Woodhead Publishing, Cambridge, UK.

De Lucia, C and Bartlett, M (2011), 'Implementing a biofuels economy in the EU: lessons from the SUSTOIL project', selected paper presented at the 18th Annual Conference of the European Association of Resource and Environmental Economists, Rome, 29 June–2 July.

Dufey, A (2006), 'Biofuels production, trade and sustainable development', Sustainable market discussion paper No. 2, IIED.

Enguídanos, M, Soria, A, Kavalov, B and Jensen, P (2002), 'Techno-economic analysis of bio-alcohol production in the EU: a short summary for decision-makers. European Commission', Joint Research Centre (DG JRC), Report EUR 20280 EN, Brussels.

Escobar, JC, Lora, ES, Venturini, OJ, Yanez, EE, Castillo, EF and Almazan, O (2009), 'Biofuels: environment, technology and food security', *Renewable and Sustainable Energy Reviews*, **13**, 1275–87.

European Biodiesel Board (2009), 'Restoring a level-playing field with Argentine biodiesels producers', Press release, 18 December 2009, available at: http://www.ebbeu.org/EBBpressreleases/Restoring%20Level%20Playing%20Field%20with%20Argentine%20biodiesels%20producers.pdf [Accessed 28 April 2011].

European Commission, (2009), 'Commission Directive 2009/28/EC of the European Parliament and of the Council of 23 April 2009 on the promotion of the use of energy from renewable sources and amending and subsequently repealing Directives 2001/77/EC and 2003/30/EC', *Official Journal of the European Union,* **L 140**, 16–62.

Faaji, APC (2006), 'Bio-energy in Europe: changing technologies choices', *Energy Policy*, **34**, 322–42.

Food and Agriculture Organisation of the United Nations (FAO) (2003), 'World Agriculture towards 2015/2030', Rome.

Food and Agriculture Organisation of the United Nations (FAO) (2009), 'The market and food security implications of the development of biofuel production'. In: FAO committee on commodity problems, sixty-seventh session, Rome, April 20–22.

Frankel, J and Romer, D (1999), 'Does trade cause growth?', *American Economic Review*, **89**(3), 379–99.

Goh, CS and Lee, KT (2010), 'Will biofuel projects in Southeast Asia become white elephants?', *Energy Policy*, **38**(8), 3847–3848.

Groom, MJ, Gray, E and Townsend, PA (2008), 'Biofuels and biodiversity: principles for creating better policies for biofuel production', *Conservation Biology*, **22**, 602–609.

Grossman, G and Helpman, E (1990), 'Comparative advantage and long-run growth', *American Economic Review*, **80**(4), 796–815.

Grossman, G and Helpman, E (1991a), 'Quality ladders in the theory of growth', *Review of Economic Studies*, **58**, 43–61.

Grossman, G and Helpman, E (1991b), *Innovation and Growth in the Global Economy,* MIT Press, Cambridge, Massachusetts and London.

Hazell, P and Pachauri, RK (2006), *Bioenergy and agriculture: promises and challenges,* International Food Policy Research Institute, Washington, DC., USA. Available from: http://www.ifpri.org/publication/bioenergy-and-agriculture [Accessed 29 April 2011].

Hill, J, Nelson, E, Tilman, D, Polasky, S and Tiffany, D (2006), 'Environmental, economic and energetic costs and benefits of biodiesels and ethanol biofuels', *Proceedings of the National Academy of Sciences*, **103**(30), 11206–10.

Hoogwijk, MM, Faaij, A, van den Broek, R, Berndes, G, Gielen, D and Turkenburg, W (2003), 'Exploration of the ranges of the global potential of biomass for energy', *Biomass and Bioenergy*, **25**, 119–33.
IMF (2007) http://www.imf.org/external/pubs/ft/survey/so/2007/RES1017A.htm [Accessed 21 April 2011].
International Monetary Fund (IMF) (2007), 'World Economic Outlook 2007'.
Junginger, M, van Dam, J, Zarrilli, S, Mohamed, FA, Marchal, D and Faaij, A (2010), 'Opportunities and barriers for international bioenergy trade', IEA Bioenergy Task 40: Sustainable international bioenergy trade.
Koh, LP (2007), 'Potential habitat and biodiversity losses from intensified biodiesels feedstock production', *Conservation Biology*, **21**, 1373–5.
Koh, LP and Wilcove, DS (2007), 'Cashing in palm oil for conservation', *Nature*, **448**, 993–4.
Koh, LP and Wilcove, DS (2008), 'Is oil palm agriculture really destroying tropical biodiversity?', *Conservation Letters*, **1**, 60–4.
Krueger, A (1997), 'Trade policy and economic development: How we learn', *American Economic Review*, **87**(1), 1–22.
Larson, ED (2008), '*Biofuel production technologies: status, prospects and implications for trade and development*'. Report No. UNCTAD/DITC/TED/2007/10, United Nations Conference on Trade and Development, New York and Geneva.
Lucas, RE Jr. (1988), 'On the mechanics of economic development', *Journal of Monetary Economics*, **22**, 3–42.
Lin, L, Cunshan, Z, Vittayapadung, S, Xiangqian, S and Mingdong, D (2011), 'Opportunities and challenges for biodiesels fuel', *Applied Energy*, **88**, 1020–31.
Luque, R, Lovett, JC, Datta, B, Clancy, J, Campeloa, JM and Romero, A (2010), 'Biodiesels as feasible petrol fuel replacement: a multidisciplinary overview', *Energy and Environmental Science*, The Royal Society of Chemistry, **3**, 1706–21.
Mitchell, D (2008), 'A note on rising food prices', *Policy Research Paper* # 4682, The World Bank.
Murphy, R, Woods, J, Black, M and McManus, M (2011), 'Global developments in the competition for land from biofuels', *Food Policy*, **36**, S52–S61.
OECD/IEA (2006), *World Energy Outlook 2006 Edition*, OECD/IEA, Paris, France.
OECD/IEA (2008), *Energy Technology Perspective 2008*, OECD/IEA, Paris, France.
OECD/IEA (2010a), *World Energy Outlook 2010*, OECD/IEA, Paris, France.
OECD/IEA (2010b), '*Sustainable production of second generation biofuels. Potential and perspectives in major economies and developing countries*', OECD/IEA, Paris, France.
OECD/IEA and Sims, R, Taylor, M, Saddler, J and Mabee, W (2008), '*From first to second generation biofuel technologies. An overview of current industry and RD&D activities*', OECD/IEA, Paris, France.
OPEC Fund for International Development (OFID) (2009), 'Biofuels and food security: implications for an accelerated biofuels production', OFID Pamphlet series n. 38.
Oregon Department of Energy (2003), 'Biomass energy: cost of production'. Available at: http://www.energy.state.or.us/biomass/cost.htm [Accessed 21 April 2011].
Pretty, JN (2008), 'Agricultural sustainability: concepts, principles and evidence', *Philosophical Transactions of the Royal Society B*, **363**, 447–65.
Rajagopal, D and Zilberman, D (2007), 'Review of environmental, economic and policy aspects of biofuels', *Policy Research Working Paper* WPS4341, The World Bank Development Research Group, Washington, DC, USA, September, 2007.
Renewable Fuel Association (2007), Ethanol industries statistics, Washington, DC, USA available at: http://www.ethanolrfa.org [Accessed 28 April 2011].

Ricardo, D (1817), *The principles of political economy and taxation.* John Murray, London, UK.

Rodolfi, L, Zittelli, GC, Bassi, N, Padovani, G, Biondi, N, *et al.* (2009), 'Microalgae for oil: strain selection, induction of lipid synthesis and outdoor mass cultivation in a low cost photobioreactor', *Biotechnolgy and Bioengineering*, **102**(1), 100–12.

Romer, PM (1990), 'Endogenous technological change', *Journal of Political Economy*, **98**, 71–102.

Romer, PM (1993), 'Two strategies for economic development: using ideas and producing ideas', *World Bank economic review.*

Schumpeter, J (1912), *The Theory of Economic Development*, Cambridge, Mass, USA.

Schumpeter, J (1942), *Capitalism, socialism and democracy,* Harper & Row, New York.

Searchinger, T, Heimlich, R, Houghton, RA, Dong, F, Elobeid, A, *et al.* (2008), 'Use of U.S. croplands for biofuels increases greenhouse gases through emissions from land-use change', *Science*, **319**(5867), 1238–40.

Smeets, E, Faaij, A, Lewandowski, I and Turkenburg, W (2007), 'A bottom-up assessment and review of global bioenergy potentials to 2050', *Energy and Combustion Science*, **33**, 56–106.

Smith, A (1776), *An Inquiry into the Nature and Causes of the Wealth of Nations.* University of Chicago Press.

Trostle R. (2008), 'Global agricultural supply and demand: factors contributing to the recent increase in food commodity prices', *USDA economic research service, report* WRS–0801, Washington, DC.

United Nations Environment Programme (UNEP) (2009), 'UNEP annual report. Seizing the green opportunity'.

Vogt, KA, Vogt, DJ, Patel-Weynand, T, Upadhye, R, Edlund, D, *et al.* (2008), 'Bio-methanol: how energy choices in the western United States can help mitigate global climate change', *Renewable Energy*, **34**, 233–241.

VTT (2007), *Bioenergy Technology Review*, Espoo, Finland.

Zeiler, KG, Heacox, DA, Toon, ST, Kadam, KL and Brown, LM (1995), 'Use of microalgae for assimilation and utilisation of carbon dioxide from fossil fuel-fired power plant flue gas', *Energy Conversion and Management*, **36**(6–9), 707–12.

4
Feedstocks for advanced biodiesel production

S. PINZI and M. PILAR DORADO, Universidad de Córdoba, Spain

Abstract: This chapter presents the most frequent vegetable-based feedstocks for biodiesel production. It introduces first generation biodiesel and focuses on second generation biodiesel with emphasis on low-cost raw materials. Vegetable oils from non-edible plants, frying oils and animal fats constitute the most extensively studied raw materials for second generation biodiesel. In this chapter, other raw materials such as soapstocks or oils for green diesel production are reviewed. Finally, feedstocks for advanced biodiesel production, such as microbial oils (also named single cell oils) or microdiesel are studied, as they are considered to be among the promising raw materials for third generation biodiesel production. The chapter finishes with a brief overview about the biorefinery concept.

Key words: biorefinery, hydrotreated vegetable oils, low cost biodiesel, microalgae, second generation biofuel.

4.1 Introduction

In 2004, accessible crude oil resources were estimated to be about 171.1 billion tonnes. By extrapolating from current consumption of about 11.6 million tonnes of crude oil per day, it can be estimated that the entire resources will only suffice for a short time (Rottig *et al.*, 2011; Vasudevan *et al.*, 2005). Furthermore, the massive emissions of greenhouse gasses caused by the combustion of fossil resources are causing an irreparable change to the global climate. Consequently, to decrease our dependence on crude oil and slow down the increasing destruction of the environment, a sustainable and competitive alternative based on renewable and abundant feedstock, like biomass (Hanna *et al.*, 2005), or any other regenerative source (Rottig *et al.*, 2011), is needed.

Edible fatty oils derived from rapeseed, soybean, palm, sunflower and other oils, are currently being used for the commercial production of biodiesel. However, biodiesel from edible oils is controversial. Some non-governmental organizations and social movements pinpoint the making of biofuels from edible raw materials as the main cause of increased global food market prices. Another claim against the use of biofuels is the possible depletion of ecological resources owing to intensive agricultural practices used in crop cultivation.

Economic viability is also a concern. Several studies have identified the price of feedstock oils as one of the most significant factors affecting the economic viability of biodiesel manufacture (Dorado *et al.*, 2006; Krawczyk, 1996; Zhang *et al.*, 2003). Approximately 70–95% of the total biodiesel production cost arises from the cost of the raw material (Krawczyk, 1996). The price of feedstock is therefore a key factor in producing competitively priced biodiesel (Dorado, 2008).

Alternative feedstocks, generally from non-edible origin, including waste vegetable oils and fats, non-food crops, biomass and other technologies, have been developed in an attempt to overcome the major shortcomings of first generation biodiesel (Luque *et al.*, 2008). It is believed that this so-called second generation biodiesel could solve previous problems, supplying fuel in a sustainable and reasonably priced way. Although several advances have been made in the last few years, the majority of second generation biodiesel processing technologies are not yet available on a fully commercial scale, but it is expected to enter the market in coming years. This chapter focuses on the raw materials used to produce second generation biodiesel.

4.2 Feedstocks for first generation biodiesel

Since feedstock price is a substantial expense in biodiesel production, the choice of feedstock affects profitability. Some plant oils, like soybean and palm oils, have a highly fluctuating price and in some years their use can be shifted from making biodiesel to the food market, owing to high international prices. In Europe, increasing demand for rapeseed oil had a strong impact on the oil price, which is approximately 25% higher than the price of soybean oil in the USA (Haas, 2005). Between 2002 and 2007, the cost of rapeseed increased by 63% (Jank *et al.*, 2007). In the USA, the amount of crude soybean oil needed to create 1 litre of biodiesel ranges in price from US$0.40 to US$0.48 (Balat, 2011). Palm oil is traded internationally and has potential for import in the short term. At present, Malaysia exports palm oil to over 100 countries (China is among the main consumers) and exports palm oil-based biodiesel to Korea, Germany and Japan. The local prices of net palm oil and palm oil-based biodiesel production are US$0.39 and US$0.60 per litre respectively and the commercial price of diesel fuel is US$0.26 per litre (Kalam *et al.*, 2008).

4.3 Feedstocks for second generation biodiesel

4.3.1 Non-edible vegetable oils

Non-edible plant oils could potentially produce biodiesel. The demand for edible oils as food and their resulting high price, prevents their use as fuel, particularly in developing countries (Pramanik, 2003). Non-edible oil plants are naturally available worldwide (Demirbas, 2009).

The production of biodiesel from non-edible oilseed crops has been extensively investigated in recent years. Non-edible oleaginous crops used in biodiesel production, including Bahapilu, castor, cotton seed, cuphea, *Jatropha curcas*, karanja seed, linseed, mahua, nagchampa, neem, rubber seed, tonka bean, castañola, cardoon and milkweed, are not discussed here, but have been covered in depth by the authors in previous works (Dorado, 2008; Pinzi and Dorado, 2010; Pinzi *et al.*, 2009).

Jatropha curcas

Biodiesel produced from *J. curcas* oil meets all the requirements stipulated by EU standard EN 14214 (Kumar and Sharma, 2008). For this reason, the complete sequence of *Jatropha curcas* chloroplast genome has recently been uncovered (Asif *et al.*, 2010). Owing to its tremendous potential for biodiesel production, in the last few years jatropha has been the subject of several reviews (Nazir *et al.*, 2009; Jain and Sharma, 2011a, 2011b; Kaushik *et al.*, 2007). Recently, Zanette *et al.* (2011) studied several heterogeneous catalysts (resins, zeolites, clays, hydrotalcites, aluminas and niobium oxide) in order to optimize and carry out a kinetic study of waste oil biodiesel transesterification.

Pongamia pinnata *(karanja)*

Several scientists have investigated and confirmed karanja oil as a potential source of biodiesel (Azam *et al.*, 2005; Meher *et al.*, 2004, 2006, Meher *et al.*, 2004; Raheman and Phadatare, 2004; Karmee and Chadha, 2005). Most researchers have conducted the transesterification of *P. pinnata* oil using methanol and potassium hydroxide (Vivek, 2004; Meher *et al.*, 2004; Karmee and Chadha, 2005). Because of its high free fatty acids (FFA) content, some researchers have proposed esterification with H_2SO_4 prior to transesterification with NaOH (De and Bhattacharyya, 1999).

Madhuca indica *(mahua)*

Several approaches to producing biodiesel can be found in literature (Dorado, 2008). With the exception of water content, the fuel properties of mahua biodiesel fall within the limits specified by ASTM D 6751–02 and EN 14214 standards (Ghadge and Raheman, 2005). Besides calorific value, all other fuel properties of mahua biodiesel were found to be better than those of petro-diesel fuel (Raheman and Ghadge, 2008).

Ricinus communis *(castor)*

The oil contains up to 90% ricinoleic acid, which is not suitable for nutritional purposes owing to its laxative effect. The hydroxycarboxylic acid is responsible for the extremely high viscosity of castor oil – almost a hundred times the value observed in other fatty materials (Mittelbach and Remschmidt, 2004).

4.1 Ricinus communis L. (photo courtesy of Eric Winder).

Furthermore, the cetane number of methyl ricinoleate does not meet the minimum requirements for biodiesel standard specifications. The oxidative stability of methyl ricinoleate is significantly lower than that of its non-hydroxylated counterpart (methyl oleate) and even lower compared to methyl linoleate (Knothe, 2008). If this biodiesel is to become an alternative to diesel fuel, more research is needed (Fig. 4.1).

4.3.2 Waste oils

Waste cooking oils offer significant potential as low-cost raw materials for biodiesel production. The price of waste cooking oil is 2.5–3.5 times cheaper than new vegetable oils, thus it has the potential to reduce the total manufacturing cost of biodiesel significantly. Large amounts of waste cooking oil are illegally dumped into rivers and landfills, causing environmental pollution, so recycling is key to reducing environmental contamination (Balat, 2011).

Used oils can be recycled through conversion into soap by saponification, or reused as lubricating oil or hydraulic fluid. Nevertheless, biofuel production seems to be the most attractive alternative for waste oil treatment. Used oils can only supply a small percentage of diesel fuel demand, so this source will not solve the energy problem, but it may decrease the dependence on fossil oil while reducing an environmental issue (Dorado, 2008).

For economic reasons, used frying oil is an interesting feedstock for biodiesel production. Nye *et al.* (1983) described the transesterification of used frying oil with excess alcohol under both acidic and basic conditions. Tests were carried out using frying margarine and partially hydrogenated soybean oil. Later, Mittelbach and Enzelsberger (1999) investigated the use of waste oils in producing biodiesel and found that the increased number of polymers during heating indicates its suitability for biodiesel production. For used olive oil, better results were also achieved using KOH and methanol instead of NaOH and ethanol, which decreased transesterification rates (Dorado *et al.*, 2004).

Several parameters can influence conversion from waste oils into biodiesel, including heating conditions, FFA composition and water content. It has been found that heating over a long period leads to a significantly higher FFA content, which can reach values of up to 10% and has a detrimental effect on the transesterification process. Nevertheless, in most cases, simple heating and filtering of solid impurities is sufficient for further transesterification (Mittelbach *et al.*, 1992; El Sabagh *et al.*, 2010). The methyl and ethyl esters of fatty acids produced by alcoholysis of triglycerides seem to be excellent fuels (Mittelbach *et al.*, 1985). Dorado *et al.* (2002) compared biofuels from waste vegetable oils from several countries (different fatty acid methyl esters –FAME– composition) including Brazil, Spain and Germany. The transesterification process was carried out in two steps, using methanol and KOH. Results revealed that an FFA value of less than 3% is needed to complete the reaction. The two-step transesterification process (without a costly purification step) was found to be an economical method of biofuel production using waste vegetable oils. A two-step transesterification using ferric sulfate and KOH with methanol has also been developed in order to reduce FFA content (Wang *et al.*, 2006). Acid-catalysed pretreatment to esterify FFA before transesterification with an alkaline catalyst was also proposed (Canakci and Van Gerpen, 2001) and the kinetics of the process were studied (Jain *et al.*, 2010). Enzyme catalysed transesterification of waste oils has also been proposed. Charpe and Rathod (2010) produced biodiesel from waste sunflower frying oil by transesterification using *Pseudomonas fluorescens* as a catalyst in a batch reactor.

An extensive review study was carried out on the combustion and emission properties of waste oil biodiesel (Enweremadu and Rutto, 2010). It was found that waste oil fatty acids usually exhibit the same properties as biodiesel from fresh vegetable oil. So, the influence of biodiesel on engine performance and combustion characteristics is probably more closely related to both its oxygenated nature (which is almost constant for every biodiesel) and its higher viscosity and lower calorific value, which have a major bearing on spray formation and initial combustion. Nevertheless, Ozsezen and Canakci (2010) tested waste frying palm methyl esters in an unmodified diesel engine. Findings showed that when biodiesel was used in the test engine, the fuel line pressure increased while air fuel equivalence ratio and ignition delay decreased, compared to petro-diesel fuel. This behaviour affected the combustion phenomena of biodiesel, causing a 57% reduction in carbon monoxide (CO) emission, about 40% in unburned hydrocarbon (HC) emission and about 23% in smoke opacity, when compared with petro-diesel.

4.3.3 Animal fats

An important group of feedstocks for second generation biodiesel production consists of fats derived from animals. Animal fats used to produce biodiesel include tallow (Oner and Altun, 2009), lard (Lu *et al.*, 2007) and chicken fat (Guru *et al.*, 2010). Compared to plant crops, these fats frequently offer an economic advantage because of their low price for conversion into biodiesel (Oner and Altun, 2009). Biodiesel production from animal fat is currently among the cheapest options, costing US$0.4–US$0.5 per litre, while traditional transesterification of vegetable oil, at present costs around US$0.6–US$0.8 per litre (Balat, 2011).

Despite drawbacks such as poor cold properties and high viscosity, animal fat methyl ester has several advantages, including its high cetane number, non-corrosive nature, clean alternative and renewable properties (Guru *et al.*, 2009). Animal fats tend to be low in FFA and water, but availability is limited, meaning they would never be able to meet world fuel needs (Balat, 2011).

Tallow

Food grade tallow generally has a low free fatty acid (FFA) content, making alkaline transesterification reactions feasible. Chung *et al.* (2009) optimized duck tallow methanolysis allowing a FAME yield of 97%. Ma *et al.* (1998) found that low quantities of NaOH used as a catalyst result in completed methanolysis of beef tallow in 15 min. However, in a recent study, Liu *et al.* (2010) reduced the time of reaction to 5 min using radio frequency heating. Araujo *et al.* (2010) carried out alkaline methanolysis of beef tallow with a high content of FFA (3.6%) by heating and preliminary formation of a microemulsion. Muniyappa *et al.* (1996) found that transesterification of beef tallow produced a mixture of esters with a high concentration of saturated fatty acids, but with physical properties somewhat similar to those of soybean oil esters. Nebel and Mittelbach (2006) found that n-hexane was the most suitable solvent for extracting fats from meat and bone meal. The extracted material was converted into fatty acid methyl esters through a two-step process.

Food grade tallow is costly and the low cold properties of methyl esters mean they are unlikely to compete with rapeseed oil in the EU. Low grade tallow is comparatively inexpensive, but its high FFA content makes it unsuitable for the one-step base catalysed methanolysis currently in use in most biodiesel plants in Europe (Fröhlich *et al.*, 2010). These authors have proposed two methods of transesterifying low grade tallow with a high FFA content:

1 a one-step reaction, neutralizing excess FFA with alkaline catalyst, significant excess of methanol and some water to promote phase separation;
2 a two-step reaction with a previous esterification reaction.

Both methods gave good conversion rates of low-grade tallow to methyl ester and yields above 90 wt% were achieved. Mbaraka *et al.* (2006) also synthesized propylsulfonic acid-functionalized mesoporous silica materials for methanol esterification of FFA in beef tallow, as a pretreatment step for alkyl ester production.

Lard

Lard is pork fat with soft grain. Lard was commonly used in many cuisines as a cooking fat, shortening, or as a spread similar to butter. Pigs that have been fed on different diets will yield lard with significantly different fatty acid contents and iodine values (D'Arrigo *et al.*, 2002).

Berrios *et al.* (2009) obtained a 90 wt% FAME from lard with very low FFA content (0.13 wt%). Dias *et al.* (2009) transesterified waste lard with an acid catalyst, achieving a low biodiesel yield (65 wt%). However, they achieved a significant improvement in yield when lard was blended with soybean oil (purity of 99.8 wt% and a yield of 77.8 wt%), showing that blending might be an interesting alternative for recycling such wastes. Lee *et al.* (2002) have also performed a three-step transesterification using a porous substance, such as silica gel, to produce biodiesel from lard and restaurant grease.

Chicken fat

Chicken fat is extracted from feather meal, which is prepared from chicken wastes such as feathers, blood, offal and trims. Feather meal contains substantial amounts of chicken fat, varying from 2–12%, depending on feather type (Kondamudi *et al.*, 2009).

Mattingly (2006) produced biodiesel from chicken fat with 2.3% FFA. This author concluded that a pretreatment was needed in order to get high biodiesel yields from chicken fat. Schulte (2007) achieved biodiesel yields of up to 91 wt% using supercritical methanol. Recently, Alptekin and Canakci (2010) produced biodiesel from chicken fat with a high FFA content (13.45 wt%), optimizing the pretreatment reaction by using different acid catalysts to reduce the FFA level (below 1 wt%). The maximum methyl ester yield was found to be 87.4 wt%. Moreover, Feddern *et al.* (2010) characterized chicken skin fat and found it possesses both a low FFA content and a fatty acid profile suitable for transesterification and biodiesel production.

Diesel blends of FAME of lard, beef tallow and chicken fat showed lower NO_x emission levels (3.2–6.2%) than soybean oil methyl ester blended with diesel fuel

(Wyatt *et al.*, 2005). However, cold-flow properties of fat-based fuels were found to be less desirable than those of soybean oil-based biodiesel, with comparable lubricity and oxidative stability (Canakci and Van Gerpen, 2003b).

4.3.4 Grease

Greases can be obtained from vegetable or animal sources and mainly contain triglycerides (TG), diglycerides (DG), monoglycerides (MG) and FFA (8–40%) (Fan and Burton, 2009). A grease containing 8–12 wt% FFA is categorized as a yellow grease, while grease containing above 35 wt% FFA is categorized as brown grease (Fan and Burton, 2009). Canakci and Van Gerpen (2001) extended the two-step process to yellow and brown grease and scaled it successfully to a pilot plant (Canakci and Van Gerpen, 2003a). The biodiesel produced from yellow grease was further tested in a four-cylinder turbocharged diesel engine and significant reductions in particulates, CO and HC were observed compared with No. 2 diesel fuel (Canakci and Van Gerpen, 2003b). Cao *et al.* (2008) used a continuous membrane reactor to produce biodiesel from different feedstocks, including yellow and brown grease, which produced high purity biodiesel.

4.4 Other feedstocks for biodiesel production

4.4.1 Soapstocks

Soapstock is a by-product of vegetable oil refinement and another low value feedstock for biodiesel production. Soapstock contains a substantial amount of water, which can be emulsified with the lipid constituents and is hard to remove. In addition, the presence of both FFA and acylglycerols makes the transesterification reaction more difficult. Alkaline catalysis is impossible owing to its high FFA level (Canakci and Van Gerpen, 2001). Haas *et al.* (2000) developed a simple, highly efficient method for synthesis of biodiesel from soapstock. The process involved two steps, the first one consisting of alkaline hydrolysis of all bounded fatty acids (to saponify all glycerides and phosphoglycerides) and the second step comprising acid-catalysed esterification of the resulting fatty acid sodium salts. After water removal, the resulting FFA sodium salts were rapidly converted into fatty acid methyl esters (FAME) by incubation with methanol and sulfuric acid at 35°C and ambient pressure. Further economic analysis by Haas (2005) suggested that the production cost of soapstock biodiesel would be approximately US$0.41/l, giving a 25% reduction on the estimated cost of biodiesel produced from soybean oil.

Other soapstocks apart from that from soybean oil are also used to produce biodiesel, thus increasing the potential supply of this raw material. Usta *et al.* (2005) used a mixture of hazelnut soapstock and waste sunflower oil to produce biodiesel. The process involved two steps, including sulfuric acid and sodium hydroxide catalysis. The mixture was first heated to 100°C to remove water, then allowed to cool to 35°C before two-step catalysis.

Another alternative is provided by the soapstock that is produced by the milk transformation industry: residual butter and related fat, named scum, are washed and collected in effluent treatment plants. Scum usually comprises a mixture of fats, lipids, proteins and packing materials. A large dairy, processing 500 000 litres of milk per day, will produce approximately 200–350 kg of effluent scum per day, which makes disposal difficult (Sivakumar *et al.*, 2010). Most dairies remove scum to solid waste disposal sites or by incineration. However, this is economically wasteful and generates pollutants. Sivakumar *et al.* (2010) transesterified triglycerides from scum using alkaline homogeneous catalyst, achieving an ester yield of 96.7 wt%.

Keskin *et al.* (2008) used cotton oil soapstock to produce biodiesel. The cotton oil soapstock biodiesel was blended with diesel fuel and tested in a single cylinder direct injection diesel engine. The engine power output decreased with respect to petro-diesel fuel performance, but showed a reduction in particulate (PM) emissions compared to petro-diesel.

4.4.2 Hydrotreated vegetable oils: green diesel

Researchers are working on different processing routes for converting vegetable oils into high quality diesel fuel that is fully compatible with fossil diesel fuel. The isoparaffin rich diesel known as 'green diesel' is produced from renewable feedstocks containing triglycerides and fatty acids by catalytic saturation, hydrodeoxygenation, decarboxylation and hydroisomerization (Naik *et al.*, 2010). This technology could be applied to any type of oleaginous feedstock to produce an isoparaffin-rich diesel substitute. Green diesel is an aromatic and sulfur-free diesel fuel with a high cetane value. Cold flow properties can be adjusted to meet climate-specific cloud point specifications, because they depend on the degree of saturation of hydrocarbon chains (Kalnes *et al.*, 2007).

Feedstocks used for green diesel production are the same as those used for conventional biodiesel (fatty raw materials of different origins). However, the fatty acid profile of the raw materials plays a significant role, since the level of saturation of the fatty acid chains influences the amount of hydrogen consumed. Highly saturated feedstocks such as palm oil and animal fats are thus preferred because they reduce hydrogen requirements and subsequently production costs (the expenses for hydrogen may be significant), especially for products with long hydrocarbon chains. This aspect may play a significant role in selecting palm oil and animal fats as feedstocks for producing renewable diesel (Knothe, 2009). However, the focus changes when products with good low-temperature properties are needed; aviation fuels, for example. Camelina oil is particularly suitable for producing aviation fuels, because it contains a relatively high amount of unsaturated fatty acid chains (27–40% of C18:3, necessary for the low-temperature requirements of aviation fuels), thus offering more cleavage sites for producing hydrocarbons (Gunstone and Harwood, 2007). A renewable diesel fuel blended with aviation fuel has been produced primarily (84%) from a camelina oil feedstock (Knothe, 2009). Some oils from algae can also contain high amounts of unsaturated chains (Hu *et al.*, 2008) and fuel derived from this feedstock has also been reported to be suitable for use in aviation (Knothe, 2009). Linseed oil (around 47% of C18:3) may also be appropriate for the same reason.

4.5 Feedstocks for advanced biodiesel production

4.5.1. Microorganisms for triglycerides production

Microbial oils, also named single cell oils (SCO), are produced by many microorganisms with the ability to accumulate lipids. Those with a lipid content in excess of 20% are classified as oleaginous (Ratledge and Wynn, 2002). These microbe species can transform renewable material into microbial oil, which can be used to produce biodiesel by transesterification (Ma *et al.*, 2008). These microorganisms belong to different families, such as microalgae, fungi (moulds and yeast) and bacteria (Subramaniam *et al.*, 2011). Table 4.1 shows oil production of microorganisms from different families that are suitable for triglycerides production. The first commercial production of SCO appeared in 1995 but, because it was not economically sound, only lasted for 6 years (Ratledge, 2004).

Microalgae

Algae are simple organisms that are mainly aquatic and microscopic. Microalgae are unicellular photosynthetic microorganisms, living in saline or freshwater environments, which convert sunlight, water and carbon dioxide to biomass (Demirbas and Fatih Demirbas, 2010). They are categorized into four main classes: diatoms, green algae, blue-green algae and golden algae. There are two main populations: filamentous and phytoplankton algae. Both species, in particular phytoplankton, increase in number rapidly to form algae blooms (Demirbas, 2010). Like higher plants, microalgae store lipids in the form of triacylglycerides (TAG). Many species exhibit rapid growth and high productivity and many microalgal species can be induced to accumulate substantial quantities of lipids, exceeding 60% of their dry biomass (Lin *et al.*, 2011). In fact, the oil productivity of many microalgae greatly surpasses that of the best producing oil crops (Chisti, 2010). The relative composition of algal lipids depends greatly on the species, medium and environmental conditions in which the cells are cultured and harvested (Greenwell *et al.*, 2010). In general, algal cells synthesize triglycerides when the energy input, through carbon assimilation, exceeds the immediate metabolic needs of the cell. However, it has been also shown that microalgae

Table 4.1 Oil content of oleaginous microorganisms

Microorganism	Oil content (% dry wt)	Microorganisms	Oil content (% dry wt)
Microalgae		Yeast	
Botryococcus braunii	25–75 (Meng *et al.*, 2009; Chisti, 2010)	*Candida curvata*	58 (Meng *et al.*, 2009)
Cylindrotheca sp.	16–37 (Meng *et al.*, 2009)	*Cryptococcus albidus*	65 (Meng *et al.*, 2009)
Chlorella sp.	28–32 (Chisti, 2010)	*Lipomyces starkeyi*	64 (Meng *et al.*, 2009)
Nitzschia sp.	45–47 (Meng *et al.*, 2009)	*Rhodotorula glutinis*	72 (Meng *et al.*, 2009)
Schizochytrium sp.	50–77 (Meng *et al.*, 2009)		
Crypthecodinium cohnii	20 (Chisti, 2010)		
Nannochloropsis sp.	31–68 (Chisti, 2010)		
Neochloris oleoabundans	35–54 (Chisti, 2010)		
Nannochloris sp.	20–35 (Chisti, 2010)		
Bacterium		Other fungi	
Arthrobacter sp.	>40 (Meng *et al.*, 2009)	*Aspergillus oryzae*	57 (Meng *et al.*, 2009)
Acinetobacter calcoaceticus	27–38 (Meng *et al.*, 2009)	*Mortierella isabellina*	86 (Meng *et al.*, 2009)
Rhodococcus opacus	24–25 (Meng *et al.*, 2009)	*Humicola lanuginosa*	75 (Meng *et al.*, 2009)

increase the proportion of triglycerides produced upon nutrient starvation and other environmental stresses, such as temperature and essential nutrients (Greenwell *et al.*, 2010).

Microalgae are the fastest-growing photosynthesizing organisms, able to complete an entire growth cycle every few days. Approximately 46 t of oil/ha/year can be produced from diatom algae. Specially bred mustard varieties can produce reasonably high oil yields and have the added benefit that the biomass left over once the oil has been pressed out acts as an effective and biodegradable pesticide (Demirbas, 2010). Furthermore, algae can be grown almost anywhere, even on sewage or saltwater and do not require fertile land or food crops.

The production cost of algal oil depends on many factors, including the biomass yield from the culture system, oil content, the scale of the production system and the cost of recovering oil from biomass (Balat, 2011). Biodiesel production from algae is therefore still far more expensive than petroleum–diesel fuels. The production of algae-generated oil currently costs US$20–30 per gallon (US$5.3–8.0 per litre), with some estimates soaring to US$60. Conventional diesel fuel costs less than US$3.5 per gallon (US$0.9 per litre) (Whipple, 2009).

Most commercial microalgae production is based on R-select species such as *Chlorella* sp., or extremophile species, such as *Arthrospira* sp. They are grown in shallow fertilized ponds or raceways (Sheehan *et al.*, 1998). Raceways typically consist of independent closed-loop recirculation channels in which paddle wheel-generated flow is guided around bends by baffles placed in the flow channel (Greenwell *et al.*, 2010). Closed microalgae bioreactors can be used to culture some species heterotrophically at high densities, offering theoretical advantages in terms of avoiding contamination, yielding higher culture densities and providing closer control over physico-chemical conditions (Greenwell *et al.*, 2010).

Moulds: filamentous fungi

Some oleaginous moulds (filamentous fungi) can store up to 80% of their biomass as lipids. Different fungi species able to accumulate lipids have been studied in recent years, including *Aspergillus oryzae*, *Mortierella isabellina*, *Humicola lanuginosa*, *Mortierella vinacea* and *Mucor circinelloides* (Li *et al.*, 2008; Liu and Zhao, 2007a). They have mainly been investigated in terms of the production of specific lipids such as docosahexaenoic acid, linolenic acid, eicosapentaenoic acid and arachidonic acid (Li *et al.*, 2008).

Vicente *et al.* (2009) studied the filamentous fungi *Mucor circinelloides* as a potential feedstock for biodiesel production, because it can accumulate around 25% of lipids using a wide range of carbon sources, even in large-scale growth. According to Azocar *et al.* (2010) unlike microalgae, microorganisms such as bacteria and fungi can be grown in conventional bioreactors, which improves biomass yield while reducing biomass and oil production costs.

Yeast

Oleaginous yeasts have fast growth rates, high oil contents and their TAG fraction is similar to that of plant oils. These organisms can grow on several carbon sources (glucose, xylose, arabinose, mannose, glycerol and other agricultural and industrial residues) (Subramaniam *et al.*, 2011). Most oleaginous yeasts can accumulate lipids above 40% of their dry weight and 70% under nutrient-limiting conditions (Beopoulos *et al.*, 2009). However, lipid contents and fatty acid profiles differ between species (Beopoulos *et al.*, 2009; Meng *et al.*, 2009). Some yeasts with high oil content are *Rhodotorula glutinis*, *Cryptococcus albidus*, *Lipomyces starkeyi* and *Candida curvata* (Meng *et al.*, 2009). Myristic, palmitic, stearic, oleic, linoleic and linolenic acids are the main fatty acids found in the lipids (Li *et al.*, 2007; Liu and Zhao, 2007b). Patil (2010) investigated the effect of the C/N ratio on lipid production and on lipid fatty acid composition in *L. starkeyi*, cultivated under various conditions. This author observed an increment of lipids in cells from 19–30% when the C/N ratio decreased from 20 to 61. The nature of the substrate affects the composition of fatty acids in yeast lipids. Patil observed that cultivation of *L. starkeyi* using starch as substrate resulted not only in a higher fraction of lipids in cells, but also in more oleic acid in cellular lipids than when cells were cultivated using glucose as a carbon source.

Bacteria

Biosynthesis and intracellular accumulation of TAG has been reported in Gram-negative as well as Gram-positive bacteria. However, this capability seems to be restricted to only a few groups of prokaryotic microorganisms. To date, TAG biosynthesis has been detected only in both aerobic heterotrophic bacteria and cyanobacteria (Alvarez and Steinbüchel, 2002). Owing to their higher growth rate and the simplicity of the culture methods, bacteria could present some advantages for biodiesel production compared with microalgae (Meng *et al.*, 2009). Investigations of the use of oleaginous bacteria as raw material for biodiesel production have therefore recently been performed (Lu *et al.*, 2008). The composition of bacterial TAG varies considerably depending on the microorganism and the carbon source (Alvarez and Steinbüchel, 2002).

4.5.2. Microdiesel

Kalscheuer *et al.* (2006) devised a method for synthesizing ethyl esters of fatty acids (FAEE, called microdiesel), based on the analysis of triacylglycerides and synthesis of wax esters by microorganisms. An effective ethanol synthesis pathway was established by introducing two genes from *Zymomona mobilis* into *Escherichia coli* (Rottig *et al.*, 2011). Synthesis of fatty acids from sugars by genetically engineered *E. coli* has also been suggested (Wackett, 2008).

Several microorganisms, plants, insects and mammalian species possess the ability to synthesize FAEE, for example, many yeasts and filamentous fungi such as *Rhizopus arrhizus* and *Neurospora* sp. For some beetles and bees, like *Trogoderma garnarium* or *Bombus terrestris*, FAEE serve as pheromones. Furthermore, they are synthesized in mammals as a consequence of ethanol consumption (Park *et al.*, 2009). FAEE are also used as aroma additives in food, beverages and cosmetics (Liu *et al.*, 2004) and serve as dietary supplements for increasing uptake of constitutional polyunsaturated fatty acids (Armenta *et al.*, 2007).

Strobel *et al.* (2008) described a newly isolated *Gliocladium roseum* strain that produces a mixture of hydrocarbons (named mycodiesel) identical to those found in diesel fuel. The strain *G. roseum* was discovered in Northern Patagonia (Chile) during the search for natural producers of volatile antibiotics. Hydrocarbon synthesis was possible under microaerophilic conditions on oatmeal or cellulose-based media. Significantly, *G. roseum* was able to utilize cellulose –a renewable resource– for the production of hydrocarbons. Unfortunately, the yield was lower than that found on the oatmeal medium, probably because of a low rate of cellulose digestion. This new microorganism could be a valuable source of information for genetic or metabolic engineers, who may perhaps find ways to improve yield, or transfer the ability to other more efficient microorganisms (Adamczak *et al.*, 2009).

4.6 Future trends: the biorefinery concept

The biorefinery system is based on processing biomass (feedstocks) for the manufacture of multiple bio-based products. According to the US National Renewable Energy Laboratory (NREL), a biorefinery is a facility that integrates biomass conversion processes and equipment to produce fuels, chemical, value-added products and power (Naik *et al.*, 2010). The power of the biorefinery is supported by economies of scale and efficient use of bioresources. The imbalance between commodity chemical needs and transportation fuels is key to the biorefinery concept (Ragauskas *et al.*, 2006).

Several authors have provided overviews of the biorefinery concept (FitzPatrick *et al.*, 2010; Cherubini, 2010; Kamm *et al.*, 2006). There are many research works that associate the idea of biorefineries with the use of different feedstocks, such as crops and crop residues, forest residues, green grasses, lignocellulosic biomass and industrial waste (Fitzpatrick *et al.*, 2010; Kamm *et al.*, 2006; Marinova *et al.*, 2009; Luo *et al.*, 2010; Kaparaju *et al.*, 2009). Renewable carbon-based raw materials for biorefineries are available from four different sectors: (a) agriculture (dedicated crops and residues), (b) forestry, (c) industry (process residues and leftovers) and households (municipal solid waste and wastewaters) and (d) aquaculture (algae and seaweeds). A further distinction can be made between feedstocks derived from dedicated crops and residues from agricultural, forestry and industrial activities, which can be available without upstream concerns. The main biomass feedstocks can be grouped into three wide categories: carbohydrates and lignin, triglycerides and mixed organic residues (Cherubini, 2010).

In order to develop biorefineries comparable to petroleum refineries, it is necessary to produce a broad variety of bio-based products in an efficient construction set system. Each biorefinery refines and converts biological raw materials into a multitude of valuable products. The change from petroleum hydrocarbons to bio-based ones will create notable opportunities for the chemical processing industry. For example, the use of carbohydrates as chemical raw materials will eliminate the need for several capital-intensive oxidative processes used in the petroleum industry (Ragauskas *et al.*, 2006). The product palette of a biorefinery includes many that are not made in petroleum refineries (Kamm and Kamm, 2004).

As humankind progresses from the oil refinery to the biorefinery, the challenges associated with the separation process will change but not diminish in importance. In the petroleum industry, distillation is the unit operation that dominates the refinery separation scheme. For chemicals derived from biomass, this dominance will be transferred to solvent-based extraction. This is a result of the non-volatile nature of most biomass components (Ragauskas *et al.*, 2006).

Future biorefinery operations will first extract high value-added chemicals already present in biomass, such as fragrances, flavouring agents, food-related products and high value nutraceuticals that provide health and medical benefits. The biorefinery will then focus on processing plant polysaccharides and lignin into feedstocks for bio-derived materials and fuels. This requires the development of innovative separation and depolymerization processes (Ragauskas *et al.*, 2006). Although future trends are leading the scientific community to develop and implement sustainable energy systems like the biorefinery concept, the best policy to preserve the environment will remain better energy and resource conservation through a reduction in energy dependence.

4.7 Acknowledgements

This research was supported by the Spanish Ministry of Education and Science (ENE2010–15159) and the Andalusian Research, Innovation and Enterprise Council, Spain (TEP–4994).

4.8 References

Adamczak, M., Bornscheuer, U. T. and Bednarski, W. (2009) The application of biotechnological methods for the synthesis of biodiesel. *European Journal of Lipid Science and Technology*, **111**, 800–13.

Alptekin, E. and Canakci, M. (2010) Optimization of pretreatment reaction for methyl ester production from chicken fat. *Fuel*, **89**, 4035–9.

Alvarez, H. M. and Steinbüchel, A. (2002) Triacylglycerols in prokaryotic microorganisms. *Applied Microbiology and Biotechnology*, **60**, 367–76.

Araujo, B. Q., Nunes, R. C. D., De Moura, C. V. R., De Moura, E. M., Cito, A. and Dos Santos, J. R. (2010) Synthesis and characterization of beef tallow biodiesel. *Energy and Fuels*, **24**, 4476–4480.

Armenta, R., Vinatoru, M., Burja, A., Kralovec, J. and Barrow, C. (2007) Transesterification of fish oil to produce fatty acid ethyl esters using ultrasonic energy. *Journal of the American Oil Chemists' Society*, **84**, 1045–1052.

Asif, M. H., Mantri, S. S., Sharma, A., Srivastava, A., Trivedi, I., *et al.* (2010) Complete sequence and organisation of the Jatropha curcas (Euphorbiaceae) chloroplast genome. *Tree Genetics and Genomes*, **6**, 941–952.

Azam, M., Waris, A. and Nahar, N. M. (2005) Prospects and potential of fatty acid methyl esters of some non-traditional seed oils for use as biodiesel in India. *Biomass and Bioenergy*, **29**, 293–302.

Azocar, L., Ciudad, G., Heipieper, H. J. and Navia, R. (2010) Biotechnological processes for biodiesel production using alternative oils. *Applied Microbiology and Biotechnology*, **88**, 621–36.

Balat, M. (2011) Potential alternatives to edible oils for biodiesel production – A review of current work. *Energy Conversion and Management*, **52**, 1479–92.

Beopoulos, A., Cescut, J., Haddouche, R., Uribelarrea, J. L., Jouve, C. M. and Nicaud, J. M. (2009) *Yarrowia lipolytica* as a model for bio-oil production. *Progress in Lipid Research*, **48**, 375–87.

Berrios, M., Gutiérrez, M. C., Martìn, M. A. and Martìn, A. (2009) Application of the factorial design of experiments to biodiesel production from lard. *Fuel Processing Technology*, **90**, 1447–1451.

Canakci, M. and Van Gerpen, J. V. (2001) Biodiesel production from oils and fats with high free fatty acids. *Transaction of the ASAE*, **44**, 1429–1436.

Canakci, M. and Van Gerpen, J. V. (2003a) A pilot plant to produce biodiesel from high free fatty acid feedstocks. *Transaction of the ASAE*, **46**, 945–954.

Canakci, M. and Van Gerpen, J. H. (2003b) Comparison of engine performance and emissions for petroleum diesel fuel, yellow grease biodiesel, and soybean oil biodiesel. *Transactions of the ASAE*, **46**, 937–44.

Cao, P. G., Dubé, M. A. and Tremblay, A. Y. (2008) High-purity fatty acid methyl ester production from canola, soybean, palm and yellow grease lipids by means of a membrane reactor. *Biomass and Bioenergy*, **32**, 1028–1036.

Charpe, T. W. and Rathod, V. K. (2010) Biodiesel production using waste frying oil. *Waste Management*, **31**, 85–90.

Cherubini, F. (2010) The biorefinery concept: Using biomass instead of oil for producing energy and chemicals. *Energy Conversion and Management*, **51**, 1412–1421.

Chisti, Y. (2010) Biodiesel from microalgae. *Biotechnology Advances*, **25**, 294–306.

Chung, K.-H., Kim, J. and Lee, K.-Y. (2009) Biodiesel production by transesterification of duck tallow with methanol on alkali catalysts. *Biomass and Bioenergy*, **33**, 155–158.

D'Arrigo, M., Hoz, L., Lopez-Bote, C. J., Cambero, I., Pin, C., *et al.* (2002) Effect of dietary linseed oil and a-tocopherol on selected properties of pig fat. *Canadian Journal of Animal Science*, **82**, 339–46.

De, B. K. and Bhattacharyya, D. K. (1999) Biodiesel from minor vegetable oils like karanja oil and nahor oil. *Lipid Letters*, **101**, 404–406.

Demirbas, A. (2009) Potential resources of non-edible oils for biodiesel. *Energy Sources Part B*, **4**, 310–314.

Demirbas, A. (2010) Use of algae as biofuel sources. *Energy Conversion and Management*, **51**, 2738–2749.

Demirbas, A. and Fatih Demirbas, M. (2010) Importance of algae oil as a source of biodiesel. *Energy Conversion and Management*, **52**, 163–170.

Dias, J. M., Alvim-Ferraz, M. C. M. and Almeida, M. F. (2009) Production of biodiesel from acid waste lard. *Bioresource Technology*, **100**, 6355–6361.

Dorado, M. (2008) Raw materials to produce low cost biodiesel. *Biofuels Refining and Performance.* Nag, A. (ed). McGraw Hill Professional, New York, USA, chapter 4.

Dorado, M. P., De Almeida, J. A., Schellert, C., Ballesteros, E., Löhrlein, H.-P. and Krause, R. (2002) An alkali-catalyzed transesterification process for high free fatty acid feedstocks. *Transactions of ASAE*, **45**, 525–9.

Dorado, M. P., Ballesteros, E., Mittelbach, M. and Lopez, F. J. (2004) Kinetic parameters affecting the alkali-catalyzed transesterification process of used olive oil. *Energy and Fuels*, **18**, 1457–1462.

Dorado, M. P., Cruz, F., Palomar, J. M. and Lopez, F. J. (2006) An approach to the economics of two vegetable oil-based biofuels in Spain. *Renewable Energy*, **31**, 1231–1237.

El Sabagh, S. M., Keera, S. T. and Taman, A. R. (2010) The characterization of biodiesel fuel from waste frying oil. *Energy Sources Part a–Recovery Utilization and Environmental Effects*, **33**, 401–9.

Enweremadu, C. C. and Rutto, H. L. (2010) Combustion, emission and engine performance characteristics of used cooking oil biodiesel--A review. *Renewable and Sustainable Energy Reviews*, **14**, 2863–2873.

Fan, X. and Burton, R. (2009) Recent development of biodiesel feedstocks and the applications of glycerol: a review. *The Open Fuels and Energy Science Journal*, **2**, 100–109.

Feddern, V., Kupski, L., Cipolatti, E. P., Giacobbo, G., Mendes, G. L., *et al.* (2010) Physico-chemical composition, fractionated glycerides and fatty acid profile of chicken skin fat. *European Journal of Lipid Science and Technology*, **112**, 1277–1284.

FitzPatrick, M., Champagne, P., Cunningham, M. F. and Whitney, R. A. (2010) A biorefinery processing perspective: Treatment of lignocellulosic materials for the production of value-added products. *Bioresource Technology*, **101**, 8915–8922.

Fröhlich, A., Rice, B. and Vicente, G. (2010) The conversion of low grade tallow into biodiesel-grade methyl ester. *JAOCS*, **87**, 825–833.

Ghadge, S. V. and Raheman, H. (2005) Biodiesel production from mahua (Madhuca indica) oil having high free fatty acids. *Biomass and Bioenergy*, **28**, 601–605.

Greenwell, H. C., Laurens, L. M. L., Shields, R. J., Lovitt, R. W. and Flynn, K. J. (2010) Placing microalgae on the biofuels priority list: a review of the technological challenges. *Journal of the Royal Society Interface*, **7**, 703–26.

Gunstone, F. D. and Harwood, J. (2007) Occurrence and characterisation of oils and fats. In *The Lipid Handbook.* Gunstone, F. D., Harwood, J. L. and Dijkstra, J. L. (eds). CRC Press. Boca Raton, FL.

Guru, M., Artukoglu, B. D., Keskin, A. and Koca, A. (2009) Biodiesel production from waste animal fat and improvement of its characteristics by synthesized nickel and magnesium additive. *Energy Conversion and Management*, **50**, 498–502.

Guru, M., Koca, A., Can, O., Cinar, C. and Sahin, F. (2010) Biodiesel production from waste chicken fat based sources and evaluation with Mg based additive in a diesel engine. *Renewable Energy*, **35**, 637–643.

Haas, M. J. (2005) Improving the economics of biodiesel production through the use of low value lipids as feedstocks: vegetable oil soapstock. *Fuel Processing Technology*, **86**, 1087–96.

Haas, M. J., Bloomer, S. and Scott, K. (2000) Simple, high-efficiency synthesis of fatty acid methyl esters from soapstock. *JAOCS*, **77**, 373–379.

Hanna, M. A., Isom, L. and Campbell, J. (2005) Biodiesel: Current perspectives and future. *Journal of Scientific and Industrial Research*, **64**, 854–857.

Hu, Q., Sommerfeld, M., Jarvis, E., Ghirardi, M., Posewitz, M. and Seibert, M. (2008) Microalgal triacylglycerols as feedstocks for biofuel production: perspectives and advances. *Plant Journal*, **54**, 621–39.

Jain, S. and Sharma, M. P. (2011a) Biodiesel production from Jatropha curcas Oil. *Renewable and Sustainable Energy Reviews*, **14**, 3140–7.

Jain, S. and Sharma, M. P. (2011b) Prospects of biodiesel from Jatropha in India: A review. *Renewable and Sustainable Energy Reviews*, **14**, 763–71.

Jain, S., Sharma, M. P. and Rajvanshi, S. (2010) Acid base catalyzed transesterification kinetics of waste cooking oil. *Fuel Processing Technology*, **92**, 32–8.

Jank, M. J., Kutas, G., Amaral, L. F. and Nassar, A. M. (2007) *EU and US Policies on Biofuels: Potential Impacts on Developing Countries*. Study, G. (ed). The German Marshall Fund, Washington, DC.

Kalam, M. A., Hassan, M., Hajar, R., Yusuf, M. S., Umar, M. R. and Mahlia, I. (2008) Palm oil diesel production and its experimental tests on a diesel engine. In *Handbook of Plant-Based Biofuels*. Pandey, A. E. (ed). Taylor and Francis LLC, Boca Raton, FL.

Kalnes, T., Marker, T. and Shonnard, D. R. (2007) Green diesel: A second generation biofuel. *International Journal of Chemical Reactor Engineering*, **5**, 748.

Kalscheuer, R., Stolting, T. and Steinbuchel, A. (2006) Microdiesel: Escherichia coli engineered for fuel production. *Microbiology*, **152**, 2529–2536.

Kamm, B. and Kamm, M. (2004) Principles of biorefineries. *Applied Microbiology Biotechnology*, **64**, 137–145.

Kamm, B., Gruber, P. R., Kamm, M. (2010) *Biorefineries – Industrial Processes and Products Status Quo and Future Directions*, Wiley-VCH Verlag GmbH & Co. KGaA, Weinheim, Germany.

Kaparaju, P., Serrano, M., Thomsen, A. B., Kongjan, P. and Angelidaki, I. (2009) Bioethanol, biohydrogen and biogas production from wheat straw in a biorefinery concept. *Bioresource Technology*, **100**, 2562–8.

Karmee, S. K. and Chadha, A. (2005) Preparation of biodiesel from crude oil of *Pongamia pinnata*. *Bioresource Technology*, **96**, 1425–1429.

Kaushik, N., Kumar, K. and Kumar, S. (2007) Potential of *Jatropha curcas* for biofuels. *Journal of Biobased Materials and Bioenergy*, **1**, 301–314.

Keskin, A., Gürü, M., Altiparmak, D. and Aydin, K. (2008) Using of cotton oil soapstock biodiesel–diesel fuel blends as an alternative diesel fuel. *Renewable and Sustainable Energy Reviews*, **33**, 553–7.

Knothe, G. (2008) 'Designer' biodiesel: optimizing fatty ester composition to improve fuel properties. *Energy and Fuels*, **22**, 1358–64.

Knothe, G. (2009) Biodiesel and renewable diesel: A comparison. *Progress in Energy and Combustion Science*, **36**, 364–73.

Kondamudi, N., Strull, J., Misra, M. and Mohapatra, S. K. (2009) A green process for producing biodiesel from feather meal. *Journal of Agricultural and Food Chemistry*, **57**, 6163–6166.

Krawczyk, T. (1996) *Biodiesel. International News on Fats, Oils and Related Materials.* American Oil Chemists Society Press, Champaign, Illinois.

Kumar, A. and Sharma, S. (2008) An evaluation of multipurpose oil seed crop for industrial uses (Jatropha curcas L.): A review. *Industrial Crops and Products*, **28**, 1–10.

Lee, K. T., Foglia, T. A. and Chang, K. S. (2002) Production of alkyl ester as biodiesel from fractionated lard and restaurant grease. *JAOCS*, **2**, 191–195.

Li, Q., Du, W. and Liu, D. (2008) Perspectives of microbial oils for biodiesel production. *Applied Microbiology and Biotechnology*, **80**, 749–756.

Li, Y., Zhaob, Z. and Bai, F. (2007) High-density cultivation of oleaginous yeast Rhodosporidium toruloides Y4 in fed-batch culture. *Enzyme and Microbial Technology*, **41**, 312–317.

Lin, L., Cunshan, Z., Vittayapadung, S., Xiangqian, S. and Mingdong, D. (2011) Opportunities and challenges for biodiesel fuel. *Applied Energy*, **88**, 1020–1031.

Liu, B. and Zhao, Z. (2007a) Biodiesel production by direct methanolysis of oleaginous microbial biomass. *Journal of Chemical Technology and Biotechnology*, **82**, 775–80.

Liu, B. and Zhao, Z. B. (2007b) Biodiesel production by direct methanolysis of oleaginous microbial biomass. *Journal of Chemical Technology and Biotechnology*, **82**, 775–80.

Liu, S. Q., Holland, R. and Crow, V. L. (2004) Esters and their biosynthesis in fermented dairy products: a review. *International Dairy Journal*, **14**, 923–45.

Liu, S. Y., Wang, Y. F., Oh, J. H. and Herring, J. L. (2010) Fast biodiesel production from beef tallow with radio frequency heating. *Renewable Energy*, **36**, 1003–1007.

Lu, J., Nie, K., Xie, F., Wang, F. and Tan, T. (2007) Enzymatic synthesis of fatty acid methyl esters from lard with immobilized Candida sp. *Process Biochemistry*, **42**, 1367–1370.

Lu, X. F., Vora, H. and Khosla, C. (2008) Overproduction of free fatty acids in *E. coli*: Implications for biodiesel production. *Metabolic Engineering*, **10**, 333–9.

Luo, G., Talebnia, F., Karakashev, D., Xie, L., Zhou, Q. and Angelidaki, I. (2010) Enhanced bioenergy recovery from rapeseed plant in a biorefinery concept. *Bioresource Technology*, **102**, 1433–9.

Luque, R., Herrero-Davila, L., Campelo, J. M., Clark, J. H., Hidalgo, J. M., *et al.* (2008) Biofuels: a technological perspective. *Energy and Environmental Science*, **1**, 542–564.

Ma, B., Li, Q., Niwa, T., Oura, T., Du, W., *et al.* (2008) Improvement of lipid accumulation in an oleaginous yeast. *Journal of Biotechnology*, **136**, S415–S415.

Ma, F., Clements, L. D. and Hanna, M. A. (1998) The effects of catalyst, free fatty acids and water on transesterification of beef tallow. *Transactions ASAE*, **41**, 1261–4.

Marinova, M., Mateos-Espejel, E., Jemaa, N. and Paris, J. (2009) Addressing the increased energy demand of a Kraft mill biorefinery: The hemicellulose extraction case. *Chemical Engineering Research and Design*, **87**, 1269–75.

Mattingly, B. G. (2006) *Production of Biodiesel from Chicken Fat Containing Free Fatty Acids*. PhD Thesis, Department of Chemical Engineering, University of Arkansas, Arkansas, USA.

Mbaraka, I. K., Mcguire, K. J. and Shanks, B. H. (2006) Acidic mesoporous silica for the catalytic conversion of fatty acids in beef tallow. *Industrial and Engineering Chemistry Research*, **45**, 3022–3028.

Meher, L. C., Naik, S. N. and Das, L. M. (2004) Methanolysis of *Pongamia pinnata* (karanja) oil for production of biodiesel. *Journal of Scientific and Industrial Research*, **63**, 913–918.

Meher, L. C., Dharmagadda, V. S. S. and Naik, S. N. (2006) Optimization of alkali-catalyzed transesterification of *Pongamia pinnata* oil for production of biodiesel. *Bioresource Technology*, **97**, 1392–7.

Meng, X., Yang, J., Xu, X., Zhang, L., Nie, Q. and Xian, M. (2009) Biodiesel production from oleaginous microorganisms. *Renewable Energy*, **34**, 1–5.

Mittelbach, M. and Enzelsberger, H. (1999) Transesterification of heated rapeseed oil for extending diesel fuel. *JAOCS*, **76**, 545–550.

Mittelbach, M. and Remschmidt, C. (2004) *Biodiesel: The Comprehensive Handbook*, Martin Mittelbach, Graz, Austria.

Mittelbach, M., Tritthart, P. and Junek, H. (1985) Diesel fuel derived from vegetable oils. II. Emission tests using rape oil ester. *Energy Agriculture*, **4**, 207–215.

Mittelbach, M., Pokits, B. and Silberholz, A. (1992) Diesel fuels derived from vegetable oils, IV: Production and fuel properties of fatty acid methyl esters from used frying oil. *Liquid Fuels from Renewable Resources. Proceedings of an Alternative Energy Conference*. Nashville, Tennessee (USA).

Muniyappa, P. R., Brammer, S. C. and Noureddini, H. (1996) Improved conversion of plant oils and animal fats into biodiesel and co-product. *Bioresource Technology*, **56**, 19–24.

Naik, S. N., Rout, G. V., Vaibhav, P. K. and Dalai, A. K. (2010) Production of first and second generation biofuels: A comprehensive review. *Renewable and Sustainable Energy Reviews*, **14**, 578–97.

Nazir, N., Ramli, N., Mangunwidjaja, D., Hambali, E., Setyaningsih, D., *et al.* (2009) Extraction, transesterification and process control in biodiesel production from Jatropha curcas. *European Journal of Lipid Science and Technology*, **111**, 1185–200.

Nebel, B. A. and Mittelbach, M. (2006) Biodiesel from extracted fat out of meat and bone meal. *European Journal of Lipid Science Technology*, **108**, 398–403.

Nye, M. J., Williamson, T. W., Deshpande, W., Schrader, J. H., Snively, W. H., *et al.* (1983) Conversion of used frying oil to diesel fuel by transesterification: preliminary tests. *JAOCS*, **60**, 1598–1601.

Oner, C. and Altun, S. (2009) Biodiesel production from inedible animal tallow and an experimental investigation of its use as alternative fuel in a direct injection diesel engine. *Applied Energy*, **86**, 2114–2120.

Ozsezen, A. N. and Canakci, M. (2010) The emission analysis of an IDI diesel engine fueled with methyl ester of waste frying palm oil and its blends. *Biomass and Bioenergy*, **34**, 1870–78.

Park, Y., Shaffer, C. and Bennett, G. (2009) Microbial formation of esters. *Applied Microbiology and Biotechnology*, **85**, 13–25.

Patil, S. (2010) *Lipid Production from Glucose and Starch Using* Lipomyces starkeyi. PhD Thesis, Chemical Engineering Department, UL Lafayette, Lafayette, LA, USA.

Pinzi, S. and Dorado, M. P. (2010) Vegetable-based feedstocks for biofuels production. In *Handbook of Biofuels Production: Processes and technologies*. Luque, R., Campelo, J. M. and Clark, J. H. (eds). Woodhead Publishing, Cambridge, UK.

Pinzi, S., Garcia, I. L., Lopez-Gimenez, F. J., Luque De Castro, M. D., Dorado, G. and Dorado, M. P. (2009) The ideal vegetable oil-based biodiesel composition: a review of social, economical and technical implications. *Energy and Fuels*, **23**, 2325–41.

Pramanik, K. (2003) Properties and use of jatropha curcas oil and diesel fuel blends in compression ignition engine. *Renewable Energy*, **28**, 239–248.

Ragauskas, A. J., Williams, A., Davison, B. H., Britovsek, G., Cairney, J., *et al.* (2006) The path forward for biofuels and biomaterials. *Science*, **311**, 484.

Raheman, H. and Ghadge, S. V. (2008) Performance of diesel engine with biodiesel at varying compression ratio and ignition timing. *Fuel*, **87**, 2659–2666.

Raheman, H. and Phadatare, A. G. (2004) Diesel engine emissions and performance from blends of karanja methyl ester and diesel. *Biomass and Bioenergy*, **27**, 393–7.

Ratledge, C. (2004) Fatty acid biosynthesis in microorganisms being used for single cell oil production. *Biochimie*, **86**, 807–815.

Ratledge, C. and Wynn, J. P. (2002) The biochemistry and molecular biology of lipid accumulation in oleaginous microorganisms. *Advances in Applied Microbiology*, **51**, 1–44.

Rottig, A., Wenning, L., Broker, D. and Steinbuchel, A. (2011) Fatty acid alkyl esters: perspectives for production of alternative biofuels. *Applied Microbiology and Biotechnology*, **85**, 1713–1733.

Schulte, W. B. (2007) *Biodiesel Production from Tall Oil and Chicken Fat via Supercritical Methanol Treatment*. PhD Thesis, University of Arkansas, USA.

Sheehan, J., Dunahay, T., Benemann, J. and Roessler, P. (1998) *A Look Back at the US Department Of Energy's Aquatic Species Program—Biodiesel from Algae*. National Renewable Energy Laboratory, Golden, CO, USA.

Sivakumar, P., Anbarasu, K. and Renganathan, S. (2010) Bio-diesel production by alkali catalyzed transesterification of dairy waste scum. *Fuel*, **90**, 147–151.

Strobel, G. A., Knighton, B., Kluck, K., Ren, Y., Living-House, T., *et al.* (2008) The production of myco-diesel hydrocarbons and their derivatives by the endo- phytic fungus *Gliocladium roseum* (NRRL 50072). *Microbiology*, **154**, 3319–28.

Subramaniam, R., Dufreche, S., Zappi, M. and Bajpai, R. (2011) Microbial lipids from renewable resources: production and characterization. *Journal of Industrial Microbiology and Biotechnology*, **37**, 1271–1287.

Usta, N., Öztürk, E., Can, O., Conkur, E. S., Nas, S., *et al.* (2005) Combustion of biodiesel fuel produced from hazelnut soapstock/waste sunflower oil mixture in a diesel engine. *Energy Conversion and Management*, **46**, 741–55.

Vasudevan, P., Sharma, S. and Kumar, A. (2005) Liquid fuel from biomass: An overview. *Journal of Scientific and Industrial Research*, **64**, 822–831.

Vicente, G., Bautista, L. F., Rodrìguez, R., Gutièrrez, F. J., S·daba, I., *et al.* (2009) Biodiesel production from biomass of an oleaginous fungus. *Biochemical Engineering Journal*, **48**, 22–27.

Vivek, G. A. K. (2004) Biodiesel production from Karanja oil. *Journal of Scientific and Industrial Research*, **63**, 39–47.

Wackett, L. P. (2008) Biomass to fuels via microbial transformations. *Current Opinion on Chemical Biology*, **12**, 187–193.

Wang, Y., Ou, S., Liu, P., Xue, F. and Tang, S. (2006) Comparison of two different processes to synthesize biodiesel by waste cooking oil. *Journal of Molecular Catalysis A: Chemical*, **252**, 107–112.

Whipple, T. (2009) Peak oil review. ASPO (ed) *International Peak Oil Conference*. Association for the Study of Peak Oil and Gas (ASPO), Denver, Colorado, USA.

Wyatt, V. T., Hess, M. A., Dunn, R. O., Foglia, T. A., Haas, M. J. and Marmer, W. N. (2005) Fuel properties and nitrogen oxide emission levels of biodiesel produced from animal fats. *Journal of the American Oil Chemists Society*, **82**, 585–91.

Zanette, A. F., Barella, R. A., Pergher, S. B. C., Treichel, H., Oliveira, D., *et al.* (2011) Screening, optimization and kinetics of Jatropha curcas oil transesterification with heterogeneous catalysts. *Renewable Energy*, **36**, 726–31.

Zhang, Y., Dube, M. A., Mclean, D. D. and Kates, M. (2003) Biodiesel production from waste cooking oil: 2. Economic assessment and sensitivity analysis. *Bioresource Technology*, **90**, 229–40.

5
Advances in biodiesel quality control, characterisation and standards development

A. CARRERO, Universidad Rey Juan Carlos, Spain and
Á. PÉREZ, Universidad de Castilla la Mancha, Spain

Abstract: Successful commercialisation and market acceptance of biodiesel requires stringent quality assurance standards. The evaluation of biodiesel quality is achieved through the determination of its chemical composition and physical properties. Regional standards provide some guidance on the methods of analysis of biodiesel to be used. The properties and limits specified in biodiesel standards must protect performance and durability of automobile engines or combustion equipment. However, biodiesel quality specifications are dynamic and they have to be periodically reviewed by institutions like the European Committee of Standardization (CEN), the International Organization for the Standardization (ISO) and the American Society for Testing and Materials (ASTM).

Key words: biodiesel blends, biodiesel specifications, quality control, standard methods.

5.1 Introduction

The evaluation of biodiesel quality is achieved through determination of its chemical composition and physical properties. The fatty ester composition, along with the presence of contaminants and minor components, dictates the fuel properties of biodiesel. In this sense, the specifications represent the definition of the characteristics (list of terms and conditions) that biodiesel must meet. There are two types of standards for biodiesel fuel:

- the standards for classification and characteristics (requirements),
- the standards for test.

The standard requirements are to ensure that biodiesel is of adequate purity, quality and stability to ensure engine and fuel component durability and reliability. The standards for classification and characteristics include feedstock-inherent properties such as the oxidative stability or the iodine value. These so-called structure indices originally served to exclude the use of certain vegetable oils or animal fats as feedstocks. The standards for test specify how to make biodiesel parameters determinations (procedures and material needed). They are standardised in order to obtain reproducible and comparable results. The standards and the specifications are constantly changing and should be continuously kept up-to-date for different reasons: new uses, new manufacturing processes, new analytical methods, changes in regulations (environmental protection), and so on.

Biodiesel properties related to the production process, also called 'quality indexes', are indicative of the content of unreacted starting material in the biodiesel. Process-related parameters comprise the acid number as well as the glycerine, methanol, water and sodium hydroxide content. As mentioned below, the determination of water content and acid number is crucial for the quality control of the feedstock and for optimising the production process. Successful commercialisation and market acceptance of biodiesel requires stringent quality assurance standards.

Although there are several important properties of biodiesel, the three that mainly affect the engine performance are volatility, ignition quality and viscosity. The ignition quality (tendency to autoignite) is expressed by cetane number (CN), it does not directly affect the motor performance but rather acts on driving comfort factors (cold starting, noise and exhaust emissions). The biodiesel is burned in the presence of air to enable thermal engines to run or for the production of heat energy in boilers, furnaces, and so on. The mass heating value represents the quantity of energy released by unit mass of fuel during the chemical reaction for complete combustion.

Another essential biodiesel characteristic is its behaviour in cold climates. Low temperatures promote precipitation of saturated compounds that can plug the fuel filter. Other parameters are related to control of emissions, like sulphur, for environmental reasons. In addition, phosphorous and metals like Na, K, Ca and Mg affect the efficiency and durability of diesel particulate filter (DPF) and selective catalytic reduction (SCR) systems. Finally, properties related to the storage, handling and distribution of biodiesel are oxidative stability, water content, acid value, and so on. Research into understanding the fundamental chemistry of biodiesel oxidative stability is being performed worldwide.

Depending on its application, biodiesel can be used for two energy purposes, that is, motor fuels and heating fuels. The major biodiesel standards for motor fuels, which commonly serve as a reference for other standards, are EN 14214 and ASTM D 6751. In addition, EN 14213 defines the minimal requirements for biodiesel used as heating oil or as a blending component for heating oil.

Biodiesel producers propose the running of certification and accreditation programmes to control biodiesel quality. To ensure product traceability, the German Association for Quality Management of Biodiesel (AGQM) has produced a multi-level quality assurance system that monitors the quality of the biodiesel from its members (producers and sellers). In the USA, the National Biodiesel Accreditation Program is a cooperative and voluntary programme for the accreditation of producers and marketers of biodiesel fuel called BQ–9000®. The programme is a unique combination of the ASTM standard for biodiesel, ASTM D6751 and a quality system programme that includes storage, sampling, testing, blending, shipping, distribution and fuel management practices. BQ–9000® is open to any biodiesel manufacturer, marketer or distributor of biodiesel and biodiesel blends in the USA and Canada. The National Biodiesel Accreditation Commission imposes the necessary standards required for quality assurance checks for biodiesel and biodiesel mixes. The Iowa Central Fuel Testing Laboratory was the first to reach this accreditation.

5.2 Characterisation, analysis techniques and quality control of feedstocks

Refined vegetable oils destined for food use are an expensive substrate for biofuel development. Although there is interest in developing speciality oil crops, some producers have turned to less desirable or non-edible materials such as animal fats and recycled vegetable oils. A key issue in the production of biodiesel is the amount of intermediates, by-products, unreacted substrates and a variety of contaminants, which may affect engine performance. Regional standards provide some guidance on the methods of analysis of biodiesel to be used, but do not address the analysis of the feedstock. In this sense, oil quality may be assessed using the Official Methods and Recommended Practices of the American Oil Chemists' Society (AOCS).

The AOCS Analytical Guidelines Ck 1–07 list those methods which may be used to assess the quality of oils and fats used in the production of biodiesel. These Analytical Guidelines are restricted to triglyceride feedstocks for biodiesel manufacture, including vegetable oils (soybean oil, rapeseed oil, palm oil, etc.), animal fats (tallow, lard, etc.) and triglyceride greases (yellow grease, etc.). According to AOCS Ck 1–07 the principal parameters to be determined in an oil or fat to be used as biodiesel feedstock are briefly described below.

Sampling: In general, feedstock sample preparation should be carried out using an environment that minimises any changes in the properties of the lipid fraction. If lipid oxidation is a problem, it is important to preserve the sample by using a nitrogen atmosphere, cold temperature, low light or adding antioxidants. Fats products are frequently solids or semi-solids at room temperature and should be completely liquefied and blended prior to testing. Temperature during melting should not exceed the melting point of the sample by more than 10°C.

Insoluble impurities: The feedstock should be free of trash or other foreign material. The insoluble impurities test measures the amount of solids that are insoluble in kerosene and petroleum ether. These solids may consist of sand, dirt and seed fragments in the case of vegetable oil and small particles of bones and gums in the case of animal fats or used cooking oil. The determination of insoluble impurities in feedstocks can be performed following the instructions of AOCS Method Ca 3a–46. This procedure consists of dissolving the residue from the moisture and volatile matter experiment in kerosene and petroleum ether to allow all the non-polar substances to dissolve, leaving behind all the small insoluble particles. The amount of insoluble impurities in oils and fats is primarily related to the extraction and purification methods utilised, and therefore it is not related to other oil and fat characteristics.

Moisture and volatile matter: Moisture is a minor component found in biodiesel feedstocks. Moisture can react with basic catalysts during transesterification and can lead to undesirable soap formation and emulsions (Knothe *et al.*, 2005). In addition, the presence of volatile matter in a feedstock may lead to fatty acid methyl ester yield loss by reacting with the catalyst or by diluting the feedstock.

The feedstock moisture can be measured in accordance with AOCS Method Ca 2e–84 for water content of fats and oils by titration with Fisher reagent, which reacts quantitatively with water. It is applicable to fats and oils that do not react with and are soluble in the reagents and that do not contain impurities leading to secondary reactions. Such impurities are alkaline compounds and peroxides which react with the reagent and therefore show high results.

The test for moisture and volatile matter may be included in fat and oil specifications. AOCS Method Ca 2b–38 involves heating a known amount of feedstock to a certain temperature and recording the weight loss. A possible drawback of the AOCS moisture and volatile method is the precision. If the feedstock needs to be less than 0.50 wt% moisture, this method may not be able to measure moisture at or below 0.50 wt%.

On the other hand, AOCS Modified Method Ca 2f–93 determines the moisture and volatile matter in fats and oils by the addition of acetone to the fat or oil, followed by heating at 100°C. The moisture and volatile matter are removed during the evaporation of the acetone and are determined by the loss in weight of the original sample. This method can be completed in one hour without the use of elaborate equipment, permitting a rapid determination.

Usually, if the feedstock moisture was above 0.050 wt%, it was dried using heat and vacuum to reduce the moisture before further conversion to minimise effects from emulsions during transesterification.

Unsaponifiable matter consists of organics, which do not react with base to form soaps. These include sterols, higher aliphatic alcohols, pigments, waxes and hydrocarbons. Since these components are non-polar there may be a possibility that they remain in the biodiesel after the transesterification reaction. The determination of unsaponifiable matter is described in AOCS Method Ca 6a–40, which is applicable to normal animal and vegetable fats and oils. This method is not suitable for fats and oils containing an excessive amount of unsaponifiable matter, such as marine oils. In addition, this method does not apply to feed grade fats.

For fats and oils containing higher levels of unsaponifiable matter than usually found in normal tallow and greases, AOCS Method Ca 6b–53 must be used. It is specially suited for marine oils, but is also applicable to vegetable oil deodorised distillates and sludges.

Polar compounds in frying fats include monoglycerides, diglycerides and free fatty acids in unused fats, as well as polar transformation products formed during frying of foodstuffs and/or heating. Polar compounds formed during the heating of fats give information about the levels of deterioration found in used frying fats. Polar compounds in oils and fats are determined by column chromatography under the conditions specified in AOCS Method Cd 20–91. Frying oils and fats are separated by column chromatography into non-polar and polar compounds, followed by the elution of the non-polar compounds. The polar compounds are determined by calculating the difference between the weight of the test sample added to the column and that of the non-polar fraction eluted.

Determination of polymerised triglycerides by gel-permeation HPLC can be done according to AOCS Method Cd 22–91. The separation is based on the relative retention of solubilised polymer molecules in terms of their molecular size by gel permeation chromatography.

Sulphur: Generally, biodiesel contains lower sulphur than fossil diesel. However, owing to the diversity of biodiesel feedstocks, questions have been raised about their sulphur content and the sulphur content of the biodiesel made from them, which is limited to 15 ppm maximum by ASTM D6751. Sulphur in oilseeds, seed meals, oils and fats and biodiesel can be determined by ASTM D7039.

The samples can also be analysed conforming to ASTM D5453. This method is capable of measuring sulphur content in liquid and solid samples. In this method, a hydrocarbon sample is placed in a sample boat. The sample boat is then inserted into a high temperature combustion tube where the sulphur is oxidised to sulphur dioxide (SO_2) in an oxygen rich atmosphere. Water produced during the sample combustion is removed. The sample combustion gases are next exposed to UV light. The SO_2 absorbs the energy from the UV light and is converted to excited SO_2. The fluorescence emitted from the excited SO_2 as it returns to a stable state, SO_2, is detected by a photomultiplier tube and the resulting signal is a measure of the sulphur contained in the sample. The detection limit is in the ppb (0.001 ppm) range and the upper limit is 1% (or 10 000 ppm) by weight. The sample weight was in the range 30–50 mg and the oven temperature was set at 1000°C.

Phosphorous is a minor oil or fat component typically associated with phospholipids and gums that may act as emulsifiers or cause sediment, lowering yields during the transesterification process. ASTM D6751 requires phosphorous in biodiesel be limited to 10 ppm (0.001% mass maximum).

There are several methods for the determination of phosphorous content in fats and oils. AOCS Methods Ca 12–55 and Ca 12a–02 are based on spectrophotometric and colorimetric measurements respectively. Alternatively, AOCS Method Ca 12b–92 requires the vaporisation of the oil or fat in a suitable graphite furnace, connected to an atomic absorption spectrometer, previously calibrated using standard solutions of soy lecithin and measurement of the phosphorus content from the observed absorption at a wavelength of 213.6 nm. In line with this, AOCS Method Ca 20–99 explains the Analysis for Phosphorus in Oil by Inductively Coupled Plasma Optical Emission Spectroscopy (ICP-OES). Finally, AOCS Method Ca 19–86 describes the determination of Phospholipids in

Vegetable Oils by a Nephelometric Method, which measures turbidity in oil–acetone mixtures due to phospholipids. The turbidity is correlated to the phosphorus level.

Sodium and calcium: ASTM D6751 requires a combined amount of alkaline metal (Na + K) to be lower than 5ppm. In the same way, calcium and magnesium in biodiesel are limited to 5 ppm. The determination of Na and Ca is based on AOCS Method Ca 15b–87, applicable to unemulsified fats and oils. Sodium and calcium (soaps) present in a fat or oil sample are acidified and then extracted into a known volume. The sodium and calcium concentrations are measured using flame atomic absorption spectrophotometry (AAS) and can be directly measured after calibration of the instrument.

Alkalinity: Fats and oils can contain an alkaline constituent either naturally (e.g. calcium soaps from bone) or accidentally (e.g. sodium soaps in imperfectly refined fats and oils). The AOCS Method Cd 3e–02 determines the alkalinity (proportion of alkali) in a fat or oil by titration with hydrochloric acid. The alkalinity of animal and vegetable fats and oils is determined without distinguishing between the various constituents. This method is not applicable to dry melted animal fats, nor to oils and fats with an acidity greater than 60% (mass fraction).

Trace elements: AOCS Method Ca 17–01 describes the determination of Calcium, Copper, Iron, Magnesium, Nickel, Silicon, Sodium, Lead and Cadmium in Oil by Inductively Coupled Plasma Optical Emission Spectroscopy. The oil sample is energised to make elements emit or absorb a quantifiable amount of energy, which indicates the element's concentration in the fuel.

Almost all forms of vegetable oils may be analysed, from crude oil to degummed, refined, bleached, deodorised and hardened, depending on the dilution solvent. This procedure is suitable only when the elements are present in a solubilised form. When they are present as fine particles (bleaching earth, catalysts, or rust), inductively coupled plasma optimal emission spectroscopy (ICP-OES) analysis results in poor recovery owing to nebulisation and atomisation problems. The only suitable non-ashing direct method for these samples is graphite furnace AAS.

Oxidation stability index (OSI): The OSI determination conforming to AOCS Method Cd 12b–92 consists in passing a stream of purified air through a test sample of oil or fat which is held in a thermostatic bath. The effluent air from the oil or fat test sample is then bubbled through a vessel containing deionised water. The effluent air contains volatile organic acids, swept from the oxidising oil that increase the water conductivity as the oxidation proceeds. The OSI is defined as the point of maximum change of the rate of oxidation. The OSI result should specify the OSI time, with the analysis temperature (100–140°C) reported immediately afterwards. Feedstock oxidation stability is primarily influenced by the presence of hydrogen atoms next to carbon–carbon double bonds and the presence of natural antioxidants that can prevent oxidation of the triglyceride molecules.

Peroxide value: Fat stability is also measured by this parameter which can be measured by AOCS Method Cd 8b–90. This method determines all substances, in terms of milliequivalents of peroxide per 1000 g of test sample that oxidise potassium iodide under the conditions of the test. The substances are generally assumed to be peroxides or other similar products of fat oxidation. This method is applicable to all normal fats and oils, including margarine. This method is highly empirical and any variation in the test procedure may result in erratic results.

p-Anisidine value is defined by convention as 100 times the optical density measured at 350 nm in a 1 cm cuvette of a solution containing 1.00 g of the oil in 100 mL of a mixture of solvent and reagent according to the procedure described in AOCS Method Cd 18–90.

Acidity: The acid value is the number of milligrams of KOH necessary to neutralise the free acids in one gram of test sample. With test samples that contain virtually no free acids other than fatty acids, the acid value may be directly converted by means of a suitable factor to per cent free fatty acids. The acid value can be determined by AOCS Method Cd 3d–63.

Free fatty acid (FFA): Interaction of FFA in the feedstock with base catalysts like sodium methoxide may form emulsions which make separation of the biodiesel more difficult, possibly leading to yield loss. Emulsions can also increase biodiesel cost by introducing extra cleaning steps and replacement of filters. The FFA determination was performed following two methods: ASTM D664, Method A, was first used to determine TAN in the samples, after this, the FFA values were calculated using the mathematical formulas found in the AOCS Method Ca 5a–40.

Saponification value is defined as the amount of potassium hydroxide (KOH) in milligrams required to saponify one gram of fat or oil under the conditions specified (AOCS Method Cd 3–25 and AOCS Method Cd 3c–91). The method includes refluxing the known amount of fat or oil with a fixed but excess amount of alcoholic KOH. The amount of KOH remaining after hydrolysis was determined by back titrating with standardised 0.5 N HCl and the amount of KOH consumed during saponification was calculated.

According to the literature (Sanford *et al.*, 2009), the saponification value for the majority of the feedstocks is in the range 185–210 mg KOH/g. This range is typical for feedstocks that have predominately fatty acids with a chain length between C16 and C18 (Knothe *et al.*, 2005) Based on the length of the fatty acids present in the triacylglycerol molecule, the weight of the triacylglycerol molecule changes, which in turn affects the amount of KOH required to saponify the molecule. Hence, the saponification value is a measure of the average molecular weight or the chain length of the fatty acids present.

Fatty acid profile: To determine the fatty acid profile of a biodiesel feedstock it is necessary to begin with the preparation of methyl esters of fatty acids (AOCS Method Ce 2–66). This method implies the reaction of fats and oils with a methanolic solution of sodium hydroxide or potassium hydroxide if acid value is lower than 2, while fatty acids are reacted with a specified amount of

BF3-methanol reagent. Then, methyl esters are recovered and analysed by packed column gas chromatography (GLC) according to AOCS Method Ce 1–62.

Iodine value gives a measure of the average degree of unsaturation expressed as the grams of iodine absorbed per 100 g of oil. It can be determined by the 'Wijs method', AOCS Cd 1–25 where the oil reacts with a known excess of iodine chloride.

$$\text{R–CH} = \text{CH–R} + \text{ICl}_{\text{excess}} \rightarrow \text{R–CHI–CHCl–R} + \text{ICl}_{\text{remaining}} \quad [5.1]$$

The amount of ICl that has reacted is determined by measuring the amount of ICl by the addition of excess potassium iodide to the solution to liberate iodine and then titrating with a sodium thiosulphate ($Na_2S_2O_3$) solution in the presence of starch to determine the concentration of iodine released:

$$\text{ICl}_{\text{remaining}} + 2\text{KI} \rightarrow \text{KCl} + \text{KI} + \text{I}_2 \quad [5.2]$$

$$\text{I}_2 + \text{starch} + 2\text{Na}_2\text{S}_2\text{O}_3\ (\text{blue}) \rightarrow 2\text{NaI} + \text{starch} + \text{Na}_2\text{S}_4\text{O}_6\,(\text{colourless}) \quad [5.3]$$

The concentration of C=C in the original sample can therefore be calculated by measuring the amount of sodium thiosulphate needed to complete the titration. In samples containing conjugated double bonds, the iodine value obtained is empirical since the reagent does not react stoichiometrically with conjugated unsaturation.

5.3 Characterisation, analysis techniques and quality control of biodiesel

The evaluation of biodiesel quality is achieved by the determination of chemical composition and physical properties of the fuel. Several factors such as composition of feedstock (oil or fat), production process (reaction and purification steps), storage and handling can influence biodiesel fuel quality. The fatty ester composition, along with the presence of contaminants and minor components, dictate the fuel properties of biodiesel. In this sense, important properties of biodiesel (low temperature operability, oxidative and storage stability, kinematic viscosity, exhaust emissions and cetane number), which will also influence its storage and oxidation, are directly influenced by fatty ester composition and the presence of contaminants and minor components (Monteiro *et al.*, 2008; Knothe, 2005). These and other contaminants of biodiesel can lead to severe operational and environmental problems. Therefore, standards that limit the amount of contaminants in biodiesel fuel are necessary. Knothe (2005) stated that the successful introduction and commercialisation of biodiesel has been accompanied by the development of standards to ensure high product quality and user confidence.

EN14214 and ASTM D 6751 biodiesel standards were already established in Europe and USA, respectively and their specifications are illustrated in Table 5.1.

Table 5.1 Specifications for biodiesel (B100)

Property	ASTM D6751 Limits	UNE-EN 14214 Limits
Acid value	≤0.5 mg KOH/g	≤0.5 mg KOH/g
Carbon residue	≤0.05 wt% (on 100% sample)	≤0.3 wt% (on 10% distillation residue)
Cetane number	≥47	≥51
Cloud point	According to climate zone (°C)	–
Cold filter plugging point	–	According to climate zone (°C)
Cold soak filterability test	≤360 s	–
Copper strip corrosion (3 h, 50°C)	≤3 classification	1 classification
Density at 15°C	–	860–900 kg m^{-3}
Kinematic viscosity at 40°C	1.9–6 mm^2 s^{-1}	3.5–5 mm^2 s^{-1}
Distillation temperature	≤360°C	
Flash point	≥93°C	≥101 (°C)
Free glycerol	≤0.02 wt%	≤0.02 wt%
Total glycerol	≤0.240 wt%	≤0.25 wt%
Group I metals (Na + K)	–	–
Group II metals (Ca + Mg)	≤5.0 mg kg^{-1}	≤5.0 mg kg^{-1}
Iodine value	–	≤120 g I_2 g^{-1}
Methanol content	≤0.2 wt% or flash point ≥130°C	≤0.2 wt%
Methyl ester content	–	≥96.5 wt%
Linolenic acid methyl ester	–	≤12.0 wt%
Polyunsaturated	–	≤1 wt%
Monoglyceride content	–	≤0.8 wt%
Di, triglyceride content	–	≤0.2 wt%
Oxidation stability, 110°C	≥3 h	≥6 h
Phosphorus content	≤10 mg kg^{-1}	≤4.0 mg kg^{-1}
Sulphated ash content	≤0.02 wt%	≤0.02 wt%
Sulphur content	≤15 mg kg^{-1}	≤10 mg kg^{-1}
Total contamination	–	≤24 mg kg^{-1}
Water and sediment	≤0.05% volume (+ sediment)	≤500 mg kg^{-1}

Both standards identify specifications that B100 must meet before being blended with petroleum diesel typically in blends up to B7 or B20 according to EN 590 and ASTM D7467, respectively. In general, the European Union EN 14214 specification is similar to ASTM D6751 but is somewhat more stringent in some elements of the specification.

In Europe, biodiesel B100 may be directly used as heating oil. A comparison of the two biodiesel standards, one for vehicle use (EN 14214) and one for heating oil use (EN 14213) in Europe, reveals that most specifications are the same or very similar. The cetane number, a diesel engine-specific parameter, is, for obvious reasons, not included in the heating oil standard. Since heating oil burners and/or the conditions under which biodiesel is used as heating fuel are less sensitive to some components or contaminants, the heating oil standard does not include the specifications of Group I (Na + K) and Group II (Ca + Mg) metals as well as methanol and phosphorus. Also, the requirement for oxidative stability is more lenient in the heating oil standard. In addition, there is no restriction on the amount of linolenate in the biodiesel fuel, although the iodine value is set at 130, slightly higher than for vehicle use. The heating oil standard contains a pour point specification not found in the standard for vehicle use. The cold filter plugging point (CFPP) is listed in EN 14213 but without limits, but it is contained in EN 14214 with geographical and seasonal limits. The heating oil standard also contains, for obvious reasons, a heating value specification. The heating value specification would let virtually all biodiesel fuels be suitable for heating oil purposes (Knothe, 2005). Alternatively, ASTM D396 includes specifications for fuel oil with 5% biodiesel which must satisfy the requirements of ASTM D6751 before inclusion with the heating fuels.

The next section describes the parameters, measurement methods and limit values of the specifications used in EN 14214 and ASTM D6751 standards. These parameters are classified attending to biodiesel origin (feedstocks characteristics), production process (catalytic reaction and subsequent product purification) and biodiesel final use (fuel and physical properties).

5.3.1 Biodiesel parameters related to feedstock characteristics

The reason for restrictions on the FA profile, contained mainly in EN 14214, is to exclude components of biodiesel with less desirable properties, for example, with respect to oxidative stability. In practice, this amount can exclude certain feedstocks for biodiesel production, as described below:

(i) *Linolenic acid methyl ester content*: The content of methyl linolenate is restricted (12%) in EN 14214 because of the propensity of methyl linolenate to oxidise. As explained above, the EN 14103 Standard Test Method describes the determination of the ester content of fatty acid methyl esters (FAME) of biodiesel and it also allows determination of the linolenic acid methyl ester content. With this method it is possible to verify that the linolenic acid content is between 1% and 15%. This method is suitable for FAME which contain methyl esters between C14 and C24.

(ii) *Content of polyunsaturated (≥4 double bonds) fatty acid (PUFA) methyl esters*: This specification serves to eliminate fish oil, with their higher content of methylene-interrupted double bonds fatty acids, as biodiesel feedstock because PUFA are very prone to oxidation. EN 15779 Standard Test Method specifies the determination of the polyunsaturated content of FAME as a whole between 0.6% (m/m) and 1.5% (m/m). The method covers the predominant four polyunsaturated fatty acid methyl esters of eicosatetraenoic acid (C 20:4 (*n*–6)), eicosapentaenoic acid (C 20:5 (*n*–3)), docosapentaenoic acid (C 22:5 (*n*–3) and docosahexaenoic acid (C 22:6 (*n*–3)). Based on the linearity of results, PUFA methyl esters can be determined in FAME in the range between 0.3% (m/m) to 3.0% (m/m). This European Standard is based on AOCS Official Method Ce 1b–89, which determines the fatty acid composition of marine oils and marine oil esters, by capillary column gas-liquid chromatography.

(iii) *Iodine value* is an indicator of unsaturation in an organic compound. It is also an indicator of stability, whereby a high iodine value is related to the level of polymerisation, which may lead to injector fouling. Biodiesels with low iodine values and therefore relatively low levels of unsaturation, are less sensitive to oxidation processes. EN 14214 specifies a maximum of 120 mg iodine/100 mg sample measured through standard method EN 14111 based on the classic wet chemical titrimetric method (Wijs) previously described in Section 5.2.

(iv) *Oxidative stability* is an indication of the degree of oxidation, potential reactivity with air and can determine the need for antioxidants. Biodiesel fuels with longer OSI times are more stable to oxidation than samples with shorter values. Oxidative stability of biodiesel is determined by measurement of the OSI by the Rancimat method EN 14112 Standard Test Method. The AOCS official method Cd 12b–92 explained previously is nearly identical to EN 14112 and provides essentially interchangeable data. Products that ultimately form by oxidation of lipids may include aldehydes, shorter chain fatty acids, other oxygenated species (such as ketones) and polymers. For this reason, oxidative degradation has a negative impact on acid value and kinematic viscosity, both of which are specified in ASTM D6751 and EN 14214.

(v) *Ester content and distillation temperature*: These parameters are important tools for determining the presence of other substances and in some cases meeting the legal definition of biodiesel (i.e. mono-alkyl esters). Low values of pure biodiesel samples may originate from inappropriate reaction conditions or from various minor components within the original fat or oil source. A high concentration of unsaponifiable matter such as sterols, residual alcohols, partial glycerides and unseparated glycerol can lead to values below the limit. In general, ester content is usually met if other parameters are met. The standard reference method EN 14103 is indicated for determining ester content in EN 14214. Alternatively, ASTM D 6751 prescribes the use of ASTM D 1160 to determine the distillation curve of the fuel and establishes values below 360°C.

Phosphorus and sulphur can be carried over from vegetable oils, for example, from phospholipids present in all vegetable oils. It must be ensured that they are not found in other biodiesel sources, such as used frying oils or animal fats, which can come into contact with extraneous materials containing these elements.

(vi) *Sulphur*: Although most biodiesel fuels inherently contain little or no sulphur, the specification is important to show that biodiesel will not negatively affect automotive catalyst systems. In ASTM D 6751, sulphur is determined through ASTM D 5453 standard method by UV fluorescence of the sample. The standard method ISO 20846 in EN 14214 uses the same approach. The alternative method ISO 20884 in EN 14214 uses wavelength-dispersive X-ray fluorescence spectrometry. These methods specifically mention their applicability to neat biodiesel or B100 (ASTM) or in blends up to 5% (ISO).

(vii) *Phosphorus*: Traces of phosphorus, resulting from phospholipids, can remain in vegetable oils after refining. Phosphorus can poison catalysts used for reduction of exhaust emissions. Catalytic converters are becoming more common in diesel-powered equipment as emissions standards are tightened, so low phosphorus levels will be of increasing importance. ASTM D 4951 (in ASTM D6751) and EN 14107 (in EN 14214) both use ICP emission spectrometry, differing only in the wavelengths specified.

5.3.2 Biodiesel parameters related to the production process

During the transesterification process, intermediate mono- and di-acylglycerol (MAG and DAG) are formed, small amounts of which can remain in the final biodiesel product. In addition these partial glycerols, unreacted triacylglycerol (TAG) as well as unseparated glycerol, FFA, residual methanol and catalyst can contaminate the final product. The contaminants can lead to severe operational problems when using biodiesel, such as engine deposits, filter clogging, or fuel deterioration, as described below:

(i) *Mono-, di- and tri-acylglycerols*: the EU standard specifies individual limit values for mono-, di- and tri-acylglycerides (see Table 5.1); however the standards for USA do not provide explicit limits for the content of partial acylglycerides. EN14105 European Standard Method determines residual mono-, di- and triglyceride contents in FAME. Mono-, di- and triglyceride contents are important in describing the quality of biodiesel to ensure that the fuels can be used over a wide range of conditions. In this sense, the presence of saturated fatty acids bound to glycerol have a deleterious effect on low-temperature operability of biodiesel owing to its very low solubility in methyl esters. As a result, high temperatures are required to keep them from crystallising.

(ii) *Free and total glycerol* are present in insufficiently purified biodiesel. Both ASTM D6751 and EN 14214 limit the maximum allowable levels of free and total glycerol (see Table 5.1), because glycerol is suspected of contributing to engine deposit formation during combustion (Knothe *et al.*, 2005). Bound glycerol in biodiesel results from incomplete conversion of triglyceride into FAME during the transesterification reaction and may cause carbon deposits on the fuel injector tips and piston rings of diesel engines during combustion (Knothe *et al.*, 2005). The presence of bound glycerol in biodiesel may also influence low-temperature operability and kinematic viscosity. EN14105 and ASTM D6584 Standard Methods determine free and total glycerol by GC according to EN 14214 and ASTM D6751, respectively.

(iii) *Flash point*: The flash point specification serves to restrict the amount of alcohol in the biodiesel fuel. The flash point is the lowest temperature at which fuel emits enough vapour to ignite. Biodiesel has a high flash point, usually more than 150°C, while conventional diesel fuel has a flash point of 55–66°C. If methanol, with its flash point of 12°C is present in the biodiesel, the flash point can be lowered considerably. The prescribed methods for flash point determination are ASTM D 93 in ASTM D 6751 and ISO 3679 in EN 14214. These methods consist of the controlled heating of the biodiesel in a closed cup, introducing an ignition source and observing if the heated biodiesel flashes. The temperature at which the biodiesel flashes is recorded as the flash point. For biodiesel, a flash point of below 93°C in ASTM D6751 or 101°C in EN 14214 is considered to be out of specification (Table 5.1). If the biodiesel has not flashed at 160°C, the test is finished and the result is reported as >160°C.

(iv) *Methanol content*: If biodiesel is contaminated with methanol, it will fail to meet the minimum flash point specifications. Methanol contamination normally results from insufficient purification of biodiesel following the transesterification reaction. As explained above, methanol in biodiesel is indirectly measured through flash point determination following ASTM D93 or by EN 14110 Standard Test Methods. EN 14110, contained in EN 14214, is a GC-based method applied to mixtures containing from 0.01–0.5% methanol.

5.3.3 Standards related to biodiesel properties and use

Standards related to biodiesel properties and use are described below:

(i) *Density* is only specified in EN 14214 and the purpose of its determination is to exclude extraneous material as biodiesel feedstock. Biodiesel density can be determined according to EN ISO 3675 Standard Test Method using a glass hydrometer at 15°C and also by EN ISO 12185 Standard Test Method using an oscillation U-tube densitometer, within the range 600–100 kg m^{-3}.

(ii) *Kinematic viscosity*: The reduction in viscosity is the major reason why alkyl esters of vegetable oils—biodiesel—are used as fuel and not the neat oil. Thus, the limits of this property are in the range of most, but not all, common vegetable oil (methyl) esters and serve to exclude vegetable oils as fuel. The higher viscosity of the neat oil causes operational problems such as engine deposits. In fact ASTM D 6751 prescribes the use of ASTM D 445 and EN 14214 utilises ISO 3104/ISO 3105, with ISO 3105 being the specification and operating instructions for the viscometers used in ISO 3104. This physical property can also be used to restrict the FA profile. For example, shorter-chain FA are excluded by the relatively high minimum value for kinematic viscosity in EN 14214 (3.5 mm^2 s^{-1}). Although the minimum value for kinematic viscosity prescribed in ASTM D 6751 (1.9 mm^2 s^{-1}) overlaps most diesel fuels, the high minimum kinematic viscosity value for biodiesel prescribed in EN 14214 (5 mm^2 s^{-1}) is higher than that of many diesel fuels, underscoring the feedstock-restrictive nature of the EN 14214 limit.

(iii) *Cetane number*: Both ASTM D 6751 and EN 14214 specify ASTM D613 and ISO 5165 Standard Test Methods using a cetane engine, an engine specifically modified for testing cetane number. Cetane number is one of the primary indicators of diesel fuel quality. It is related to the ignition delay time that a fuel experiences once it has been injected into a diesel engine's combustion chamber. Generally, shorter ignition delay times result in higher cetane numbers and vice versa. Hexadecane, also known as cetane (trivial name), which gives the cetane scale its name, is a high-quality reference standard with a short ignition delay time and an arbitrarily assigned CN of 100. The compound 2,2,4,4,6,8,8-heptamethylnonane is the low quality reference standard with a long ignition delay time and an arbitrarily assigned CN of 15 (Knothe, 2006). In general, cetane number is influenced by chain length and higher values are obtained for longer chain compounds.

(iv) *Cloud point and cold-filter plugging point (CFPP)*: The behaviour of automotive diesel fuel at low temperatures is an important quality criterion, as partial or full solidification of the fuel may cause blockage of the fuel lines and filters, leading to fuel starvation and problems in starting, driving and engine damage owing to inadequate lubrication. The melting point of biodiesel products depends on chain length and degree of unsaturation, with long chain saturated fatty acid esters displaying particularly unfavourable cold temperature behaviour.

Cloud point is defined as the temperature at which crystal growth is large enough (diameter ≥ 0.5 μm) to be visible to the naked eye. At temperatures below the cloud point, larger crystals fuse together and form agglomerations that eventually become extensive enough to prevent pouring of the fluid. ASTM D 2500 and ISO 3015 Standard Test Methods are prescribed for cloud point determination. The cold-filter plugging point is defined as the lowest temperature at which a given volume of biodiesel completely flows under vacuum through a wire mesh filter screen within 60 s. CFPP is generally considered to be a more reliable indicator of low-temperature operability than cloud point since the fuel will contain solids of sufficient size to render the engine inoperable owing to fuel filter plugging once the CFPP is reached (Moser, 2009). CFPP can be determined by ASTM D6371 and EN 116 Standard Test Methods. Each country using EN14214 can specify certain temperature limits for different times of year depending on climate conditions. The cloud point is not limited in ASTM D6751; rather a 'report' is specified. The reason is that the climate conditions in the USA vary considerably and therefore the needs of biodiesel users vary accordingly.

(v) *Carbon residue*: Carbon residue formed by decomposition and subsequent pyrolysis of the fuel components can clog the fuel injectors The carbon residue tests (ASTM D 4530 in ASTM D 6751; ISO 10370 in EN 14214) are designed to indicate the extent of deposits that result from the combustion of a fuel. The amount of carbon residue formed after evaporation and pyrolysis of the weighed petroleum sample is determined and the results correspond to the Conradson Carbon Residue test (ASTM D 189). For an expected test result of less than 0.10%, the sample can be distilled to give the remaining 10% of the original sample, which is the material that will then be tested.

(vi) *Sulphated ash* tests (ASTM D 874 in ASTM D 6751; ISO 3987 in EN 14214) are designed to determine sulphated ash from lubricating oils containing various metal-containing additives. Metals that are covered include Ba, Ca, Mg, Na, K and Sn, although S, P and Cl can be present in combined form. To carry out this test, the sample is burned fully with only ash and carbon remaining. This residue is treated with sulphuric acid and heated until oxidation of carbon is complete. The ash is cooled, treated again with sulphuric acid and heated to constant weight. The determination of residual Na or K from the catalyst can obviously be applied to biodiesel.

(vii) *Total contamination*: EN 12662, contained in EN14214, is a method for determining contamination as the content of undissolved substances in middle distillates (in mg kg^{-1}). It applies to liquid petroleum products with a kinematic viscosity <8 mm^2 s^{-1} at 20°C or <5 mm^2 s^{-1} at 40°C.This European Standard specifies a method for determining contamination as the content of undissolved substances in diesel fuel containing up to 5% (V/V) FAME and in 100% (V/V) FAME from 6 mg kg^{-1} to 30 mg kg^{-1}. Excessive contamination in a fuel system can give rise to premature blocking of filters and/or hardware failure and is therefore undesirable. This standard applies to liquid petroleum products that have a kinematic viscosity not exceeding 8 mm^2 s^{-1} at 20°C, or 5 mm^2 s^{-1} at 40°C, for example diesel fuel, as specified in EN 590 or light fuel oils.

(viii) *Sodium, potassium, calcium and magnesium*: These may be present in biodiesel as abrasive solids or soluble metallic soaps. Abrasive solids can contribute to injector, fuel pump, piston and ring wear, as well as to engine deposits. Soluble metallic soaps have little effect on wear, but they may contribute to filter plugging and engine deposits. High levels of Na, K, Ca and Mg may also be collected in exhaust particulate removal devices; they are not typically removed during passive or active regeneration; they can create increased backpressure and reduced period to service maintenance as explained in ASTM D6751 standard. This specification is also contained in the EN 14214 standard. Both standards prescribe EN 14538 standard method for Na, K, Ca and Mg determination by optical emission spectral analysis with inductively coupled plasma (ICP OES).

(ix) *Copper strip corrosion*: The copper strip corrosion test (ASTM D 130 in ASTM D 6751; ISO 2160 in EN 14214) consists of dipping a strip of copper into the fuel for a specified time and defined temperature and observing the corrosive action of the fuel. It is a test for corrosive acids or sulphur compounds in the fuel. The degree of tarnish on the corroded strip correlates with the overall corrosiveness of the fuel sample.

(x) *Water or water and sediment*, is a major source of fuel contamination. While fuel leaving a production facility may be virtually free of water, once it enters the existing distribution and storage network, it will come into contact with water as a result of environmental humidity (Knothe *et al.*, 2005). Water in biodiesel causes three serious problems: corrosion of engine fuel system components, promotion of microbial growth and hydrolysis of FAME. Consequently, both ASTM D6751 and EN 14214 contain strict limits on water content in biodiesel (see Table 5.1). Water may be present in biodiesel as either dissolved or free water. Dissolved water, which is measured by the Karl Fisher titration method in EN 14214 according to EN ISO 12937 Standard Test Method, is water that is soluble in biodiesel. Free water, which is measured by a centrifugation method (ASTM D2709) Standard Test Method in ASTM D6751, arises after biodiesel becomes saturated with water, resulting in a separate water phase.

5.4 Development of standards for biodiesel quality and use

European standard EN 14214 describes biodiesel being used as a stand-alone diesel fuel or as a blending component in conventional hydrocarbon-based diesel fuel. Currently biodiesel is mixed according to EN 14214 and EN 590 with up to 7% v/v of fossil diesel fuel (B7). With a further increase of biodiesel fraction to 10% (B10 blend), there will be some uncertainties regarding the undisturbed long-life behaviour owing to changes in FAME physicochemical properties compared to commercial diesel fuel.

At 100% concentration, biodiesel is applicable to fuel in diesel engines vehicles designed or subsequently adapted to run on 100% FAME. In Germany, the biodiesel producers association (VDF), oil seed promoters (UFOP) and the quality control agency AGQM provide lists of Commercial Vehicle Manufacturer approvals on using biodiesel as B100 or B30 (UFOP *et al.*, 2010). In general, these vehicles are heavy duty engines (trucks, large buses, industrial machinery, agricultural, marine and mining applications, etc.) and usually perform as captive fleets. The product information from the commercial vehicle manufacturers requires only use of biodiesel that complies with DIN EN 14214 for problem-free engine operation and warranty assurances. Engine manufacturers also recommend users to obtain quality assurances from biodiesel suppliers.

Biodiesel can also be used as heating oil. European Standard EN 14213 specifies requirements and test methods for marketed and delivered FAME to be used either as heating fuel for diesel engines at 100% concentration, or as a blending component for the production of heating fuel. At 100% concentration, it is applicable to fuel for use in heating devices designed or subsequently adapted to run on 100% FAME.

In USA ASTM D6751 establishes standard specifications for biodiesel fuel blend stock (B100) for middle distillate fuels. There has been substantial effort since 2003 to develop and formally approve specifications for the finished blend of biodiesel and conventional fuel. B5 has been accepted into the conventional diesel specifications for on/off road diesel fuel (ASTM D975) and fuel oil burning equipment (ASTM D396). Furthermore, a new stand alone specification covering biodiesel blends with diesel fuel oil between 6% and 20% has been approved (ASTM 7467). The main characteristics of these standards are listed by Nadkarni (2010).

In May 2006, New Holland (a major engine manufacturer in USA) informed the National Biodiesel Board, that it approved the use of up to B20 blends and announced full support for use of high quality B20 in all equipment having New Holland engines. Mercedes-Benz approved the use of B5 according to ASTM D975 (standard ultra low sulphur diesel) in all common rail injection (CDI) and BlueTEC diesel engines. Diesel fuels containing a higher percentage of biodiesel (B6-B20) according to ASTM D7467 as well as straight biodiesel (B100 /ASTM D 6751) may cause severe damage to the engine/fuel system and are not approved (car passenger and light duty truck applications equipped with diesel particulate traps).

5.4.1 Improving test methods and proficiency testing programmes

There are a number of test methods specified for biodiesel that up to now have not had the precision obtained for petroleum products. This may be due to different reasons: the lack or the unsuitability of reference materials, interference from some constituents of the biodiesel matrix, the lack of sensibility of the established detection limits in the standards, and so on. One measure of how well a test method is working or how well a laboratory or a group of them is performing can be deduced from the performance of that entity in an ILCP (interlaboratory comparison programme) conducted by national institutes of metrology and standardisation.

The Institute for Interlaboratory Studies (IIS, the Netherlands) has organised a Proficiency Test for FAME used as B100 every year, since 2001. In the penultimate interlaboratory comparison programme held in September 2010, 51 laboratories from 23 different countries participated. In this case a rapeseed biodiesel sample was analysed by each laboratory and in the final report 10G05 (IIS, 2010) details of the sample preparation, quality system, tests methods used in accordance with the requirements of EN14214 and/or ASTM D6751 and evaluation results were given. These evaluation results were discussed for each test method by the participating laboratories and compared with the previous proficiency tests.

For many tests there was not good compliance by the laboratories with the relevant standards. The problematic tests were acid number, carbon residue, kinematic viscosity, sulphur (ASTM D5453), methanol, di-glycerides, tri-glycerides, free glycerol, total glycerol, total ester and total contamination. In most of them the calculated reproducibility, after rejection of the statistical outlier, was not in agreement with the requirements of the assayed standard test. The performance of the determinations of the proficiency tests was compared against the requirements of the respective standards. Sulphated ash, sulphur (ASTM D5453), methanol, total glycerol and total contamination were categorised as '– –' (group performance was much worse than the standard).

ASTM's Biodiesel Proficiency Test in cooperation with AOCS provides laboratories who need it with a statistical quality assurance tool that will enable them to improve and maintain a high level of performance in conducting routine ASTM biodiesel analyses by comparing their results with those of other laboratories worldwide doing the same testing. The programme is described by Nadkarni and McKlindon (2009). Test performance index (TPI) is a term used for the efficacy of a test method or a laboratory and is defined in ASTM D6792, as the ratio of test method reproducibility divided by the robust reproducibility of proficiency testing programme data. A TPI >1.2 means the performance of the group providing data is probably satisfactory relative to the corresponding ASTM published precision. TPI values in the range of 0.8–1.2 means the performance may be marginal and each laboratory should consider reviewing the test method procedures to identify opportunities for improvement. TPI <0.8 means the performance is not consistent with ASTM published precision and laboratory method performance improvements and should be investigated by all the laboratories. The subcommittee responsible for the test should also review if the published precision reflects reality. Whether an improvement in the test methods employed in PTP is needed or not can be judged from the TPIs. Some TPIs are satisfactory, but many are not, indicating that (a) an improvement in the laboratory performance is needed; (b) a test method improvement is needed; (c) alternate test methods need to be investigated; and/or (d) product specification limits need to be reviewed in order to conform with the expected test variability. The TPIs calculated for the tests used in 2008 biofuels ILCPs are summarised by Nadkarni and McKlindon (2009) and Nadkarni (2010). There are very few tests for which satisfactory TPIs are obtained. This clearly reinforces the conclusion that for a number of test methods used in biofuels specifications, further work in development of appropriate test methods is necessary before they could be simply extrapolated from their usage in petroleum products analyses.

On the other hand, the accuracy and precision of any quantitative analysis begins with the quality of the calibration standards used. Reference materials for biofuels with well-characterised property values are essential for the development and validation of measurement methods. Also, these materials are an important tool in quality assurance of the day-to-day measurements, that is, in obtaining reliable, traceable measurement results. The aim of the BIOREMA PROJECT (VSL *et al.*, 2010) was to demonstrate the feasibility of preparing and characterising reference materials for biofuels. The focus is on providing SI-traceable reference values. Further, the project established the current state of measurement quality by means of interlaboratory comparisons using these test materials. The organisations involved in this project were: VSL (the Netherlands) project coordinator, IRMM (European Commission), NPL (United Kingdom), INMETRO (Brazil), NIST (USA) and LGC (United Kingdom). The information supplied on the applied methodology and the performance characteristics can be used for the development of harmonised standard methods. An additional aim was to provide guidance and material for producing reference materials for the different test methods required by the ultimate global biofuel blend specifications.

The main discussion points dealt with were:

- the difficultly of measuring specific parameters such as FAME content in biodiesel;
- the need with respect to 'fit for purpose' Certified Reference Materials for biofuel measurements;
- the requirements in the ISO/IEC 17025 standard for the accreditation of testing laboratories relative to the quotation of uncertainty estimates in measurement results and to the frequency of participation in ILCs.

5.4.2 Biodiesel specification standards: global harmonisation efforts

European fuel quality specification EN 14214 was set in 2003 by the European Committee for Standardization (CEN). By request of the European Commission, it was a FAME specification for both 100% fuel and for blends up to 5% in diesel fuel (7% at present). At that time it was already known that some test methods had problems in terms of precision, application or even were not available at all. Meanwhile, a group of experts are updating the test methods and have suggested alternatives to the CEN community.

Because usual feedstock, market use, car park and regular diesel in the USA differ from those in Europe, the biodiesel specification shows many differences. Since the beginning of 2007, Brazilian, American and European authorities combined their initiatives towards an increase in bio-energy. The European Council established an independent EU commitment to reduce greenhouse gases by at least 20% by 2020 and President Bush gave the same targets in his State of the Union message. Together with India, China and South Africa, an International Biofuels Forum was set up in March 2007. To make biofuels a success story, they agreed that joint standards were needed. For these 'International Harmonisation Efforts' they established two task forces of biofuel standardisation experts. The goal of the task forces was to submit a White Paper to the Brazilian, EC and USA authorities by the end of 2007. The report was circulated to the standardisation bodies of the International Biofuels Forum members so that further evolution of the current standards could take into account closer alignment with other members' standards, rather than diverging from them. This White Paper was the basis of further work at ISO level to establish global biofuel standards.

Summaries, analyses, observations and conclusions of this report are compiled below (Tripartite Task Force, 2007). The tripartite task force classified the various specifications into three categories:

1. Specifications that are already similar: sulphated ash, alkali and alkaline earth metal content, free glycerol content, copper strip corrosion, methanol and ethanol content and acid number.

2. Specifications with significant differences between parameters and methods, but which might be aligned by work on documentary standards and measurement standards: total glycerol content, phosphorous content, carbon residue, ester content, distillation temperature, flash point, total contamination and water and sediment.
3. Specifications with fundamental differences, perhaps due to emissions or environmental regulations within one or more regions, which are not bridgeable in the foreseeable future: sulphur content, cold climate operability, cetane number, oxidation stability, mono, di- and tri-acylglycerides, density, kinematic viscosity, iodine number, linolenic acid methyl ester content and polyunsaturated methyl ester content.

The current standards (EN 14214, ASTM D6751) established to govern the quality of biodiesel on the market are based on a variety of factors which vary from region to region, including characteristics of the existing diesel fuel standards, the predominance of the types of diesel engines most common in the region and the emissions regulations governing those engines. Europe, for example, has a much larger diesel passenger car fleet, while the USA and Brazilian markets are mainly comprised of heavier duty diesel engines. It is therefore not surprising that there are some significant differences between the three sets of standards.

Other sources of regional differences in biodiesel standards arise from the following factors. The biodiesel standards in Brazil and the USA are applicable for both FAME and fatty acid ethyl esters (FAEE), whereas the current European biodiesel standard is only applicable for FAME. Also, the standards for biodiesel in Brazil and the USA are used to describe a product that represents a blending component in conventional hydrocarbon-based diesel fuel, while the European biodiesel standard describes a product that can be used either as a stand alone diesel fuel or as a blending component in conventional hydrocarbon-based diesel fuel.

It should also be noted that some specifications for biodiesel are feedstock neutral and some have been formulated around the locally available feedstocks. The diversity in these technical specifications is primarily related to the origin of the feedstock and the characteristics of the local markets. Although this could imply some significant divergence in specifications and properties of the derived fuels – which could be perceived as an impediment to trade – in most cases it is possible to meet the various regional specifications by blending the various types of biodiesel to the desired quality and specifications.

Key factors influencing issues of alignment include the following:

- Biodiesel, both FAME and FAEE, are two chemically different mixtures making it a challenge to develop a common standard that can address the complex fuel and engine requirements.
- Further alignment of existing standards and specifications where necessary from both a technical and trade perspective will help establish Internationally

> Compatible Standards for Biofuels that should facilitate trade, improve efficiency of biofuels production, and promote innovative energy resources, and associated economic security.

At that time the European Commission began discussions with CEN to initiate the development of a European Standard for FAEE to be used as a fuel for diesel engines (Commission Mandate M/393, 2006). It should be noted that this mandate was submitted to the CEN/Technical Board, along with Commission Mandate M/394 (2006) – Mandate to CEN on the revision of EN 590 to increase the concentration of FAME to 10% v/v.

Technically and chemically, FAEE production is very comparable to FAME production. Therefore, existing biodiesel methyl ester plants have the technical possibility to produce FAEE without major retrofitting problems. Practical experience in the production of FAEE is limited mainly on funded studies and a few demonstration projects. Brazil, as producer of the cheapest ethanol, has some experiences with FAEE production and at least one biodiesel factory is claimed to be capable of producing FAEE. The fuel quality parameters are expected generally to be in line with the European biodiesel standard EN14214, except that the analytical methods would have to be adopted for ethyl ester application. Application of FAEE offers generally larger benefits in terms of well-to-wheel energy efficiency than in terms of greenhouse gas emissions. This is because methanol production is rather energy intensive, but as methanol is rich in hydrogen the greenhouse gas penalty is limited.

5.4.3 Current biodiesel standards development

International standardisation is a multi-stakeholder process involving industry, science, academia, consumers and governments. Standards are developed through precise processes including several pre-defined stages in order to create an industry-wide consensus.

The properties and limits specified in biodiesel standards must protect performance and durability of automobile engines or combustion equipment. However, as with any fuel specification, biodiesel quality specifications are dynamic and several aspects have to be periodically reviewed by institutions like the European Committee of Standardization (CEN), the International Organization for the Standardization (ISO) and the American Society for Testing and Materials (ASTM). They work for the development and continuous improvement of standards for biodiesel applications through their corresponding technical committees: CEN/TC19, ISO/TC28 and ASTM/D02. In addition, these technical committees have different working groups mainly focused on extending the scope of existing standards to unconventional components and products and developing new standards or modifying existing ones to meet lower detection limits (lowering acceptable levels of some components may often be imposed by legislation).

Despite the particular work of each technical committee, in recent years, there has been a continued move toward reducing duplication of efforts, so the development of standards is minimised where they exist; market relevant standards already fulfil the needs of the international community. This effort is characterised by the Memorandum of Understanding (MoU) developed in 2007 between ISO/TC 28, CEN/TC 19 and ASTM International Committee D02.

Next, standards related to biodiesel characterisation which are now under development are reviewed. ASTM refers to Work Items (WK) as new standards or a revision of existing standards that are under development by a committee. ASTM publishes Work Item descriptions in the interest of openness, as well as to solicit input from interested stakeholders who may not be members of the committee. The standards are:

- ASTM WK20385: This method determines the total glycerine in biodiesel fuel (range TBD) by liquid chromatography. Mono- di- and triacylglycerols are eluted as single peaks for each compound class and therefore could be reported separately as well as together as total glycerine. Also, monoglyceride content is of particular interest as a component in biodiesel fuel.
- ASTM WK13125: Subcommittee D02.04.0H is developing a methodology to obtain the boiling point distribution of biodiesel and biodiesel mixtures (B5-B20, etc.).These mixtures are difficult to characterise by glass distillation (i.e. ASTM D1160) due to severe foaming yielding poor precision. The proposed method will use gas chromatography to obtain the boiling range. Boiling point distribution is required specification for biodiesel in D6751. Since gas chromatographic boiling point distributions methods have been shown over the years to have been reliable methodologies for obtaining the boiling point distribution, the technique will be applied to biodiesel. Subcommittee E has requested that D02–04-H undertake the development of a method to obtain boiling point characterisation of biodiesel with better precision than using glass vacuum distillation. It will be used as a method for determining boiling point range as required by D6751, that is, to control the biodiesel quality. The users will be providers and producers of biodiesel.
- ASTM WK22309 describes an anion exchange chromatography procedure for determining free and total glycerine content of biodiesel (B100) and blends (B0-B20) with diesel fuel oils defined by Specifications D 6751 and D 975 Grades 1-D, 2-D and low-sulphur 1-D and 2-D. It is intended for the analysis of biodiesel blend samples containing between 0.5 and 50-mg kg^{-1} total glycerine. As explained before (in Section 5.3) high levels of free glycerine can cause injector deposits as well as clog fuel systems. Low levels of total glycerine ensure that high conversion of the oil or fat into its monoalkyl ester has taken place. High levels of mono-, di- and triglycerides can cause injector deposits and may adversely affect cold weather operation and filter plugging.

- prEN 16091 describes the determination of oxidation stability by a rapid small scale oxidation method. In biodiesel and its blends with diesel, oxidation of fuel can result in acid formation and polymer solids that plug filters and leave problematic deposits throughout the new test methods for evaluation of oxidation stability. This method utilises a new rapid small scale oxidation tester (RSSOT) named PetroOXY. Within the RSSOT's test chamber, a 5 mL sample of fuel is combined with oxygen, at 700 kPa and heated to 140°C. This initiates a fast oxidation whose results are determined by the time needed to achieve a pressure drop of 10% in the system. Testing confirmed that the time needed to achieve the 10% drop in pressure is directly related to the sample's oxidation stability. The RSSOT-method is quicker than other actual methods of evaluating the oxidation stability of biodiesel, petroleum middle-distillates and blends.
- prEN 16294 describes the determination of phosphorus content in FAME by optical emission spectral analysis with inductively coupled plasma (ICP OES).
- prEN 590 rev collect the requirements and test methods for blending up to 10% (v/v) of FAME in diesel fuels. This initiative started in 2007 with the European Union mandate M/394, but the revised normative is not expected to be available before 2013. The CEN underlined the problem of reaching an agreement between all stakeholders that made the B10 standardisation process slow. CEN mentioned that a political decision was required, too, which was beyond the scope of the CEN. Regarding car manufacturers there is some controversy because some of them are already indicating that they welcome B10 or even higher blends, while others are already not comfortable with actual B7. In addition, the push for further work on the B10 mandate may come from policy makers in the EU and Member States, especially those concerned about the fulfilment of renewable energy targets. This is because today Member States have the possibility to allow higher biodiesel volumes in diesel than 7vol%, or for other biocomponents with no limits (e.g. HVO); they are unwilling, however, to allow products for which there are no technical standards or which are not in compliance with the standards.

Apart from these prenormatives, CEN/TC19 is also involved in the development of a standard about the determination of the iodine value through a calculation method using GC data, together with a Round Robin report on EN 116 revision work regarding the cold filter plugging point. This report includes the results of the inter-laboratory study performed under CEN/TC 19/WG 14 to determine the precision of the newly developed procedures for EN 116.

Also, the European Commission published a Mandate M/393 about 'fatty acid ethyl esters (FAEE) for use in diesel engines and heating fuels'. The reason for giving a mandate to CEN for development of standards for FAEE biodiesel was to achieve the targets of the biofuels directive, because a stable and reliable supply and demand of bioethanol and biodiesel is mandatory. Therefore, there is a need to develop appropriate standards for fatty acid ethyl esters for diesel engines. This European Standard should be developed based on EN 14214 and EN 590. In addition, a European standard for fatty acid ethyl esters to be used as heating fuel, should be developed on the basis of EN 14213.The production of FAEE will also facilitate the energy security of the European Union and contribute significantly to meeting the Kyoto objectives. However FAEE is not on the market currently so the certification work has been suspended.

5.5 Future trends

In the future there is expected to be an increase to the blending limit of FAME up to B30. Car manufacturers advise about cold flow, oxidation stability and post treatment durability issues, even though most of the actual cars can run on biodiesel blends up to 30 vol% FAME, as a consequence of changes in many of the engine components over the last 10 years.

In 2008, when B7 was introduced in the European Union, there was an increasing demand for biodiesel and another jump is expected with the introduction of B10. From the point of view of producers, there will be many difficulties in biodiesel quality control when new feedstocks are used. In this sense, impurities are the main challenge for quality control at the producer's level because FAME is a vegetable product containing many different components.

For these reasons, new requirements will appear beyond those stipulated in EN 14214. These new requirements and the modification of previous ones mean that the point of view of biodiesel suppliers, car manufacturers, oil companies, customers, standardisation committees and so on, must be considered. Areas of interest will include understanding the effect of the wide variability of fuel chemistry resulting from the very broad range of potential feedstocks, challenges of fuel storage resulting from instability and reactivity to air. It will be necessary to reduce water content, determine minor components and their effects over filterability and cold flow properties. At the same time, standards must be improved, checking their accuracy, doing interlaboratory comparatives (ILCs) favouring worldwide biodiesel harmonisation and developing new reference materials. This approach will maximise the benefits of the pre-normative research to be carried out.

5.5.1 Point of view of biodiesel producers and distributors and of car manufacturers

To ensure product traceability, the German Association for Quality Management of Biodiesel (AGQM) has produced a multi-level quality assurance system that monitors the quality of the biodiesel from its members (producers and sellers). This QM concept has been gradually aligned over the years to the increasing requirements of the DIN EN 14214:2010–04 European Standard on minimum standards and has been extended to meet the AGQM's own requirements. These additional requirements are specified in Table 5.2 (AGQM, 2010).

The first edition of biodiesel guidelines from the Worldwide Fuel Charter Committee was presented in March 2009 (WWFC, 2009). The recommended limits in these guidelines were established for 100% biodiesel (B100) blend stock intended for blending with petroleum-based diesel fuel to make a blend containing a maximum of 5% biodiesel by volume (B5) suitable for use in vehicles with compression ignition engines. The parameters, limits and test methods coincide with EN 14214:2008+A1:2009. However, this proposal is more restrictive in the oxidation stability induction period (10 hours, minimum) and in the flash point (100°C, minimum). Higher level blends may require different specifications, labelling and additional controls to protect biodiesel users. Additionally, final biodiesel blends and petroleum-based diesel fuel have to meet the requirements of the appropriate diesel fuel category in the WWFC.

The Japan Automobile Manufacturers Association (JAMA) Fuels and Lubricants Sub-committee (Ogura, 2009; JAMA, 2009) supports WWFC guidelines. If FAME-blended diesel with more than 5% FAME has to be prepared, the use of hydro-treated vegetable oil (HVO) or biomass to liquid (BTL) as blendstocks is strongly recommended. Therefore, in order to assure materials compatibility with FAME-blended diesel, its quality control reference to oxidation stability, total acid number and water content is extremely important. One of the

Table 5.2 Additional requirements of biodiesel specified by AGQM

Parameter	Value
Clear and bright	Appearance: at 15°C, at least clear and bright of undissolved water and visible contamination
Water content	max. 300 mg kg^{-1}
CFPP	During the period 16 Nov–28 Feb (29), it is assured that a CFPP of –20°C is achieved using suitable flow improvers
Cloud point	April–28 Feb (29) max. 5°C Rem. time: max. –2°C
Pour point	16 Nov–28 Feb (29) max. –6°C and min. –18°C (with additives)
Total contamination	max. 20 mg kg^{-1}
Saturated monoglycerides	max. 0.12 wt%
Sterol glucosides (SG) non-acylated	To be determined
Filter blocking tendency	To be determined
Oxidation stability	min. 8 h

classical problems associated with the use of FAME is infiltrating and remaining impurities caused by the shortcomings of the reaction and refining process. To ensure FAME quality, it is essential to reduce the total impurities that negatively affect vehicle and engine operation as much as possible. When biodiesel is not completely refined, unreacted tri-glyceride can remain in the products and induces defective operation of the moving parts of the fuel injection systems and/or clogging of fuel filters. Recent findings indicate that fuel filters are also clogged by mono-glyceride and sterol glycoside. Metallic residues from reaction catalysts can promote deposits on the fuel injection system and this is why control of these impurities is extremely important. In the case of insufficient washing, methanol that has not been removed may induce softening or swelling of rubber and plastic materials. As methanol is intensely corrosive, infiltration of methanol must be avoided as much as possible.

Based on the experience with captive fleets, PSA Peugeot Citroën (Crepeau, 2010) showed the requirements that it is necessary to look at closely during the validation of an engine for use with higher biodiesel blends (B10–B30). These are:

- Oil changing interval and lube specification.
- Oxidation stability: it is necessary to have sufficient protection which takes into account the impact of long periods of vehicle inactivity and the faster ageing of the fuel. EN590: 28 h min (tested with EN 15751); acid number after ageing at 115°C: 0.3 mg KOH/g max.
- Cold flow properties: suitable engine protection and the need to include in the EN 14214 standard a saturated monoglyceride content specification (when adequate methods are available). In addition, a new relevant filterability method based on IP 387 (with or without cold soak filtration test ASTM D7501) must be added to EN590.
- Contaminants EN14214: need to reduce the P, Na, K, Ca, Mg contents to protect the post treatment systems at 160 000 km and respect the emissions standard.

5.5.2 View of CONCAWE

CONCAWE (conservation of clean air and water in Europe) was established in 1963 by a small group of leading oil companies to carry out research on environmental issues relevant to the oil industry. Full membership of CONCAWE is currently open to companies that own crude oil refining capacity within the European Union, Iceland, Norway, Switzerland and Turkey. In addition, a status of associate member is available to refining companies operating in other European countries.

CONCAWE prepared the last guidelines for handling and blending FAME by November 2009, according EN 14214 and EN 590 (www.CONCAWE.org).

These guidelines focus on the production, blending, distribution and supply of diesel containing up to 10% v/v FAME as well as the storage and handling of neat FAME. It is not clear if a new B10 diesel standard could be incorporated into EN 590 or developed as a separate specification. Anyway, only FAME fully meeting requirements of EN 14214 should be used. The potential future production and use of FAEE in diesel fuel was also discussed.

Jansen (2009) from KPI Oil Company presented the status and issues for biofuel specifications at the 8th CONCAWE Symposium. The principal issues under study were:

- Diesel fuel stability:
  1. Monitor biodiesel oxidation of B10 in six vehicles under long-term storage conditions. 2. Relate oxidation in vehicles to accelerated storage conditions in the laboratory.
- Particulate matter, filterability and microbial contamination:
  1. Compatibility of current total contamination method for FAME testing.
  2. Options for improving filterability performance (Filter Blocking test IP 387 and Cold Soak test ASTM D7501).
  3. Microbiological contamination could be another source of filter blocking problems. Very important problems with biodiesel, both fungi and yeast, can occur anywhere in the distribution system.
- Ash and metal contents (potential impact after treatment). Conventional diesel after treatment systems use a discrete NO_X control unit followed by a diesel particulate filter (DPF) for PM control.

According to CONCAWE, European standard EN 14214 will be revised mainly in the following aspects:

- An increase in Rancimat oxidation stability from 6 to 8 hours minimum.
- Climate requirements:
  1. CFPP measurements will be carried out before adding CFPP improvers.
  2. Cloud point requirement will be added.
  3. Three classes of biodiesel will be proposed for summer, winter and extreme winter.
  4. Alternatively, climate-dependent requirements for mono-glycerides (from 0.8 to 0.3% m/m).
- Development of equivalent FAEE standard has lower priority.

5.5.3 Role of CEN in European normalisation

The European Committee for Standardization (CEN) provides a platform for the development of European Standards and other technical specifications. In addition to ENs, CEN produces other documents for reference that can be developed quickly and easily such as Draft European Standards (prEN), CEN Workshop

Agreements (CWA) which are often used in fast-evolving technologies and the creation of new markets, Technical Specifications (TS) and Technical Reports (TR).

There are several difficulties in the development of both EN 590 for B10 and a new revised version of EN14214, mainly in those aspects that are related to cold flow properties. According to Mittelbach (2010) EN 14103 and EN 14105 standard methods are being improved. However, the determination of sterol glycosides and the replacement of total contamination limits by limitation of metals content are still under discussion.

Table 5.3 shows the progress of EN 14214 standard according to the proposals reported by Mittelbach (2010) and Bondioli (2010). Developments include:

(i) *Contaminants*: the changes made to tighten the quality parameters for phosphorus, sodium, potassium, calcium and magnesium are being analysed. These elements are the subject of constant discussion owing to their negative interactions with currently employed emissions treatment systems (SCR and DPF). Test methods for determining the trace element content in biodiesel by means of optical emission spectroanalysis with inductively coupled plasma (ICP OES) are proposed elsewhere (Fuelpro Biofuel Metals Analyzer/Teledyne Leeman Labs).

(ii) *Oxidation stability* is being analysed by CEN TC19/TC 307 JWG, oxidation stability subgroup (Guanziroli and Vignati, 2010). Alternative or complementary methods to EN 15571 Rancimat are under development. For example, a Petroxy test and a long-term test similar to EN 12205, carried out at 115°C with a final determination of the change in acid number (ΔAN). The Petroxy method is more exact than Rancimat EN 15571, except if oxidation stability is greater than 40 h. The presence of ignition improvers may lead to lower oxidation stability results by the Petroxy test method. In practice it has been observed that the addition of EHN (2-ethylhexylnitrate) can reduce the measured oxidation stability values.

(iii) *Minor components and filterability*: the presence of impurities consisting mainly of sterol glucosides (SGs) and saturated monoglycerides (SMG) have been found to have a significant impact on the low temperature storage stability and filter blocking tendency (Imperial Oil, 2009).

Table 5.3 Evolution of EN 14214 standard

Parameter	EN 14214:2008+A1:2009	prEN 14214:20XX
Phosphorus content	≤4 mg kg^{-1}	2
Group I metals (Na + K)	≤5 mg kg^{-1}	3.5
Group II metals (Ca + Mg)	≤5 mg kg^{-1}	3.5
Oxidation stability, 110°C	6 h min	8
Sterol glycosides	–	Test method
Saturated monoglycerides	–	Test method
Filterability	–	In discussion
Cold flow properties	CFPP (according climate zones)	CFPP + CP

Owing to their high melting point (240°C) and insolubility in biodiesel, the presence of these sterol glucosides could contribute to filterability problems. The impact of SGs on filter plugging is being thoroughly investigated (Lee *et al.*, 2007; AGQM-OVID, 2009; Chupka *et al.*, 2011; AGQM, 2011). Even at relatively low levels (35 ppm or higher), sterol glucosides may promote the formation of aggregates in biodiesel, exacerbating problems caused by saturated monoglycerides and other known cold crystallising components. The presence of SGs at double-digit parts per million (ppm) levels may cause the formation of a cloud-like haze in biodiesel, even at room temperature. Under the microscope, the cloud-like substance appears as agglomerates of various sizes composed of discrete particles of 10–15 μm.

Fourier transform infrared (FTIR) spectra and gas chromatography flame ionisation detector (GC-FID) chromatograms indicated that sterol glucosides are the major cause of precipitate formation in soybean oil biodiesel; while for poultry fat biodiesel, the precipitates are due to monoglycerides. However, precipitates from cotton seed oil biodiesel are due to both sterol glucosides and monoglycerides (Tang *et al.*, 2008).

High performance liquid chromatography (HPLC) and mass spectrometry (MS) methods were applied to 24 field receipt samples of solids from biodiesels. The normal phase and reverse phase methods described could be used to analyse SG quantitatively from biodiesel precipitates with an HPLC instrument equipped with either an evaporative light-scattering detector (ELSD) or more commonly an UV detector operating at 205 nm (Moreau *et al.*, 2008). Nuclear magnetic resonance and mass spectrometry were used as fast and reliable identification methods, without the need for a prior hydrolysis of the glucosidic bond. A GC method, including optimised sample preparation, was developed for the quantification of the free sterol glucoside (FSG) in biodiesel as well as in filter residues. The FSG concentrations in biodiesel produced by different processes ranged between 55 and 275 mg kg^{-1} for palm, and from not detectable to 158 mg kg^{-1} for soy biodiesel (Van Hoed *et al.*, 2008).

In Europe and under the supervision of CEN, several institutions like ITERG/SSOG and ASG (Bondioli *et al.*, 2008; Bondioli, 2010) are working on the development of analytical methods for SG determination. It is also necessary to establish the maximum limits allowed for SG and to propose a suitable reference material (cholesteryl-D-glucopyranoside standard).

Unexpected filter plugging issues occurred in vehicles and dispensing filters during the winter of 2005/06 in Minnesota. Analysis of the plugged filters identified a preponderance of saturated monoglycerides in the organic component. In response to these problems, a cold soak filtration test was developed and incorporated into the ASTM D6751 standard. During the winter of 2007, Sweden experienced cold temperatures for long periods of time which resulted in filter blocking in vehicles and precipitates in customer above ground tanks at temperatures above the cloud point of the fuel. Lab tests showed that solubility of

SMG in Swedish diesel to be low (<50 ppm at 4°C). The monoglyceride specification was reduced to 0.3 wt% in Sweden versus 0.8 wt% in EN 14214. Infineum has also reported precipitation of SMG above the cloud point. This was demonstrated by SMG add-back experiments using IP 387 (Davis and Denecker, 2008). The filter blocking tendency test (ASTM D2068) was found to correlate fairly well with the SMG content of the fuel (Imperial Oil, 2009).

The limits of saturated monoglycerides (SMG) allowed in B10 blends imply that modification of EN 14105 standard (Bondioli, 2010) is needed. The Shell Company proposes a limit of SMG of 200 mg kg^{-1} maximum, but the actual analytical method does not have enough accuracy and precision. Some compounds like squalene can interfere. Fischer proposed calculating the SMG content from total monoglycerides determined by EN 14105 standard and the percentage of saturated fatty acids determined by EN14103 standard. Table 5.4 shows a proposal by CONCAWE to limit the maximum of SMG (mg kg^{-1}) according to the following climate related requirements (Del Manso, 2010):

(iv) *Cold flow properties* are under study and the most accepted proposals come from CONCAWE, DIN and EBB (European Biodiesel Board). Table 5.5 collects the proposal of CONCAWE (Bondioli, 2010) relating cold flow properties to monoglycerides content. Del Manso (2010, 2011) reports the proposals of DIN and EBB.

Table 5.4 CONCAWE proposal to limit saturated monoglycerides (SMG) content in B10

Class	A	B	C	D	E	F
Suggested region	Sweden	Nordic	Northern European winter	Northern European summer	Southern European winter	Southern European summer
B10 max SMG	225	275	550	950	700	950

Table 5.5 CONCAWE proposal for cold flow properties and monoglycerides content

Property	CP (°C) max EN 23015	12	7	3	0	−3
	CFPP (°C) max EN 116	10	5	0	−5	−10
MG % (m/m) max EN14105	0.8	A1	B1	C1	D1	E1
	0.5	A2	B2	C2	D2	E2
	0.3	A3	B3	C3	D3	E3

Finally, the CEN/TC19/WG 24 is evaluating the specifications for high biodiesel fuel blends (B10-B30) in diesel fuel to be used in captive fleet applications for designated vehicles (requirements and test methods). There are two topics under discussion: (1) establish the definition of captive fleets vehicles and (2) define the future diesel fuel according to EN 590, allowing higher density.

5.5.4 Oil dilution of a passenger car diesel engine in operation with blended diesel fuel B10

Currently biodiesel is mixed according to EN 14214 and EN 590 with up to 7% fossil diesel fuel. With a further increase of FAME fraction to 10% (B10 blend), there are uncertainties regarding the undisturbed long life behaviour owing to the variations in the physicochemical properties of FAME in comparison to commercial diesel fuel. It is known that an irreversible oil dilution appears in the engine lubrication system as a result of the regeneration mode of current passenger car diesel engines with particulate filters (Institute for Mobile Systems, 2009). The reason lies in the higher distillation characteristic of FAME in contrast to the fossil diesel fuel. If unburnt fuel reaches the lubricant in the oil sump via the piston-cylinder assembly during the regeneration of the particulate filter, the FAME fraction cannot evaporate from the engine oil owing to its distillation characteristic (boiling point >340°C). As a consequence of the oil dilution, an irreversible decrease of the viscosity results with the risk of increasing wear. This problem is still more important for B30 diesel blends (Institute for Mobile Systems, 2010).

Compared with diesel fuel the disadvantage of FAME becomes obvious: they all have almost the same boiling point and are not represented by a rising boiling curve in contrast to diesel fuel. A rising boiling curve is beneficial for good fuel ignition and combustion in the cylinder. Therefore FAME may not be considered to be optimal prospective fuels. Future engines will tend to have lower compression ratios and more homogeneous fuel-air charging. So, the boiling line of future biofuels should be adapted. The example of coconut methyl ester indicates the general possibility for vegetable oil to lower the boiling line, at the expense of shorter chain lengths. Of course this is a tremendous goal, but future engines may demand fuels with boiling characteristics between diesel fuel and gasoline (Dorenkamp *et al.*, 2002).

To consider FAME as an appropriate fuel, its molecule structure must be designed, probably by means of plant breeding. It is important to lower the boiling line by achieving chain lengths between 12 and 16 carbon atoms and to have in mind the oxidation stability by focussing the number and the position of double bonds (Krahl *et al.*, 2007).

Smith *et al.* (2008) from NIST (National Institute of Standards and Technology) presented the application of a distillation technique to representative batches of commercially available biodiesel fuels sold as B20 and B100. The distillation curve data presented are of sufficiently low uncertainty that they can be used for thermodynamic modelling of these complex fluids. Moreover, they not only showed the distillation curves but also a chemical characterisation of the 0.025, 10, 50 and 80% fractions and discussed the contrasts between the various mixtures.

5.5.5 Blends of paraffinic diesel from synthesis (XTL) or hydrotreatment (HVO) and FAME

Taking into account that the main drawbacks of the biodiesel-diesel blends increase with the proportion of biodiesel, it will be difficult to reach the biomass target fixed by the European Directive (2009/28/EC) on renewable energies. However, it can be reached by mixing biodiesel with HVO and/or with XTL. It is worth mentioning that HVO and XTL as biofuels can be distinguished from fossil fuels through radiocarbon (C14) measurements. ASTM D 6866–09 could be used to determine the total biomass contribution in the final product. In this case, biodiesel would be determined using an infrared spectrometry method described in the European Standard EN 14078.

Nowadays, there is a CEN workshop (WS61-*Automotive Fuels – Blends of Paraffinic Diesel from Synthesis (XTL) or Hydrotreatment (HVO) and Fatty Acid Methyl Esters (FAME) – Requirements and test methods*) analysing the physicochemical advantages of this kind of blend. Some advantages would be a higher oxidation stability of paraffinic fuels and their ability to dissolve certain FAME contaminants (i.e. monoglycerides), at low temperatures.

Krahl *et al.* (2007) reported that rapeseed oil methyl ester (RME) is currently not suitable for diesel particulate filters. The boiling characteristic of RME is an additional problem that presumably limits its further usability in Euro VI engines (2014). The phosphorus and metal contents of future biodiesel must become significantly lower than the actual specification limits. These authors also showed and anticipated the tendencies of the German biofuel share of the diesel market with regard to the passenger car release withdrawal and the probable and slow increase of second generation biofuels like XTL (biomass-to-liquid fuel) or HVO. So, biodiesel will be used at least as a blend and probably as a truck fuel and it must urgently be optimised to ensure its usability in the long-term.

5.5.6 Fuels for vegetable oil compatible combustion engines

EN 14214 is a specification suitable for all types of diesel engines, especially where the demands of light-duty, Euro V engines require analysis of the biological product. From an environmental perspective, it would be beneficial to desist from using the much optimised EN 14214 quality in less demanding (heavy duty) or even adapted engines. Until now, pure plant oil has never been adapted for existing diesel engines (except through transesterification to biodiesel), but the engine has always been adapted to the fuel. With a specification that sets specific quality requirements for the oil, its use may well be guaranteed for heavy duty and other compression ignition engines. Finally, one could even think of two classes: for regular diesel and for adapted/modified engines.

Recently, standard DIN 51605:2010–10 has been published. This standard specifies the properties, with the test methods and the limit values for using rapeseed oil as fuel in vegetable oil compatible combustion engines. The most important changes to the former pre-standard are related to the following parameters: density at 15°C (910–925 kg m^{-3}), flash point (101°C), ignition quality (DCN, 40) and total contamination (24 mg kg^{-1}). The oxide ash and carbon residue are no longer listed and therefore do not have to be analysed any more. From 01/01/2012, P, Ca and Mg content must be lower than 3, 1 and 1 mg kg^{-1}, respectively. This trend to use rapeseed oil as fuel is being also analysed by CEN Workshop WS 56.

In 2008, the DEUTZ Natural Fuel Engine® was the first series-produced industrial engine that can be operated exclusively with 100% rapeseed oil. This engine is a further development based on the proven liquid cooled DEUTZ engine series TCD 2012 and TCD 2013 and basically consists of a DEUTZ Common Rail system DCR® for optimum atomisation and combustion of the rapeseed fuel, a 2-tank system with automatic switching from biodiesel or diesel to rapeseed oil and the patented DEUTZ Fuel Management DFM®. The release is limited to rapeseed oil in accordance with DIN V 51605 (cold pressed and refined). Other vegetable oils may not be used (not even as additives). Biodiesel according to DIN EN 14214 can be used instead of normal diesel fuel according to DIN EN 590. These engines are available in the tractors produced by AGCO Fendt (FENDT 820 Vario Greentec) and DEUTZ-Fahr (Agrotron Natural Power).

5.6 References

AGQM (2010), *Recommendation on Additional Requirements for FAME as Blend Component for Diesel Fuel beyond DIN EN 14214*. Available from: http://www.agqm-biodiesel.de/index.php?msg=309 [Accessed 11 July 2011].

AGQM (2011), *Steryl Glycosides and Acylated Steryl Glycosides in Vegetable Oils and Fatty Acid Methyl Esters – Effects on the Filterability of Biodiesel*. Available from: http://www.agqm-biodiesel.de/index.php?menu_sel=27&menu_sel2=18 [Accessed 11 July 2011].

AGQM-OVID (2009), *Sterol Glucoside Content in Vegetable Oils as a Risk for the Production of Biodiesel – Study of the Technological Chain Impact*. Available from: http://www.agqm-biodiesel.de/index.php?menu_sel=27&menu_sel2=18 [Accessed 11 July 2011].

AOCS Analytical Guidelines Ck 1–07 (2009), *Analytical Guidelines for Assessing Feedstock to Ensure Biodiesel Quality*, Official Methods and Recommended Practices of the AOCS, 6th edition, AOCS Press, Champaign IL.

AOCS Method Ca 2b–38 (2009), *Moisture and Volatile Matter Hot Plate Method*, Official Methods and Recommended Practices of the AOCS, 6th edition, AOCS Press, Champaign IL.

AOCS Method Ca 2e–84 (2009), *Moisture Karl Fischer Reagent*, Official Methods and Recommended Practices of the AOCS, 6th edition, AOCS Press, Champaign IL.
AOCS Method Ca 2f–93 (2009), *Determination of Moisture and Volatile Matter in Fats and Oils*, Official Methods and Recommended Practices of the AOCS, 6th edition, AOCS Press, Champaign IL.
AOCS Method Ca 3a–46 (2009), *Insoluble Impurities*, Official Methods and Recommended Practices of the AOCS, 6th edition, AOCS Press, Champaign IL.
AOCS Method Ca 5a–40 (2009), *Free Fatty Acids*, Official Methods and Recommended Practices of the AOCS, 6th edition, AOCS Press, Champaign IL.
AOCS Method Ca 6a–40 (2009), *Unsaponifiable Matter*, Official Methods and Recommended Practices of the AOCS, 6th edition, AOCS Press, Champaign IL.
AOCS Method Ca 6b–53 (2009), *Unsaponifiable Matter*, Official Methods and Recommended Practices of the AOCS, 6th edition, AOCS Press, Champaign IL.
AOCS Method Ca 12–55 (2009), *Phosphorus*, Official Methods and Recommended Practices of the AOCS, 6th edition, AOCS Press, Champaign IL.
AOCS Method Ca 12a–02 (2009), *Colorimetric Determination of Phosphorus Content in Fats and Oils*, Official Methods and Recommended Practices of the AOCS, 6th edition, AOCS Press, Champaign IL.
AOCS Method Ca 12b–92 (2009), *Phosphorus by Direct Graphite Furnace Atomic Absorption Spectrometry*, Official Methods and Recommended Practices of the AOCS, 6th edition, AOCS Press, Champaign IL.
AOCS Method Ca 15b–87 (2009), *Sodium and Calcium by Atomic Absorption Spectrophotometry*, Official Methods and Recommended Practices of the AOCS, 6th edition, AOCS Press, Champaign IL.
AOCS Method Ca 17–01 (2009), *Calcium, Copper, Iron, Magnesium, Nickel, Silicon, Sodium, Lead and Cadmium in Oil by Inductively Coupled Plasma Optical Emission Spectroscopy*, Official Methods and Recommended Practices of the AOCS, 6th edition, AOCS Press, Champaign IL.
AOCS Method Ca 19–86 (2009), *Phospholipids in Vegetable Oils Nephelometric Method*, Official Methods and Recommended Practices of the AOCS, 6th edition, AOCS Press, Champaign IL.
AOCS Method Ca 20–99 (2009), *Analysis for Phosphorus in Oil by Inductively Coupled Plasma Optical Emission Spectroscopy (ICP-OES)*, Official Methods and Recommended Practices of the AOCS, 6th edition, AOCS Press, Champaign IL.
AOCS Method Cd 1–25 (2009), *Iodine Value of Fats and Oils Wijs Method*, Official Methods and Recommended Practices of the AOCS, 6th edition, AOCS Press, Champaign IL.
AOCS Method Cd 3–25 (2009), *Saponification Value*, Official Methods and Recommended Practices of the AOCS, 6th edition, AOCS Press, Champaign IL.
AOCS Method Cd 3c–91 (2009), *Saponification Value Modified Method Using Methanol*, Official Methods and Recommended Practices of the AOCS, 6th edition, AOCS Press, Champaign IL.
AOCS Method Cd 3d–63 (2009), *Acid Value*, Official Methods and Recommended Practices of the AOCS, 6th edition, AOCS Press, Champaign IL.
AOCS Method Cd 3e–02 (2009), *Determination of Alkalinity in Fats and Oils*, Official Methods and Recommended Practices of the AOCS, 6th edition, AOCS Press, Champaign IL.
AOCS Method Cd 8b–90 (2009), *Peroxide Value Acetic Acid–Isooctane Method*, Official Methods and Recommended Practices of the AOCS, 6th edition, AOCS Press, Champaign IL.

AOCS Method Cd 12b–92 (2009), *Oil Stability Index*, Official Methods and Recommended Practices of the AOCS, 6th edition, AOCS Press, Champaign IL.
AOCS Method Cd 18–90 (2009), *p-Anisidine Value*, Official Methods and Recommended Practices of the AOCS, 6th edition, AOCS Press, Champaign IL.
AOCS Method Cd 20–91 (2009), *Determination of Polar Compounds on Frying Fats*, Official Methods and Recommended Practices of the AOCS, 6th edition, AOCS Press, Champaign IL.
AOCS Method Cd 22–91 (2009), *Determination of Polymerised Triglycerides by Gel-Permeation HPLC*, Official Methods and Recommended Practices of the AOCS, 6th edition, AOCS Press, Champaign IL.
AOCS Method Ce 1–62 (2009), *Fatty Acid Composition by Packed Column Gas Chromatography*, Official Methods and Recommended Practices of the AOCS, 6th edition, AOCS Press, Champaign IL.
AOCS Method Ce 1b–89 (2009), *Fatty Acid Composition Of Marine Oils And Marine Oil Esters By Capillary Column Gas-Liquid Chromatography*. Official Methods and Recommended Practices of the AOCS, 6th edition, AOCS Press, Champaign IL.
AOCS Method Ce 2–66 (2009), *Preparation of Methyl Esters of Fatty Acids*, Official Methods and Recommended Practices of the AOCS, 6th edition, AOCS Press, Champaign IL.
ASTM Standard D93 (2010), *Flash Point by Pensky-Martens Closed Cup Tester*, ASTM International, West Conshohocken, PA.
ASTM Standard D130 (2004), *Standard Test Method for Corrosiveness to Copper from Petroleum Products by Copper Strip Test*, ASTM International, West Conshohocken, PA.
ASTM Standard D189- 06 (2010), *Conradson Carbon Residue of Petroleum Products*, ASTM International, West Conshohocken, PA.
ASTM Standard D396 (2008), *Specification for Fuel Oils*, ASTM International, West Conshohocken, PA.
ASTM Standard D445 (2011), *Standard Test Method for Kinematic Viscosity of Transparent and Opaque Liquids (and Calculation of Dynamic Viscosity)*, ASTM International, West Conshohocken, PA.
ASTM Standard D613 (2010), *Standard Test Method for Cetane Number of Diesel Fuel Oil*, ASTM International, West Conshohocken, PA.
ASTM Standard D664 (2009), *Standard Test Method for Acid Number of Petroleum Products by Potentiometric Titration*, ASTM International, West Conshohocken, PA.
ASTM Standard D874 (2007), *Standard Test Method for Sulfated Ash from Lubricating Oils and Additives*, ASTM International, West Conshohocken, PA.
ASTM Standard D975 (2008), *Specification for Diesel Fuel Oils*, ASTM International, West Conshohocken, PA.
ASTM Standard D1160 (2006), *Standard Test Method for Distillation of Petroleum Products at Reduced Pressure*, ASTM International, West Conshohocken, PA.
ASTM D2068 (2010), *Standard Test Method for Determining Filter Blocking Tendency*. ASTM International, West Conshohocken, PA.
ASTM Standard D2500 (2009), *Standard Test Method for Cloud Point of Petroleum Products*, ASTM International, West Conshohocken, PA.
ASTM Standard D2709 (2006), *Standard Test Method for Water and Sediment in Middle Distillate Fuels by Centrifuge*, ASTM International, West Conshohocken, PA.
ASTM Standard D4530 (2007), *Standard Test Method for the Determination of Carbon Residue*, ASTM International, West Conshohocken, PA.

ASTM Standard D4951 (2006), *Standard Test Method for Determination of Additive Elements in Lubricating Oils by Inductively Coupled Plasma Atomic Emission Spectrometry*, ASTM International, West Conshohocken, PA.

ASTM Standard D5453 (2009), *Total Sulfur in Light Hydrocarbons, Spark Ignition Engine Fuel, Diesel Engine Fuel and Engine Oil by Ultraviolet Fluorescence*, ASTM International, West Conshohocken, PA.

ASTM Standard D6371 (2005), *Standard Test Method for Cold Filter Plugging Point of Diesel and Heating Fuels*, ASTM International, West Conshohocken, PA.

ASTM Standard D6584 (2010), *Standard Test Method for Determination of Total Monoglyceride, Total Diglyceride, Total Triglyceride and Free and Total Glycerin in B 100 Biodiesel Methyl Esters by Gas Chromatography*, ASTM International, West Conshohocken, PA.

ASTM Standard D6751 (2011), *Standard Specification for Biodiesel Fuel Blend Stock (B100) for Middle Distillate Fuels*, ASTM International, West Conshohocken, PA.

ASTM Standard D6792 (2007), *Quality System in Petroleum Products and Lubricants Testing Laboratories*, ASTM International, West Conshohocken, PA.

ASTM Standard D6866 (2009), *Standard Test Methods for Determining the Biobased Content of Solid, Liquid and Gaseous Samples Using Radiocarbon Analysis*, ASTM International, West Conshohocken, PA.

ASTM Standard D7039 (2007), *Sulfur in Gasoline and Diesel Fuel by Monochromatic Wavelength Dispersive X-Ray Fluorescence Spectrometry*, ASTM International, West Conshohocken, PA.

ASTM Standard D7467 (2008), *Specification for Diesel Fuel Oil, Biodiesel Blend (B6 to B20)*, ASTM International, West Conshohocken, PA.

ASTM Standard D7501 (2009), *Standard Test Method for Determination of Fuel Filter Blocking Potential of Biodiesel (B100) Blend Stock by Cold Soak Filtration Test (CSFT)*, ASTM International, West Conshohocken, PA.

ASTM Work Item WK13125, *Determination of Boiling Point Distribution of Fatty Acid Methyl Esters (Fame) in the Boiling Range of 100°C–615°C by Gas Chromatography*. ASTM International, West Conshohocken, PA.

ASTM Work Item WK20385, *New Test Method for Determination of Total Glycerine in Biodiesel (B–100) by Liquid Chromatography*. ASTM International, West Conshohocken, PA.

ASTM Work Item WK22309, *New Test Method for Determination of Free and Total Glycerine in Biodiesel Blends by Anion Exchange Chromatography*, ASTM International, West Conshohocken, PA.

Bondioli P (2010), *Biodiesel – Evoluzione delle norme europee, Normazione tecnica SSC*, Prove Interlaboratorio Prodotti Petroliferi, Unichim riunione plenaria, 10 November, Milano, Italy. Available from: http://www.ssc.it/pdf/2010/PLENARIA_XI_2010/ID115_Bondioli.pdf [Accessed 11 July 2011].

Bondioli P, Cortesi N and Mariani C (2008), Identification and quantification of steryl glucosides in biodiesel, *Eur J Lipid Sci Technol*, **110**, 120–6.

CEN Workshop WS 56 (2010), *Fuel Quality Specification – Towards Pure Plant Oil Application in Diesel Engines*.

CEN Workshop WS 61 (2010), *Automotive Fuels – Blends of Paraffinic Diesel from Synthesis (XTL) or Hydrotreatment (HVO) and Fatty Acid Methyl Esters (FAME) – Requirements and test methods analyzing the physicochemical advantages of this kind of blends*.

Chupka G M, Yanowitz J, Chiu G, Alleman T L and McCormick R L (2011), Effect of saturated monoglyceride polymorphism on low-temperature performance of biodiesel, *Energy Fuels*, **25**, 398–405.

Commission Directive 2009/28/EC of 23 April 2009 on the *Promotion of the Use of Energy From Renewable Sources and Amending and Subsequently Repealing Directives 2001/77/EC And 2003/30/EC*.

Commission Mandate M/393 EN of 13 November 2006, *Mandate to CEN for Standards for Fatty Acid Ethyl Ester for Use in Diesel Engines and Heating Fuels*.

Commission Mandate M/394 EN of 13 November 2006, *Mandate to CEN on the Revision of EN 590 to Increase the Concentration of FAME and FAEE to 10% v/v*.

Crepeau G (2010), Engine developments and compatibility with higher biodiesel blends – biodiesel quality concerns, *The Future of Biodiesel in Europe, AGQM – HEC Meeting*, 27–28 September, Brussels, Belgium.

Davis B and Denecker V (2008), FAME cold flow properties – The challenges of measurement, *Infineum, Biofuels: Measurement and Analysis Symposium*, 15–16 April, Old Windsor, UK. Available from: http://www.energyinst.org.uk/biofuelsconference2008/presentations/12.pdf [Accessed 11 July 2011].

Del Manso F (2010), Panoramica sulle norme tecniche europee, *Evoluzione Norme EN/ISO/Nazionali*, Normazione tecnica SSC, Prove Interlaboratorio Prodotti Petroliferi, Unichim riunione plenaria, 10 November, Milano, Italy. Available from: http://www.ssc.it/pdf/2010/PLENARIA_XI_2010/ID114_DelManso.pdf [Accessed 11 July 2011].

Del Manso F (2011), *Evoluzione Norme EN/ISO/Nazionali*, Normazione tecnica SSC, Prove Interlaboratorio Prodotti Petroliferi, Unichim riunione plenaria, 13 April, Falconara, Italy. Available from: http://www.ssc.it/pdf/2011/plenaria_p_2011/ID_133_Evoluz_norme_europee_FDM.pdf [Accessed 11 July 2011].

DIN 51605 (2010), *Fuels for Vegetable Oil Compatible Combustion Engines – Fuel from Rapeseed Oil – Requirements and test methods*.

DIN EN 590 (2004), *Specifications for Common Storage of Diesel Fuel Requirements, test methods and threshold values*.

DIN EN 14214 (2010–04) *Automotive Fuels Fatty – Acid Methyl Ester (FAME) for Diesel Engines – Requirements and test methods*.

DIN V 51605 (2006), *Pre-standard for Rapeseed Oil Fuel*.

Dorenkamp R, Garbe T and Ohlendorf G (2002), Biodiesel im Zielkonflikt mit Anforderungen an den Dieselmotor, *Landbauforsch Völkenrode*, **239**, 57–62.

EN 116:1997/AC (1999), *Diesel and Domestic Heating Fuels – Determination of Cold Filter Plugging Point*.

EN 590:2009+A1 (2010), *Automotive Fuels – Diesel – Requirements and test methods*.

EN 12662 (2008), *Liquid Petroleum Products – Determination of Contamination in Middle Distillates*.

EN 14078 (2009), *Liquid Petroleum Products – Determination of Fatty Acid Methyl Ester (FAME) Content in Middle Distillates – Infrared Spectrometry Method*.

EN 14103 (2011), *Fat and Oil Derivatives – Fatty Acid Methyl Esters (Fame) – Determination of Ester and Linolenic Acid Methyl Ester Contents*.

EN 14105 (2011), *Determination of Free and Total Glycerin in B–100 Biodiesel Methyl Esters by Gas Chromatography*.

EN 14107 (2003), *Fat and Oil Derivatives – Fatty Acid Methyl Esters (FAME) – Determination of Phosphorus Content by Inductively Coupled Plasma (ICP) Emission Spectrometry*.

EN 14110 (2003), *Fat and Oil Derivatives – Fatty Acid Methyl Esters (FAME) – Determination of Methanol Content.*

EN 14111 (2003), *Fat and Oil Derivatives – Fatty Acid Methyl Esters (FAME) – Determination of Iodine Value.*

EN 14112 (2003), *Fat and Oil Derivatives – Fatty Acid Methyl Esters (FAME) – Determination of Oxidation Stability (accelerated oxidation test).*

EN 14213 (2003), *Heating Fuels – Fatty acid methyl esters (FAME) – Requirements and test methods.*

EN 14214:2008+A1 (2009), *Automotive Fuels – Fatty acid methyl esters (FAME) for Diesel Engines – Requirements and test methods.*

EN 14538 (2006), *Fat and Oil Derivatives – Fatty Acid Methyl Ester (FAME) – Determination of Ca, K, Mg and Na Content by Optical Emission Spectral Analysis with Inductively Coupled Plasma (ICP OES).*

EN 15751(2009), *Automotive Fuels. Fatty Acid Methyl Ester (FAME) Fuel and Blends with Diesel Fuel. Determination of Oxidation Stability by Accelerated Oxidation Method.*

EN 15779 (2009), *Petroleum Products and Fat and Oil Derivates – Fatty Acid Methyl Esters (FAME) for Diesel Engines – Determination of Polyunsaturated (= 4 double bonds) Fatty Acid Methyl Esters (PUFA) by Gas Chromatography.*

EN ISO 3104:1996/AC (1999), *Petroleum products – Transparent and opaque liquids – Determination of kinematic viscosity and calculation of dynamic viscosity.*

EN ISO 3675 (1998), *Crude petroleum and liquid petroleum products – Laboratory determination of density – Hydrometer method.*

EN ISO 12185 (1996), *Crude Petroleum and Petroleum Products – Determination of Density – Oscillating U-tube Method.*

EN ISO 12937 (2000), *Petroleum Products – Determination of Water – Coulometric Karl Fischer Titration Method.*

Guanziroli G and Vignati M (2010), Stabilità del biodiesel e delle miscele gasolio/biodiesel: CEN TC19/TC307 JWG, Oxidation stability subgroup: stato lavori e programmi futuri, Normazione tecnica SSC, Prove Interlaboratorio Prodotti Petroliferi, Unichim riunione plenaria, 18 March, San Donato Milanese, Italy. Available from: http://www.ssc.it/pdf/2010/PLENARIA2010/Guanziroli_id89.pdf [Accessed 11 July 2011].

IIS (2010), *Results of Proficiency Test Biodiesel 100% FAME (B100) September 2010*, News & Reports, Biodiesel B100, December 2010. Available from: http://www.iisnl.com/ [Accessed 11 July 2011].

Imperial Oil (2009), *Low Temperature Storage Test Phase 2 – Identification of Problems Species*. Available from Canadian Petroleum Products Institute: http://www.cppi.ca/userfiles/file/Low%20Temperature%20Storage%20Test%20Report%20Dec%2009%20Publication%20EN.pdf [Accessed 11 July 2011].

Institute for Mobile Systems (2009), *Oil Dilution of a Passenger Car Diesel Engine in Operation with Blended Diesel Fuel B10*. Available from: http://www.ufop.de/english_news.php [Accessed 11 July 2011].

Institute for Mobile Systems (2010), *Optimisation of the Engine-internal Late Post-injection in the Regeneration Mode of a Passenger Car Diesel Engine for Reducing the Oil Dilution during Operation with the Fuel Blends B7, B10 and B30*. 2010. Available from: http://www.ufop.de/downloads/20101210_Bericht_final_Oelverd_ENG_ohne_Ghk.pdf [Accessed 11 July 2011].

IP 387/07 *Determination of the Filter Blocking Tendency of Distillate Fuels, International Association for Stability, Handling and Use of Liquid Fuels, Inc.*

ISO 2160 (1998), *Petroleum Products – Corrosiveness to Copper – Copper Strip Test.*

ISO 3015 (1992), *Determination of Cloud Point in Petroleum Products.*

ISO 3105 (1994), *Glass Capillary Kinematic Viscometers – Specifications and operating instructions.*

ISO 3679 (2004), *Determination of Flash Point – Rapid Equilibrium Closed Cup Method.*

ISO 3987 (2010), *Determination of Sulfated Ash in Lubricating Oils and Additives.*

ISO 5165 (1998), *Determination of the Ignition Quality of Diesel Fuels – Cetane Engine Method.*

ISO 10370 (1996), *Carbon Residue of Petroleum Products.*

ISO 20846 (2004), *Sulfur by Combustion Ultraviolet Fluorescence Detection.*

ISO 20884 (2011) *Petroleum Products – Determination of Sulfur Content of Automotive Fuels – Wavelength-Dispersive X-ray Fluorescence Spectrometry.*

ISO/IEC 17025 (2005), *General Requirements for the Competence of Testing and Calibration Laboratories.*

JAMA (2009), *Quality of Biodiesel (FAME) and Use of FAME-blended Diesel.* Available from: http://www.jama.or.jp/eco/wwfc/pdf/JAMA_FQ_PositionStatement_FAME.pdf [Accessed 11 July 2011].

Jansen L (2009), CEN fuel and biofuel specifications: status and issues for CONCAWE. *8th CONCAWE Symposium*, November 2009, Brussels, Belgium.

Knothe G (2005), Dependence of biodiesel fuel properties on the structure of fatty acid alkyl esters, *Fuel Process Technology*, **86**, 1059–70.

Knothe G (2006), Analyzing biodiesel: standards and other methods, *JAOCS*, **83**(10), 823–33

Knothe G, Van Gerpen J and Krahl J (2005), *The Biodiesel Handbook*, AOCS Press, Champaign IL, 122–136.

Krahl J, Munack A and Bockey D (2007), Property demands on future biodiesel (HVO), *Landbauforschung Völkenrode*, **4**, 415–18.

Lee I, Pfalzgraf L M, Poppe G B, Powers E and Haines T (2007), The role of sterol glucosides on filter plugging, *Biodiesel Magazine*, **4**, 105–12.

Mittelbach M (2010), Future perspectives of biofuels development in Europe, *Sustainable Energy Technologies for Future Transportation – URJC Summer Course*, 12–16 July, Aranjuez, Spain.

Monteiro M R, Ambrozin A R P, Lião L M and Ferreira A G (2008), Critical review on analytical methods for biodiesel characterisation, *Talanta*, **77**, 593–605.

Moreau R A, Scott K M and Haas M J (2008), The identification and quantification of steryl glucosides in precipitates from commercial biodiesel, *Journal of the American Chemical Society*, **85**, 761–70.

Moser, B R (2009), Biodiesel production, properties and feedstocks, *In Vitro Cellular Developmental Biology: Plant*, **45**, 229–66.

Nadkarni R A K (2010), Challenge of analytical characterisation of biofuels, *Journal ASTM International,* **7** (3), 1–18.

Nadkarni R A K and McKlindon A (2009), ASTM Proficiency Testing Programs for Biofuels, *J ASTM Int*, 6 (7), 1–15.

Ogura A (2009), World wide fuel charter – recommended biofuel specifications, *7th Asian Petroleum Technology Symposium*, 18–20 February, Ho Chi Minh, Vietnam. Available from: http://www.pecj.or.jp/english/plaza/7th_asiasympo/2-3_Akira-Ogura.pdf [Accessed 11 July 2011].

prEN 590 rev *Automotive Fuels – Diesel – Requirements and test methods.*

prEN 16294, *Determination of Phosphorus Content in Fatty Acid Methyl Esters (FAME) by Optical Emission Spectral Analysis with Inductively Coupled Plasma (ICP OES).*

prEN 16091, *Liquid Petroleum Products? Middle Distillates and Fatty Acid Methyl Ester (FAME) Fuels and Blends ?Determination of Oxidation Stability by Rapid Small Scale Oxidation Method.*

Sanford S D, White J M, Shah P S, Wee C, Valverde M A and Meier G R (2009), *Feedstock and Biodiesel Characteristics Report*, Renewable Energy Group, Inc., www.regfuel.com

Smith B L, Ott L S and Bruno T J (2008), Composition-explicit distillation curves of commercial biodiesel fuels: comparison of petroleum-derived fuel with B20 and B100, *Industrial and Engineering Chemistry Research*, **47**, 5832–40.

Tang H, Salley S O and Ng K Y S (2008), Fuel properties and precipitate formation at low temperature in soy-, cottonseed- and poultry fat-based biodiesel blends, *Fuel*, **87**, 3006–17.

Tripartite Task Force (2007), White Paper on *Internationally Compatible Biofuel Standards*.

UFOP, VDB and AGQM (2010), *Commercial Vehicle Manufacturer Approvals on using biodiesel (B100)*. Available from: www.ufop.de/downloads/201006_Biodiesel_approvals_B100.pdf [Accessed 11 July 2011].

Van Hoed V, Zyaykina N, De Greyt W, Maes J, Verhé R and Demeestere K (2008), Identification and occurrence of steryl glucosides in palm and soy biodiesel, *Journal of the American Chemical Society*, **85**, 701–9.

VSL, IRMM, NPL, INMETRO, NIST and LGC (2010), *Biorema Project – Reference Materials for Biofuel Specifications*, EU Project Grant 219081.

WFCC (2009), *Biodiesel Guidelines from the Worldwide Fuel Charter Committee*. Available from: http://www.acea.be/index.php/files/b100_biodiesel_guidelines [Accessed 11 July 2011].

Part II

Advanced biodiesel production processes and technologies

6
Advances in biodiesel catalysts and processing technologies

M. MITTELBACH, University of Graz, Austria

Abstract: A large number of papers and patents have been published and filed during the last few years on new technologies for the production of biodiesel from various feedstocks. This chapter aims to give an overview of latest developments in the field of new catalysts like the use of heterogeneous catalysts or the use of lipases, alternative reaction conditions like supercritical treatment, microwave or ultrasound application, or the use of alternative reactor designs like oscillatory or micro-channel reactors. Also new technologies for separation of reaction mixtures, like the use of membranes or simultaneous derivatization of glycerol during the reaction are described. The advantages and disadvantages of individual alternative approaches are discussed and compared with traditional biodiesel production using mainly homogeneous alkaline catalysis.

Key words: biodiesel, catalysts, innovation, lipases, reactors, supercritical, technology.

6.1 Introduction

Biodiesel has been a well established biofuel all over the world for almost 20 years. Currently, there is an overcapacity of biodiesel production, especially in Europe, owing to high expectations resulting from European legislation with a binding target of 10% biofuels until 2020 (European Directive 2009/28/EC, 2009). However, until the legislation had been established into national law, which had to be done by the end of 2010, there was no urgent need for high amounts of biodiesel. Furthermore imports of cheaper biodiesel from the USA and Argentina have hindered European biodiesel production. A lack of sufficient feedstocks, like rapeseed oil, in Europe, and the unpredictable fluctuation of feedstock prices has led to a low overall production rate of biodiesel. A series of biodiesel plants have gone bankrupt within the last few years, especially those with low profitability which had high investment and running costs, due to insufficient and costly technology. This particularly affected so-called single feedstock plants, which were only capable of processing refined virgin vegetable oils and could not process low-grade, cheaper feedstocks, which led to production plant closure.

Most biodiesel production plants are currently running on traditional technology, which is based on alkaline catalyzed transesterification and, if necessary, using an acid pre-esterification step in order to reduce the amount of free fatty acids which hinder transesterification. Because there are now an excessive number of biodiesel production plants there is an urgent need for new technology that is more energy efficient and leads to higher overall yields in order to reduce production costs. Furthermore, the restrictions of the European Directive on Alternative Energy, including mandatory limits for a reduction of overall greenhouse gases, will push the introduction of new technologies that are capable of improving the overall life cycle analysis of biodiesel production.

6.2 Current production technology for biodiesel

Alkaline, or basic compounds, are by far the most commonly used catalysts for biodiesel production. Alkaline catalysts used most often for transesterification reactions are sodium or potassium hydroxide, or alcoholates. In small or medium scale biodiesel production plants the use of cheaper alkali hydroxides is preferred. Where free fatty acids represent minor components in unrefined oils they are converted into soaps and are separated together with the glycerine phase. For oils with higher amounts of free fatty acids, a combination of an acidic esterification and an alkaline transesterification leads to a conversion of almost 100% of fatty acid material.

For biodiesel production in huge industrial size plants, 100 000 tonnes or more of mostly refined vegetable oils can be used per year, which require degumming and deacidification. In this case a solution of sodium methoxide is used as a catalyst. Sodium methoxide is more expensive than sodium hydroxide, but if refined oils are used as feedstocks, less catalyst can be used and the formation of soaps as a side product is minimized.

Whereas traditional homogeneous catalysis offers a series of advantages, the major drawback is the fact that homogenous catalysts cannot be reused. The catalyst will be separated during the glycerol phase as soap. Mostly, the glycerol phase will be neutralized with inorganic acids forming the corresponding salts. These can be used as fertilizer in the case of potassium, however in the case of sodium salts, they have to be disposed of. In most biodiesel production plants, after separation of the glycerol phase the biodiesel is washed with water, dried and then directly used as the final product. Distillation of biodiesel is only carried out in some production plants where they use very low quality feedstocks in high quantities, including animal fat and used frying oils.

The majority of biodiesel production plants currently utilize quite traditional technologies, without a large number of new innovations. However, owing to the fact that there is an overcapacity of production plants leading to a large amount of competition in the market, new technologies will have to be introduced within the next decade. The main research and development (R&D) trends are listed below:

- use of alternative feedstocks, especially waste materials
- use of heterogeneous and enzymatic catalysts

- reactions under supercritical conditions without any catalyst
- microwave and ultrasound assisted reaction conditions
- micro reactors and new phase separation systems
- new biodiesel and glycerol purification methods
- simulation and theoretical models
- reactions under simultaneous modification of glycerol
- new technologies leading to non-traditional biodiesel
- fatty acid ethyl esters.

During the last few years there have been an enormous number of articles and patents published on these topics (Fig. 6.1).

It can be seen that over 1000 articles and over 500 patents relating to biodiesel have been published each year, over the past few years. A large number of research projects in this area have also received grant funding within the last few years. The main reasons for this huge interest in biodiesel are the market penetration already established in almost all countries worldwide, combined with the relatively easy process of transesterification of glyceridic oils, which can be carried out with low cost equipment in almost any laboratory.

The main focus of most of this research and development (R&D) work is the improvement of the economic and ecological sides of the biodiesel production. However, only a few of these developments are able to compete with existing technologies and have only been carried out at laboratory scale.

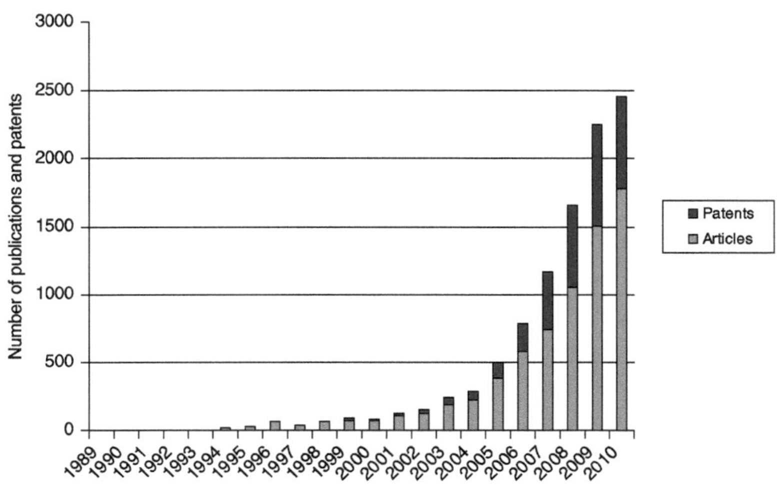

6.1 Number of scientific publications and patents on biodiesel according to SciFinder, 2011.

6.3 New types of catalysts

6.3.1 Homogenous catalysts

The distinction between homogeneous and heterogeneous catalysis is defined, respectively, by the solubility or non-solubility of the catalyst in the reaction mixture between alcohol and vegetable oil. The most important homogeneous catalysts for today's biodiesel production are still alkali hydroxides or alkoxides. The main drawback of these catalysts is the fact that free fatty acids present in the oil will react with the catalyst to form alkali soaps, which are developed further during the glycerol phase. There are three options to overcome this problem: the free fatty acids can be removed from the feedstock prior to alcoholysis, they can be esterified in an additional step, or they can be recycled out of the glycerol phase by separation and further esterified. Besides sodium and potassium hydroxide, Li, Rb and Cs hydroxides have also been investigated and compared (Arzamendi *et al.*, 2008). Performance with sunflower oil and a molar methanol/oil ratio of 12:1 at 323K and 0.1% of catalyst did not show significant differences, except for the fact that a kind of induction period could be seen in the first 60 min for LiOH. After about 90 min, the conversion with all hydroxides was about 90%. As an alternative alkaline homogeneous catalyst, organic bases such as guanidines, carnitine or arginine can be used. These catalysts can also be supported on a resin or polymer (Peter and Weidner, 2007). With guanidine carbonate, vegetable oils with a content of up to 1% of free fatty acids can be transesterified in one step; the guanidine carbonate then disintegrates into ammonia and carbon dioxide. Guanidine cannot be reused, it remains in the glycerol phase and during distillation of glycerol it disintegrates into carbon dioxide and ammonia, which can be trapped as ammonium sulfate with sulfuric acid. Acid catalysts, including concentrated sulfuric acid, can be used for both esterification and transesterification where an oil conversion of 97% is achieved after a 4 h reaction time at 80°C (Zhang *et al.*, 2003). However, reaction rates are significantly lower than for alkaline catalysis and the use of concentated sulfuric acid can lead to side reactions like oxidation and decomposition of organic compounds.

6.3.2 Heterogeneous catalysts

Intensive research has been carried out within past few years into the development of heterogeneous or solid catalysts, mainly focusing on the ease of separation of reaction products from the catalyst, easier purification of the main and side products, as well as the reuse of the catalyst. Furthermore, heterogeneous catalysts supported in columns are ideal for continuous operation. Numerous publications and patents have discussed insoluble catalysts for transesterification reactions. However, only a few have ever reached the stage of industrial application. Most of these catalysts are only active at higher temperatures and pressures, consequently leading to an enormous increase of investment and running costs. The reaction conditions mostly reach a temperature and pressure where esterification and transesterification can take place without a catalyst. In a series of publications, no control reactions have been carried out without a catalyst. Also, the main advantage of solid catalysts, their possible reuse, has not been thoroughly investigated.

Yan *et al.* (2010a) divided heterogeneous catalysts into:

- low temperature catalysts like alkali oxides, rare earth metal oxides, alkali metal exchanged zeolites, alkali metal ion-supported catalysts, clay minerals, and
- high temperature catalysts which could be acidic or basic as well as exhibiting catalytic activity in subcritical or supercritical conditions.

The term 'green catalysts' is often used without any further proof of lower environmental impact. Most of the reactions were studied only once with a fresh catalyst and with changing reaction conditions. There are also catalysts that are insoluble at room temperature, but partly soluble at elevated temperatures.

6.3.3 Low temperature catalysts

Arzamendi *et al.* (2008) have shown that catalysts like sodium and potassium carbonate, trisodium phosphate and CaO are soluble up to 50% at 323K. The most reactive catalysts of this group were potassium and sodium carbonate, whereas the corresponding bicarbonates and hydrogen phosphates showed negligible catalytic activity. A certain level of solubility seems to be responsible for the catalysis, because CaO that was calcined for 12 h at 773K almost became an inactive compound. Several researchers studied the use of calcined MgO. With a molar ratio of methanol to oil of 12:1 and a catalyst amount of 5% (m/m), a conversion over 98% could be reached (Xu *et al.*, 2010). Calcination was performed at 600°C; no leaching of the catalyst was observed, however, after four runs with the same catalyst, which has been calcined after each use, a significant loss of activity could be observed.

Naturally occurring dolomite, consisting of $CaCO_3$ and $MgCO_3$ was also successfully used as a heterogeneous catalyst (Ngamcharussrivichai *et al.*, 2010). The calcination was performed at 800°C and transesterification of palm kernel oil with a methanol/oil molar ratio of 30/1 and a methyl ester content of over 90% was reached after 3 h at 60°C. The conversion decreased slightly to about 85% after 10 repetitions, which was mainly caused by methanol leaching.

Aluminium oxide loaded with alkali metal salts was demonstrated to be an efficient solid-base catalyst. Synthesized aerogel catalysts with potassium carbonate on an alumina/silica support were tested in transesterification of sunflower oil, showing that temperatures below 100°C can lead to negligible leaching because of heterogeneous conditions (Lukic *et al.*, 2010).

Naturally occurring and commercially available hydrotalcite, a magnesium-aluminium hydroxycarbonate, could also be used as a solid catalyst after calcination, however, a lower activity than hydrotalcites synthesized in laboratory has been observed (Navajas *et al.*, 2010). When utilizing oxides in this way, the formation of soaps as side products has to be considered. In particular, traces of calcium soaps in biodiesel can lead to serious injection pump problems. Furthermore, the quantity of water in the feedstock can negatively influence the overall conversion rate. The use of modified ZnO nanoparticles, prepared from $Zn(NO_3)_2$ and $La(NO_3)_3$ with urea, has been studied by Yan *et al.*, 2010b. At 200°C the catalyst was active in both esterification and transesterification reactions and could be employed for at least 17 cycles in batch reaction and 70 days under continuous running, without any significant loss of activity. Leaching was also negligible.

A series of low temperature catalysts has been prepared from natural and renewable material like shrimp and crab shell, containing the active components $CaCO_3$ or chitin (Yan *et al.*, 2010a). Carbonization of sugar or starch and sulfonation with concentrated sulfuric acid leads to a 'green catalyst' (Toda *et al.*, 2005). However, this catalyst still showed rapid loss of activity with repeated use and the amount of acid used is far more than the amount used for direct catalysis with sulfuric acid.

Ion exchange resins, either acidic or basic, can be used for low temperature heterogeneous catalysis. Owing to the relatively high costs of the resins, mainly acidic resins are used for the esterification of free fatty acids. Some of these products, such as Amberlyst types from Rohm and Haas, are already used in pilot or small scale production for pre-esterification of highly acidic feedstocks (Pirola *et al.*, 2010). To test the lifetime of the catalyst, 90 consecutive batch deacidification runs, each lasting 6 h, were conducted using crude palm oil or soybean oil as a feedstock and Amberlyst 46 as a catalyst. At the end of the cycles, a decrease in activity of about 25% was observed.

6.3.4 High temperature catalysts

The main advantage of using solid catalysts, including alkaline catalysts such as $CaCO_3$, at higher temperatures is that free fatty acids are also converted into the corresponding fatty acid methyl esters (Suppes *et al.*, 2001).

The first and currently only commercial technology using heterogeneous catalysts like zinc oxides or zinc aluminates is being used in several commercial biodiesel production plants in both France and the USA (Stern *et al.*, 2000). The so-called Esterfip-H process was developed by the Institut Français du Pétrole (IFP) and has been designed and commercialized by Axens. The main advantages of the process are the production of high quality glycerol without the need for disposal of salts resulting from the catalyst. It is reported that the catalyst is very sensitive to water and is not capable of esterifying free fatty acids, so only fully refined vegetable oils can be used as feedstocks. A series of zinc and lanthanum catalysts were developed by Yan *et al.* (2009). Under reaction conditions of 60 min, at 200°C, with 500 psi and 36:1 molar ratio of methanol to oil and 2.3 wt% of the catalyst, refined soybean oil is converted into methyl esters with a yield of 96%. It can be shown that the catalyst is still active, even after more than 100 days.

6.3.5 Lipases as biocatalysts

Within the last few years the use of enzymes as heterogeneous catalysts for the production of biodiesel has been discussed widely within the literature (Adamczak *et al.*, 2009). Lipases are enzymes which catalyze both the hydrolytic cleavage and the synthesis of ester bonds in glycerol esters. Their application in fatty acid methyl ester (FAME) production dates back to Choo and Ong (1986), to a patent application on lipase-catalyzed methanolysis in the presence of water and to Mittelbach (1990) who reported the first water-free process for lipase-catalyzed biodiesel production.

Compared to other catalyst types, biocatalysts have several advantages. They enable conversion under mild temperatures, pressures and pH-conditions. Neither the ester product nor the glycerol phase have to be purified from basic catalyst residues or soaps. Moreover, both the transesterification of triglycerides and the esterification of free fatty acids occur in one process step. As a consequence, highly acidic fatty materials, such as palm oil or waste oils, can be used without pretreatment. Furthermore, enzymes can be easily used for the production of fatty acid ethyl esters using azeotropic technical ethanol. Although the use of enzymes seems to overcome most of the problems related to biodiesel production, the major drawback is still the price of enzymes for the production of bulk chemicals. Therefore, the enzymatic route for biodiesel production is currently still too expensive for industrial applications.

Transesterification of triglycerides takes place in the absence of water, so the lipase is not soluble in the reaction mixture and there is no need for immobilization of the enzyme. The enzyme can be filtered after each reaction step and reused for the next step. For continuous production, however, it is recommended to use a fixed bed reactor filled with immobilized enzyme. The inhibitory effect of methanol can be overcome by using stepwise addition of methanol (Shimada *et al.*, 2002). Normally it is not necessary to use a solvent for a lipase catalyzed reaction. However, in this case there is the danger of glycerol being adsorbed onto the surface of the enzyme, which necessitates an additional step of washing with polar solvents.

An interesting approach undertaken by Zhao *et al.* (2010) is the use of ionic liquids for lipase-catalyzed transesterification. Ionic liquids are well known as alternative solvents with low volatility, toxicity and flammability properties. Ionic liquids can act as a solvent as well as a stabilizer for the lipases, so they are widely used in enzymatic reactions. The first report using ionic liquids for biodiesel production was published by Ha *et al.* (2007). They studied the methanolysis of soybean oil with immobilized *Candida antarctica* lipase in 23 different ionic liquids. The production yield was stated to be eight times higher compared to a conventional solvent free system and about 15% higher than using *t*-butanol as a solvent. The highest production of FAME was obtained in commercially available 1-ethyl-3-methylimidazolium triflate. However, no data were given about economy or the potential industrial applications of this approach. A patent covering the use of ionic liquids as solvents for biodiesel production has been filed by Lozano *et al.* (2008).

Whole cells, such as immobilized *Rhizopus oryzae* cells, have been used for transesterification and esterification of fatty material. The use of a packed bed reactor protected the immobilized cells from physical damage and gave a significant increase in yield at optimum flow rates (Matsumoto *et al.*, 2001; Hama *et al.*, 2007). The main advantage of using whole cells is stated to be a reduction in the overall costs of biocatalytic biodiesel production. There is, however, still a large amount of research to be done before the first industrial application of this process.

6.4 Innovative reaction conditions

6.4.1 Production of biodiesel using supercritical conditions

The transesterification of triglycerides with lower alcohols can be undertaken in the absence of a catalyst, provided reaction temperatures and pressures are high enough (Kusdiana and Saka, 2001; Demirbas, 2006; Wen *et al.*, 2009a). Above the critical temperature, a gas cannot be liquefied. The supercritical conditions show liquid, as well as gaseous properties. Owing to this fact, this approach has been applied, in particular, to CO_2 extractions, but has also been used for carrying out chemical reactions. The advantage of not using a catalyst for transesterification is the production of high-purity esters and soap-free glycerol. The high excess of methanol, which has to be used during supercritical transesterification, seems to make the process not economically feasible. However, a two-step process has been described, where the first step is the hydrolysis of the glycerides into fatty acid with an excess of water, followed by esterification as the second step, thus requiring lower amounts of methanol (Fig. 6.2; Kusdiana and Saka, 2004).

Although the non-catalytic hydrolysis of vegetable oils is now a common practice in the industrial production of fatty acids, non-catalytic production of fatty acid methyl or ethyl esters has not yet been commercialized. The main reason for this is the high investment costs for high temperature and high pressure procedures and also the possibility of side reactions occurring, which automatically lead to lower yields and higher purification costs.

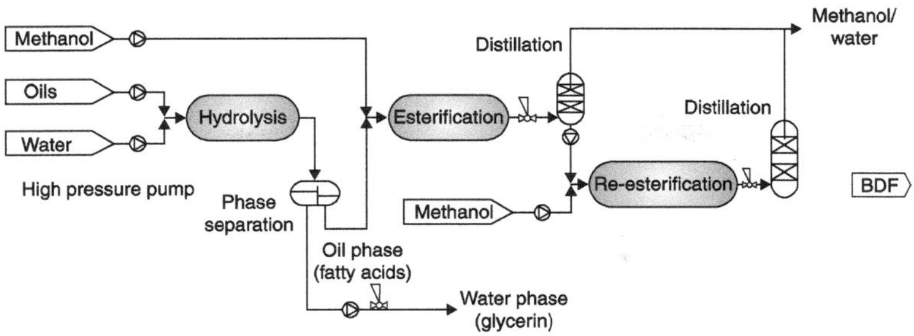

6.2 Supercritical two-step biodiesel production process (from Saka, with permission).

6.4.2 Microwave- and ultrasound-assisted biodiesel production technologies

Microwave-promoted synthesis is an area of increasing interest for both academic and industrial applications. As well as being energy efficient, microwaves can also enhance the rate of reactions and in many cases improve product yields. Microwave-promoted chemistry is based on the fact that solvents or reagents can absorb microwave energy and convert it into heat. Therefore the main applications of this in biodiesel production are for reactions which need higher temperatures, such as acid catalyzed esterification, reactions with longer chain alcohols and for reactions taking place under sub- or supercritical conditions.

Leadbeater and Stencel (2006) reported the use of microwave technology for the transesterification of triolein with methanol under atmospheric conditions with KOH as a catalyst at 50°C. Over 98% conversion was achieved with 5% KOH after 1 min. However, no reference example was given without the use of a microwave. One of the most important weaknesses in described literature using alternative methods is that the corresponding reference experiment is missing. Geuens *et al.* (2008) reported the catalyst-free transesterification of triglycerides with *n*-butanol. The microwave especially assisted reactions under supercritical conditions, which were faster because of the reduced heating time.

Stavarache *et al.* (2005) reported the use of low frequency ultrasound (28 and 40 kHz) for alkaline catalyzed biodiesel production, stating that the quantity of catalyst required is two to three times lower using this approach. Chand *et al.* (2010) studied transesterification of soybean oil with catalytic amounts of NaOH using ultrasonic energy in pulse and continuous sonication at three levels of amplitude. It was observed that ultrasonic treatment resulted in a 96% yield of biodiesel in less than 90 s, compared to 30–45 min for the unsonicated control (which produced an 83–86% yield). It has been established that the physical mechanism responsible for the beneficial action of ultrasound is the formation of a fine emulsion between the oil and the methanol that enhances the interface area for the reaction. No chemical effect, such as the formation of radicals, plays a role in this process (Kalva *et al.*, 2009).

6.4.3 Reactive distillation

Reactive distillation is a technique which combines chemical reactions and product separation in one unit. Because of the continuous removal of products formed during the reaction, the conversion limitation of equilibrium is avoided. He *et al.* (2006) used a continuous flow reactive distillation reactor for biodiesel production from rapeseed oil and methanol, with KOH as the catalyst. The upward-flowing methanol vapour served as an agitator in the reactant mixture, providing constant mixing whilst bubbling throughout the liquid phase. The main advantages were shown to be the short reaction time, the avoidance of excess methanol and the lower investment costs, because there is no need for additional separation units. Fatty acid esterification by catalytic reactive distillation with solid metal oxides as catalysts was carried out by Kiss *et al.* (2008). Further review of the process efficiency and economic analysis of this process is required to establish the impact on biodiesel production.

6.5 Advanced biodiesel reactor design

Biodiesel can be produced either in batch or continuous reactors. Most industrial biodiesel production units today use one or a series of continuous stirred tank reactors with intermediate removal of glycerine. Fonseca *et al.* (2010) employed a kinetic model to compare the behaviour of batch and continuous processes. They found that a single continuous stirred tank reactor is incapable of achieving the same productivity accomplished in a batch reactor indicating the need for continuous reactors in series.

The choice of the type of processing depends on the size of the production plant and on the type of feedstock used. With constant changes in the quality and type of feedstock used, in most cases batch reactors are preferred, where the reactants are mixed with vigorous stirring followed by separation of the glycerol phase without stirring. In most cases the reaction is carried out in two steps, the first step leading to a conversion of approximately 90% and the second step leading to conversions of up to almost 100%. One of the first two-step processes for the industrial production of biodiesel was described by Ahn *et al.* (1995). The two steps could either be performed in the same reactor, or in two different reactors. After separation of the glycerol phase, the biodiesel is washed in the same reactor, or in a different reactor, followed by separation of the water and drying the biodiesel at an elevated temperature, either under reduced pressure or by stripping with air. However, batch processes suffer several disadvantages: they are less efficient owing to their start-up and shut-down nature, they require larger reactor volumes and also the labour costs are usually higher (Fonseca *et al.*, 2010). On the other hand, this kind of reaction type is favoured when different feedstocks with different qualities are used. Using this method, the change from one feedstock to another can easily be done.

For large scale production, continuous operation is favoured. The non-soluble phases, methanol and vegetable oil, are mixed either via a static stirrer or with the help of a homogenizer. The separation of the glycerol phase can be done in a continuous settler or with the help of a centrifuge. To overcome the problem of immiscibility of the methanol and vegetable oil, the use of solvents has been experimented with, leading to a faster reaction at the beginning of the process. However, this reaction has to be stopped at the glycerol separation stage in order to avoid the equilibrium formation, which will halt conversion. Noureddini *et al.* (1998) investigated a continuous process in a pilot plant, using a variation of the mixing intensity, the stoichiometry and the catalyst concentration. They found that a motionless mixer appeared to be more effective than a high-shear mixer. Recent research has been focused on novel reactors and processing technologies (Qiu *et al.*, 2010).

6.5.1 Static mixers

Static mixers consist of specially designed motionless geometric elements, enclosed within a pipe or a column, creating the effective mixing of two immiscible liquids. Thompson and He (2007) used a closed-loop static mixer system as a continuous flow reactor for the alkaline catalyzed production of biodiesel. The most favourable conditions were found to be a temperature of 60°C, with a molar ratio of methanol to oil of 6:1 and 1.5% sodium hydroxide as the catalyst. Static mixers have the advantage of being low maintenance, having low operating costs and low space requirements because there are no moving parts.

6.5.2 Micro-channel and oscillatory reactors

Another approach suitable for continuous biodiesel production is the use of microreactor systems. A microreactor system consists of a series of parallel narrow channels, through which vegetable oil and alcohol are pumped simultaneously. Although the amount of biodiesel produced from a single micro reactor is quite small, the reactors can be connected and stacked in banks to increase the production dramatically. In the microreactor system, mass and heat transfer can be intensified owing to the small space, with a large surface area to volume ratio, providing a high conversion rate in short reaction times.

Microreactors for biodiesel production have been described under classical alkaline catalysis (Canter, 2006; Guan *et al.*, 2009; Sun *et al.*, 2008), for esterification of high fatty acid feedstocks under acidic catalysis (Sun *et al.*, 2010), as well as at sub- and supercritical conditions using ethanol (da Silva, 2010). Sun *et al.* (2008) studied transesterification of vegetable oils with KOH as a catalyst, using

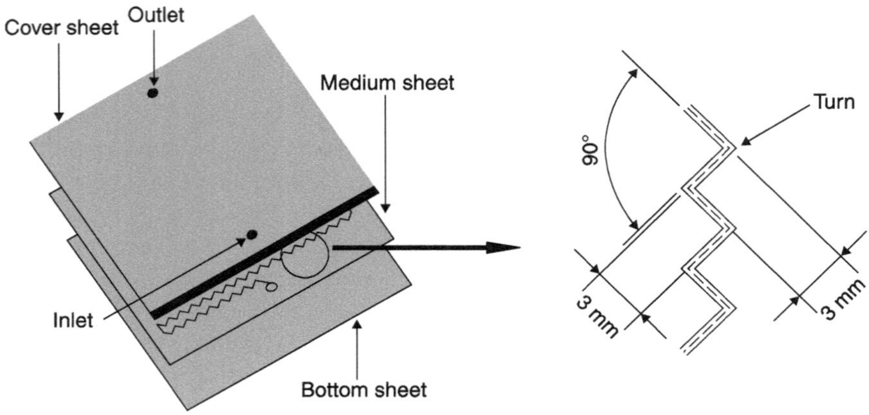

6.3 Zigzag micro-channel reactor according to Wen *et al.* (2009a).

microreactors with an inner diameter of 0.25 mm. With 1 wt% of KOH, a reaction time of 6 min and a 6:1 molar ratio of methanol to oil, a conversion of 99.5% can be reached. The zigzag microchannel reactors are described by Wen *et al.* (2009b) in Fig. 6.3. The process is intensified by the formation of smaller droplets, leading to a conversion of 99.5% within 28 s, at a temperature of 56°C, 1.2 wt% sodium hydroxide and 9:1 molar ratio of methanol to oil. Less energy is used compared to a conventionally stirred reactor, leading to lower investment and operating costs.

Oscillatory flow reactors are tubular reactors in which orifice plate baffles are equally spaced and produce oscillatory flow using a piston drive. Harvey *et al.* (2003) developed a continuous oscillatory flow reactor to produce biodiesel from rapeseed oil on a pilot scale (Fig. 6.4). Conversions of biodiesel up to 99% were

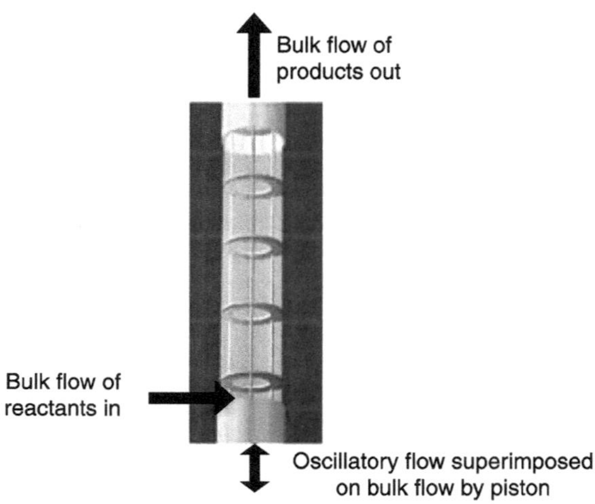

6.4 Configuration of an oscillatory flow reactor (Harvey *et al.*, 2003).

achieved after 30 min at 50°C, in the presence of sodium hydroxide. Another pilot plant was installed in 2004 by the Polymer Fluids Group at the University of Cambridge (http://www.ceb.cam.ac.uk/pages/biofuels.html), producing around 25 l/h. This technology is believed to lead to lower investment and operating costs.

6.5.3 Membrane reactors

Membrane reactors integrate reaction and membrane-based separation into a single process. They can increase the conversion of an equilibrium-limited reaction by continuous separation of the reaction stream via membranes. Dube *et al.* (2007) developed a biodiesel process using a two-phase tubular membrane reactor (Fig. 6.5). The pore size of the carbon membrane was 0.05 μm. They investigated acid and base-catalyzed transesterification of rapeseed oil with methanol and found that the microporous carbon membrane selectively permeated fatty acid methyl esters, methanol and glycerol and excluded the bigger mono-, di- and triglycerides molecules.

Cao *et al.* (2008) investigated the effect of pore size of the membrane in a semi-continuous reactor.

Another route to overcome mass-transfer limitations is the use of liquid–liquid reactors, either as a packed bed reactor (Ataya *et al.*, 2008) or as a film reactor (Narváez *et al.*, 2009). A gas–liquid reactor has been proposed, where vegetable oil is atomized and sprayed into a reaction chamber filled with methanol vapour

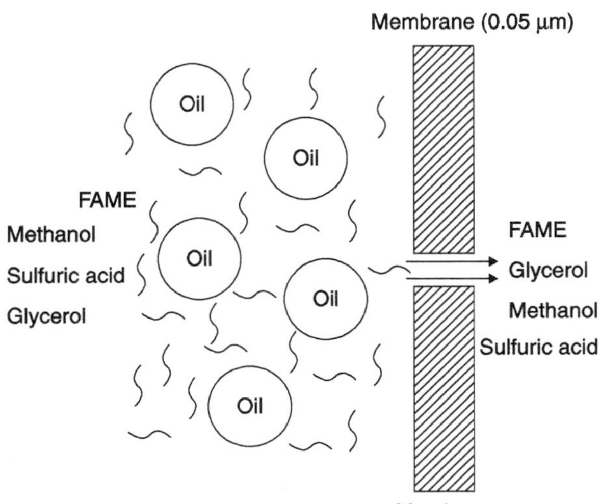

6.5 Schematic of biodiesel production in a membrane reactor (Dube *et al.*, 2007).

(Behzadi and Farid, 2009). This process produced micro-sized droplets, which increased the heat and mass transfer.

Among the techniques developed recently, hydrodynamic cavitation is a potential method for biodiesel production, which uses various different catalysts (Gogate *et al.*, 2006). Cavitation can be defined as the generation, growth and collapse of cavities resulting in very high energy densities. Cavitation can occur at millions of locations in a reactor simultaneously and can generate conditions of very high temperatures and pressures without affecting the overall environment. In the hydrodynamic cavitation method mixing of two phases is carried out under cavitation conditions produced by pressure variation (Pal *et al.*, 2010). Cavitation is generated by the flow of liquid under controlled conditions through simple geometries, such as venture tubes and orifice plates.

Kozyuk (2009) developed flow cavitation technology for esterification of fatty acids to fatty acid alkyl esters. The reactor consisted of one or more controlled flow cavitation zones with one or more orifices. Precise process control ensures repeatable droplet size distribution. The conversion of fatty acids into fatty acid alkyl esters can take microseconds, through four cavitation constrictions in series, with a 1:6 molar ratio of fatty acids to methanol. Transesterification using the principle of cavitation mixing was developed by Hydro Dynamics, Inc. and is described by Mancosky *et al.* (2007). The reaction mixture is passed through a generator's spinning cylinder. The shock waves increase the surface area so that a higher mass transfer occurs. It takes only several seconds to complete transesterification, which leads to less saponification and emulsification.

6.5.4 Rotating reactors and centrifugal contactors

Rotating or spinning tube reactors consist of one inner tube which rotates rapidly within a concentric stationary outer tube. Lodha and Jachuz (2007) produced biodiesel in a rotating tube reactor from canola oil and sodium hydroxide as a catalyst. Within 40 s a conversion of over 98% was achieved at 40–60°C.

The centrifugal contactor is a process intensification technology; it integrates a mixing and a separating zone. A rotor is rapidly rotated within a stationary cylinder, achieving simultaneously intensive mixing and quick phase separation by high centrifugal force. McFarlane *et al.* (2008) modified a conventional centrifugal contactor to produce biodiesel continuously via a base-catalyzed reaction. At 60°C more than 99% conversion was reached after 1 min, with a spinning rotor at 3600 rpm. Good product separation was reached after 3 min.

6.6 Production of biodiesel with simultaneous derivation of glycerol

The traditional transesterification of triglycerides with monohydric alcohols, also called alcoholysis, produces fatty acid alkyl esters and glycerol. The reaction can

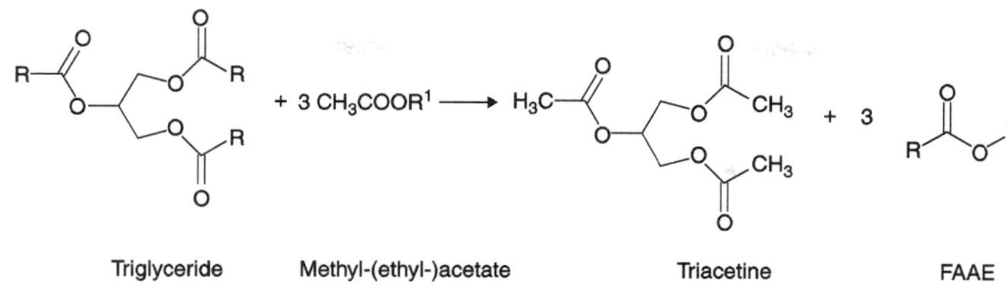

6.6 Inter-esterification of triacylglycerol with alkyl acetate (Mittelbach, 1995).

also be carried out with alkyl acetates, which produce fatty acid alkyl esters and triacetin (Fig. 6.6). This reaction could be carried out either under traditional catalysis with KOH (Mittelbach, 1995) or under supercritical conditions (Saka *et al.*, 2009). Triacetin, which is formed under these conditions, can be either separated or used directly in a mixture with biodiesel as alternative fuel. However, the overall conversion into triacetin is not complete and the existence of mono- and diacetin can lead to incomplete solubility in the resultant diesel fuel. Triacetin can be used in series of technical applications, such as in the tobacco industry as a moisture agent.

Triacetin can also be obtained from vegetable oils via acidolysis with subcritical acetic acid and supercritical methanol (Saka *et al.*, 2010). In the first step the triglycerides are split into fatty acids and triacetin and in the second step the fatty acids are esterified with methanol under supercritical conditions. This two step route has advantages over the direct reaction with supercritical acetates; it has higher reaction rates as well as better overall conversion.

Another approach for direct derivatization of glycerol during transesterification (Fig. 6.7) has been described by Ilham and Saka (2009).The reaction of triacylglycerol with dimethyl carbonate under supercritical conditions leads to the formation of fatty acid methyl esters and glycerol carbonate, which is a valuable building block for synthetic organic chemistry like, for example, the production of ethylene or propylene carbonate or for the production of glycidol, which is widely utilized in the pharmaceutical and polymer industry (Kob and Klein, 2009).

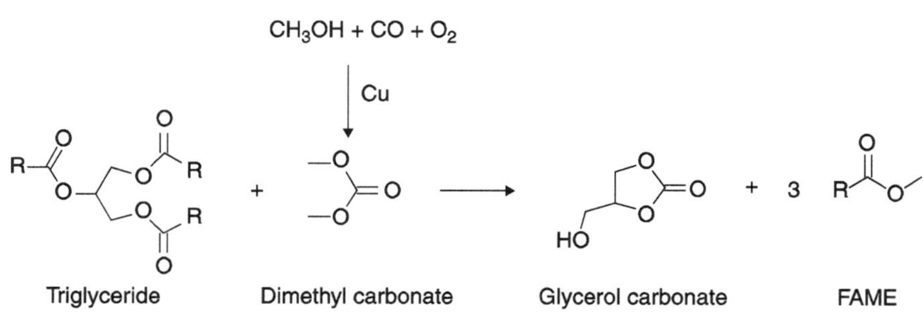

6.7 Transesterification with dimethyl carbonate (adapted from Ilham and Saka, 2009).

6.7 *In situ* transesterification

Conventional biodiesel production starts with an oil or fat, which is expelled or extracted from oil-containing seeds or fruits. With so-called *in situ* production the starting material is any oil containing biomass, which is directly transesterified with an alcohol. The main advantage of this is the direct processing of the feedstock without requiring an expelling or extraction process. The alcohol itself, with or without a catalyst, acts directly as an extraction solvent. Such methods have been called reactive extraction. No additional solvents have to be used and the overall conversion of the total lipids in the seeds is higher than conventional expelling and transesterification of the oil.

The drawback, however, is the fact that unwanted components like polar carbohydrates or proteins can be extracted. Harrington and D'Arcy-Evans (1985) studied *in situ* transesterification of a series of sunflower seeds, focusing mainly on the overall reduction of production costs of biodiesel. Kaul *et al.* (2010) studied reactive extraction of jatropha seeds. With a seed size of >2.46 mm and a seed to solvent ratio of 1:7.8, using 0.1 M NaOH as the catalyst and a reaction time of 1 hour at 65°C, the reaction reached a 98% conversion rate under optimized conditions.

The general applicability of *in situ* transesterification for the production of fatty acid esters from a variety of feedstocks was described by Haas *et al.* (2007). Soybean flakes can be transesterified directly, using an alkali-catalyzed reaction leading to almost quantitative conversion into fatty esters. However, the resulting soybean meal has to be purified from the remaining alcohol. The direct *in situ* transesterification is more favoured when using non-conventional low quality feedstocks with a lower oil content, like distillers dried grains with solubles (DDGS), a co-product of the fermentation of corn that is used to produce ethanol, or meat and bone meal, a product of the animal rendering industry containing residual fat of up to 10% (m/m). It can be shown that the FAME produced from these feedstocks does not contain unreacted triglycerides, because these remain within the feedstock. However, the content of free fatty acids could retard transesterification. The process is therefore limited to low acidic feedstocks.

Wyatt and Haas (2009) reported the *in situ* transesterification of soy flakes with sulfuric acid in methanol containing a 50% mole fraction of CO_2. Supercritical CO_2 is used as a co-solvent that is dissolved in methanol. It can be shown that the addition of CO_2 increased the rate of the acid-catalyzed reaction by as much as 2.5 times. Nitrogen analysis showed that about 85% of protein was left intact under these conditions.

As *in situ* transesterification of traditional oil seed feedstocks is not competitive for industrial application, owing to long tradition in oil seed processing, this technique could be very interesting for feedstocks like microalgae, where a cost effective extraction and processing has not yet been developed. Wahlen *et al.* (2011) studied the simultaneous extraction and conversion of total lipids from microalgae, cyanobacteria and wild mixed-cultures. This direct transesterification approach yielded more biodiesel than was expected from the triglyceride content, indicating capture of fatty acids from membrane phospholipids. An amount of 100 mg of lyophilized algal biomass was mixed with methanol containing sulfuric acid as a catalyst and the reactions were carried out in a commercial scientific microwave. The reaction mixture was extracted with chloroform and quantified by gas chromatography. The optimum conditions for dry biomass were found to be 2 ml of methanol containing 1.8% sulfuric acid (v/v) and a reaction time of 20 min at 80°C. Between 18 and 36% of FAME from the biomass and 77–82% of FAME from the extractable lipids was produced, compared to a triglyceride content of between 8 and 27% of the dry biomass. Optimum conditions were also used for the study of an *in situ*-processing of wet biomass, showing that an increase of methanol content to 4 ml led to very similar yields. With this approach the very costly drying process of microalgae can be avoided. However further studies have to be carried out in order to evaluate industrial applicability.

6.8 Conclusions

Biodiesel is an already well established biofuel in many countries worldwide. Production technologies are available from a series of commercial companies mainly based on traditional esterification and transesterification steps using acidic or base homogeneous catalysis. Although a huge number of papers have been published within the last years describing new technologies, only few have been applied on an industrial scale. The main innovations are the use of alternative catalysts like heterogeneous compounds or lipases as enzymatic catalysts, or supercritical reaction conditions without any use of catalysts. There are also technologies using microwave, cavitation or ultrasound for improved heat transfer or mixing conditions. Also better separation of side products with the help of membrane technology or the purification of the end product like distillation or adsorption have been described. Another approach could be the simultaneous derivatization of the side product glycerol, which could lead to higher value added products. However, all these technologies have the drawback of higher investment or running costs, therefore today they are not really competitive with traditional existing technologies. Perhaps the need for higher reduction of overall greenhouse gas emissions during the whole life cycle according to the European Renewable Energy Directive or similar legislation could help new innovations to become reality. The use of alternative cheap waste feedstocks or of totally different feedstocks like microalgae will require the use of alternative technologies.

6.9 References

Adamczak M, Bornscheuer U and Bednarski W (2009) The application of biotechnological methods for the synthesis of biodiesel, *Eur. J. Lipid Sci. Technol.*, **111**, 800–13.

Ahn E, Koncar M, Mittelbach M and Marr R (1995) A low-waste process for the production of biodiesel, *Sep. Sci. Technol.*, **30**, 2021–33.

Arzamendi G, Arguinarena E, Campo I, Cabala S and Gandía L (2008) Alkaline and alkaline-earth metals compounds as catalysts for the methanolysis of sunflower oil, *Catal. Today*, **133–135**, 305–13.

Ataya F, Dubé M and Ternan M (2008) Transesterification of canola oil to fatty acid methyl ester (FAME) in a continuous flow liquid–liquid packed bed reactor, *Energy Fuels*, **22**, 3551–6.

Behzadi S and Farid M (2009) Production of biodiesel using a continuous gas–liquid reactor, *Biores. Technol.*, **100**, 683–9.

Canter N (2006) Making biodiesel in a micro reactor, *Tribol. Lubr. Technol.*, **62**, 15–17.

Cao P, Dubé M and Trembly A (2008) High purity fatty acid methyl ester production from canola, soybean, palm and yellow grease lipids by means of a membrane reactor, *Biomass Bioenergy*, **32**, 1028–36.

Chand B, Chintareddy R, Verkade J and Grewell D (2010) Enhancing biodiesel production from soybean oil using ultrasonics, *Energy Fuels*, **24**, 2010–2015.

Choo Y and Ong S (1986) *Transesterification of Fats and Oils*, British Patent GB 2 188 057 A.

Demirbas A (2006) Biodiesel Production via Non-Catalytic SCF Method and Biodiesel Fuel Characteristics, *Energy Convers. Manag.*, **47**, 2271–82.

Dube M, Tremblay A and Liu J (2007) Biodiesel production using a membrane reactor, *Biores. Technol.*, **98**, 639–647.

European Directive 2009/28/EC (2009) Directive 2009/28/EC of the European Parliament and of the council on the promotion of the use of energy from renewable sources; April 2009.

Fonseca F, Vidal-Vieira J and Ravagnani S (2010) Transesterification of vegetable oils: Simulating the replacement of batch reactors with continuous reactors, *Biores. Technol.*, **101**, 8151–8157.

Geuens J, Kremsner J, Nebel B, Schober S, Dommisse R, *et al.* (2008) Microwave-assisted catalyst free transesterification of triglycerides with 1-butanol under supercritical conditions, *Energy Fuels*, **22**, 643–5.

Gogate P, Tayal R and Pandit A (2006) Cavitation: a technology on the horizon, *Curr. Sci.*, **91**, 35–46.

Guan G, Kusakabe K, Moriyama K and Sakurai N (2009) Transesterification of sunflower oil with methanol in a microtube reactor, *Ind. Eng. Chem. Res.*, **48**, 1357–1363.

Ha S, Lan M, Lee S, Hwang S and Koo Y (2007) Lipase-catalyzed biodiesel production from soybean oil in ionic liquids, *Enzyme Microb. Technol.*, **41**, 480–483.

Haas M, Scott K, Foglia T and Marmer W (2007) The general applicability of in situ transesterification for the production of fatty acid esters from a variety of feedstocks, *J. Am Oil Chem. Soc.*, **84**, 963–970.

Hama S, Yamaji H and Fukumizu T (2007) Biodiesel fuel production in a packed bed reactor using lipase-producing *Rhizopus oryzae* cells immobilized within biomass supported particles, *Biochem. Eng. J.*, **34**, 273–278.

Harrington K and D'Arcy-Evans C (1985) A comparison of conventional and in situ methods of transesterification of seed oil from a series of sunflower cultivars, *J. Am. Oil Chem. Soc.*, **62**, 1009–1013.

Harvey A, Mackley M and Seliger T (2003) Process intensification of biodiesel production using a continuous oscillatory flow reactor, *J. Chem. Technol. Biotechnol.*, **78**, 338–341.

He B, Singh A and Thompson J (2006) A novel continuous – flow reactor using reactive distillation for biodiesel production, *Trans ASAE*, **49**, 107–12.

Ilham Z and Saka S (2009) Dimethyl carbonate as potential reactant in non-catalytic biodiesel production by supercritical method, *Biores. Technol.*, **100**, 1793–1796.

Kalva A, Sivasankar T and Moholkar V (2009) Physical mechanism of ultrasound-assisted synthesis of biodiesel, *Ind. Eng. Chem. Res.*, **48**, 534–544.

Kaul S, Porwal J and Garg M (2010) Parametric study of Jatropha seeds for biodiesel production by reactive extraction, *J. Am. Oil Chem. Soc.*, **87**, 903–908.

Kiss A, Dimian A and Rothenberg G (2008) Biodiesel by catalytic reactive distillation powered by metal oxides, *Energy Fuels*, **22**, 598–604.

Kob N and Klein H (2009) Five-membered cyclic alkylene carbonates, *Spec. Chem. Mag.*, **29**, 44–45.

Kozyuk O (2010) *Apparatus and Method for Producing Biodiesel from Fatty Acid Feedstock*, US Patent 7,754,905 B2.

Kusdiana D and Saka S (2001) Methyl esterification of free fatty acids of rapeseed oil as treated in supercritical methanol, *J. Chem. Eng. Jap.*, **34**, 383–387.

Kusdiana D and Saka S (2004) Effects of Water on Biodiesel Fuel Production by Supercritical Methanol Treatment, *Biores. Technol.*, **91**, 289–295.

Leadbeater N and Stencel, L (2006) Fast, easy preparation of biodiesel using microwave heating, *Energy Fuels*, **20**, 2281–2283.

Lodhar H and Jachuz R (2007) Intensified biodiesel reaction using continuous rotating tube reactor technology, in *Proceeding of the AIChE Annual Meeting*, Salt Lake City, USA.

Lozano P, De Tiego T, Iborra J and Vaultier M (2008) *Use of Ionic Liquids for Implementing a Process for the Preparation of Biodiesel*. European Patent EP 2189535 A1.

Lukic I, Krstic J, Glisic S, Jovanovic D and Skala D (2010) Biodiesel synthesis using K_2CO_3/Al-O-Si aerogel catalysts, *J. Serb. Chem. Soc.*, **75**, 789–801.

Mancosky D, Armstead D, McGurk T, Hopkins G and Hudson K (2007) The use of a controlled cavitation reactor for bio-diesel production, in *Proceedings of the AiChE Spring Meeting*, Houston, USA.

Matsumoto T, Takahashi S and Kaieda M (2001) Yeast whole cell biocatalyst constructed by intracellular overproduction of *Rhizopus oryzae* lipase is applicable to biodiesel fuel production, *Appl. Microbiol. Biotechnol.*, **57**, 515–520.

McFarlane J, Birdwell J, Tsouris C and Jennings H (2008) Process intensification in continuous base-catalyzed biodiesel production, in *Proceedings of the AIChE Annual Meeting*, Philadelphia, PA, 16–21 November.

Mittelbach M (1990) Lipase catalyzed alcoholysis of sunflower oil, *J. Amer. Oil Chem. Soc.*, **67**, 168–170.

Mittelbach M (1995) Unpublished results resulting from diploma thesis of Pressnig M, University of Graz.

Narváez P, Sánchez F and Godoy-Silva R (2009) Continuous methanolysis of palm oil using a liquid–liquid film reactor, *J. Amer. Oil Chem. Soc.*, **86**, 343–352.

Navajas A, Campo I, Arzamendi G, Hernández W, Bobadilla L, *et al.* (2010) Synthesis of biodiesel from the methanolysis of sunflower oil using PURAL® Mg–Al hydrotalcites as catalyst precursors, *Appl. Catal. B: Environmental*, **100**, 299–309.

Ngamcharussrivichai Ch, Nunthasanti P, Tanachai S and Bunyakiat, K (2010) Biodiesel production through transesterification over natural calciums, *Fuel Proc. Technol.*, **91**, 1409–1415.

Noureddini H, Harkey D and Medikonduru V (1998) A continuous process for the conversion of vegetable oils into methyl esters of fatty acids, *J. Am. Oil Chem. Soc.*, **75**, 1775–1783.

Pal A, Verma A, Kachhwaha S and Maji S (2010) Biodiesel production through hydrodynamic cavitation and performance testing, *Renewable Energy*, **35**, 619–624.

Peter S and Weidner E (2007) Methanolysis of triacylglycerols by organic basic catalysts, *Eur. J. Lipid Sci. Technol.*, **109**, 11–16.

Pirola C, Bianchi L, Boffito D, Carvoli G and Ragaini V (2010) Vegetable oil deacidification by Amberlyst: Study of the catalyst lifetime and a suitable reactor configuration, *Ind. Eng. Chem. Res.*, **49**, 4601–4606.

Qiu Z, Zhao L and Weatherley L (2010) Process intensification technologies in continuous biodiesel production, *Chem. Eng. Process.*, **49**, 323–330.

Saka S and Isayama Y (2009) A new process for catalyst-free production of biodiesel using supercritical methyl acetate, *Fuel*, **88**, 1307–1313.

Saka S, Isayama Y, Ilham Z and Jiayu X (2010) New process for catalyst-free biodiesel production using subcritical acetic acid and supercritical methanol, *Fuel*, **89**, 1442–6.

SciFinder®, Chemical Abstract Service, American Chemical Society; Online-Search carried out in January 2011.

Shimada Y, Watanabe Y and Sugihara A (2002) Enzymatic alcoholysis for biodiesel fuel production and application of the reaction to oil processing, *J. Mol. Catal. B*, **17**, 133–142.

da Silva C, de Castilhos F, Oliveira J and Cadozo Filho L (2010) Continuous production of soybean biodiesel with compressed ethanol in a microtube reactor, *Fuel Proc. Technol.*, **91**, 1274–1281.

Stavarache C, Vinatoru M, Nishimura R and Maeda Y (2005) Fatty acid methyl esters from vegetable oil by means of ultrasonic energy, *Ultrasonics Sonochemistry*, **12**, 367–372.

Stern R, Hillion G and Rouxel J (2000) Institut Français du Pétrole, US Patent 6,147.196, 2000, http://www.axens.net/html-gb/offer/offer_processes_104.html; accessed 20 July 2011.

Suppes G, Bockwinkel K, Lucas S, Botts J, Mason M and Heppert J (2001) Calcium carbonate catalyzed alcoholysis of fats and oils, *J. Amer. Oil Chem. Soc.*, **78**, 139.

Sun J, Ju J, Ji L, Zhang L and Xu N (2008) Synthesis of biodiesel in capillary microreactors, *Ind. Eng. Chem. Res.*, **47**, 1398–1403.

Sun P, Sun J, Yao J, Zhang L and Xu N (2010) Continuous production of biodiesel from high acid value oils in microstructured reactor by acid-catalyzed reactions, *Chem. Eng. J.*, **162**, 364–70.

Thompson J and He B (2007) Biodiesel production using static mixers, *TransASABE*, **50**, 161–165.

Toda M, Takagaki A, Okamura M, Kondo J, Hayashi S, *et al.* (2005) Biodiesel made with sugar catalyst, *Nature*, **438**, 178.

Wahlen B, Willis R and Seefeldt L (2011) Biodiesel production by simultaneous extraction and conversion of total lipids from microalgae, cyanobacteria and wil mixed-cultures, *Biores. Technol.*, **102**: 2724–2730.

Wen D, Jiang H and Zhang K (2009) Supercritical fluids technology for clean biofuel production, *Progr. Nat. Sci.*, **19**, 273–284.

Wen Z, Yu X, Tu S, Yan J and Dahlquist E, (2009) Intensification of biodiesel synthesis using zigzag micro-channel reactors, *Biores. Technol.*, **100**, 3054–3060.

Wyatt V and Haas M (2009) Production of fatty acid methyl esters via in situ transesterification of soybean oil in carbon dioxide-expanded methanol, *J. Ar. Oil Chem. Soc.*, **86**, 1009–1016.

Xu Ch, Enache D, Lloyd R, Knight D, Bartley J and Hutchings G (2010) MgO catalysed triglyceride transesterification for biodiesel synthesis, *Catal. Lett.*, **138**, 1–7.

Yan S, Salley S and Ng K (2009) Simultaneous transesterification and esterification of unrefined or waste oils over ZnO–La_2O_3 catalysts, *Appl. Catal. A*, **353**, 203.

Yan S, DiMaggio C, Mohan S, Kim M, O'Salley S and Ng S (2010a) Advancements in heterogeneous catalysis for biodiesel synthesis, *Top Catal.*, **53**, 721–736.

Yan S, Mohan S, DiMaggio C, Kim M, Ng S and O'Salley S (2010b) Long term activity of modified ZnO nanoparticles for transesterification, *Fuel*, **89**, 2844–2852.

Zhang Y, Dube M, Mc.Lean D and Kates M (2003) Biodiesel production from waste cooking oil: 2. Economic assessment and sensitivity analysis, *Biores. Technol.*, **90**, 229–240.

Zhao H, Song Z, Olubajo O and Cowins J (2010), New ether-functionalized ionic liquids for lipase-catalyzed synthesis of biodiesel, *Appl. Biochem. Biotechnol.*, **162**, 13–23.

7

Biodiesel from waste oils and fats

J. IGLESIAS and G. MORALES, Universidad Rey Juan Carlos, Spain

Abstract: One of the main challenges that the biodiesel industry faces nowadays is that the production costs are still not competitive with conventional petrol fuels. Furthermore, biodiesel production is based on expensive edible biomass as raw material. In this way, innovative technologies are needed to produce biofuels from a wider range of biomass resources allowing a reduction of production costs while maintaining energy efficiency. Waste oleaginous feedstock represents an attractive alternative to edible fat and oils for the production of biodiesel fuel from different points of view: they are low-cost raw materials, their use supposes an environmentally friendly approach and they do not compete with food industry, avoiding social concerns. Thus, this chapter deals with the exploitation of this relatively abundant alternative feedstock for the production of biodiesel.

Key words: animal fat, brown grease, biodiesel, waste cooking oil, wastes, yellow grease.

7.1 Introduction

Despite the great interest awakened by biodiesel as a renewable source of energy for transportation and the excitement of its market potential, the use of virgin vegetable oil feedstock, that is basically meant for human consumption, not only makes the biodiesel product expensive but also hikes the cost of oil-related food items on the market. Added to several other important issues such as price volatility and low price of petroleum oil, sustainable water use, secondary land use and biodiversity effects, and so on, this has hindered the competitive market introduction of biodiesel. Improving the economics of the production process through the use of less expensive feedstocks, to provide a more affordable fuel, is one of the current interests of the sector. Apart from edible vegetable oils, plenty of alternative low-cost triglyceride sources have been identified and demonstrated, either in research literature or in operating demonstration plants. Over the years, researchers have focused attention on developing various catalytic processes to achieve the desired goal in this important aspect of renewable energy. In this context, the present chapter relates to the description of waste oils and fats with potential use as biodiesel feedstock. Also, the catalytic systems and technologies proposed for the exploitation of this available alternative feedstock are reviewed. Special focus has been brought onto innovative environmentally friendly technologies to produce biodiesel from a wide range of lipid wastes.

7.2 Waste oils and fats

Waste oils and fats can be considered to be some of the most important feedstocks to be used in biodiesel production, not only because the destination of these materials in the production of fuels does not affect other markets, such as the food market, as vegetable oils do (Rosillo-Calle *et al.*, 2009), but also because their rather high worldwide availability and low cost allow reduction in biodiesel production costs (Rice *et al.*, 1999); cost being one of the main drawbacks of biodiesel when compared to conventional fuels. There is a large variety of raw materials that can be considered as wastes from certain processes but which contain significant amounts of lipids. Among many others, the possibility of preparing biodiesel from raw materials such as waste cooking oil (WCO) (Wiltsee, 1998; Canakci, 2007), low-grade animal fats (Canoira *et al.*, 2008), trap greases (Mustapic and Stanisa, 2006), sewage sludge from municipal waste-water treatment plants (Mondala *et al.*, 2009), along with other special raw materials such as fish oil (Lin and Li, 2009), chicken fat (Marulanda *et al.*, 2010), tobacco seed oil (Veljkovic *et al.*, 2006) or the lipid fraction of spent coffee grounds (Kondamudi *et al.*, 2008) has been described. All these lipid wastes display potential for use in the production of biodiesel, although there are several features a lipid waste must fulfil before being considered as a feedstock with enough potential for biodiesel production (Groschen, 2002). These requirements include a wide range of factors affecting important concerns such as the profitability of the biodiesel production process, the technical feasibility for the transformation of the raw material into biodiesel, or its performance when used as biofuel in conventional combustion engines.

7.2.1 Types of waste raw materials for biodiesel production

Many raw materials containing lipids have been described as showing high potential for use in the production of biodiesel, bearing in mind the above-mentioned requirements. These include low-grade animal fats and WCOs as the most reported cases, but also other lipid wastes such as trap greases (Tyson *et al.*, 2004) and lipid fractions extracted from sewage sludges (Boocock *et al.*, 1992). All these waste lipids are usually classified into three different categories (Groschen, 2002): animal fats, yellow grease and brown grease.

Waste animal fats reported to be used in biodiesel production include a wide variety of fats obtained from different sources such as beef tallow (Ma *et al.*, 1998; Espinosa da Cunha *et al.*, 2009), poultry fat (Kondamudi *et al.*, 2009; Marulanda *et al.*, 2010), waste lard (Dias *et al.*, 2009; Berrios *et al.*, 2009), fish oil (Lin and Li, 2009) and, among many others, lipid fractions extracted from meat-and-bone meal (Nebel and Mittelbach, 2006). In the latter case, several laws, like the EU Regulation for laying down the health rules concerning animal by-products not intended for human consumption, have catalogued various animal by-products as category I grease, whose use for human or animal feeding has been banned (Commission Regulation 2002R1774), and thus this raw material appears to be a good candidate for biodiesel production. In addition, recent studies on biodiesel production using category I fats, classified as risk material owing to the presence of bovine spongiform encephalopathy prions, resulted in considering the final biodiesel as safe (Seidel *et al.*, 2007; Mittelbach *et al.*, 2007), although the solid residue still displays transmissible spongiform encephalopathy activity (Bruederle *et al.*, 2008) and hence requires secure management.

Yellow grease is usually considered to be a mixture of used vegetable oils, such as WCOs and animal fats from commercial or industrial cooking operations. Usually, these wastes show free fatty acid (FFA) levels lower than 15% (Canakci and van Gerpen, 2001). The main destination of this oil, since it is considered as not being adequate for human consumption, is animal feed as supplementary food or food additive, so it still has some commercial interest. As a consequence, there is already an existing infrastructure for the collection of yellow grease, reaching almost 95% of the yellow grease producers in metropolitan areas (Groschen, 2002).

Brown grease differs from yellow grease in the amount of FFA content, typically over 20%, and essentially it can be formed by cooking oils that have decomposed more than the norm or may be composed of greases that were overcooked at the renderers (Tyson, 2002). This material is not suitable as animal food, with the exception of being blended with low-FFA material to meet the yellow grease specifications, unless it also contains trap grease. Trap grease is a kind of waste collected in grease traps either from restaurants, food processing plants or even waste-water treatment plants (Wiltsee, 1998). It is usually included within the brown grease classification owing to its high acidity, usually displaying FFA contents ranging from 40–100% (van Gerpen *et al.*, 2003). This high level of acidity is caused by its contact with water, which hydrolyses the starting oil glycerides. Finally, among the most potentially interesting raw materials to be used as feedstocks for biodiesel production, waste-water treatment sewage sludge emerges as a good candidate (Boocok *et al.*, 1992), both from primary and secondary units.

The reasons for considering these lipid-containing wastes (animal fats, yellow grease and brown grease) as technically and economically feasible possibilities for the production of biodiesel are several. On the one hand, their collection can be accomplished in a concentrated manner since they can be harvested from household waste recycling centres, restaurants and rendering companies – in the case of yellow grease – and in municipal sewage systems or from grease traps in processing rendering plants – for sewage sludge and trap grease, respectively. On the other hand, their lower price relative to refined and semi-refined vegetable oils is a key factor allowing a reduction in the biodiesel production costs and improving the economics of the overall process (Haas, 2005). However, this advantage can be eclipsed by the greater difficulty in processing such feedstocks (Canakci and Sanli, 2008).

7.2.2 Availability of waste lipids for biodiesel production

The availability of a waste lipid is a limiting feature for its use in biodiesel production, since the preparation of this fuel must be accomplished on a scale large enough to fulfil the market requirements and take advantage of the economy of scale to reduce the production costs. Not all lipid wastes are produced in enough quantity and available data for waste lipid production is scarce. However, Table 7.1 lists the availability of different rendered products – animal fats – and the potential for the collection of different lipid wastes – yellow and trap/brown grease – based on several estimations, both in the USA and EU.

Table 7.1 Current and potential production of rendered waste fats and greases, biodiesel and diesel substitution capacity in the USA and EU in 2009

Product (tonnes/year)	USA	EU 25
Waste fats and greases – Rendered	2880 356	3963 990
Inedible tallow	1514 913[a]	89 5051[b]
Lard	143 963[a]	1056 000[b]
Poultry fat	606 646[a]	808 682[b]
Yellow grease	614 835[a]	1204 257[c]
Waste fats and greases – Potential[d]	4915 110	4445 693
Yellow grease	1057 459[e]	1685 960[c]
Brown/trap grease	1592 129[e]	n.a.
Biodiesel production (current)	1610 113[f]	9046 000[g]
From rendered wastes (current)	370 326[h]	220 779[i]
Potential from rendered fats and oils[j]	2102 659	2893 713
Potential from waste lipids[j]	3588 030	3245 356
Potential from sewage sludge[k]	725 400[l]	1405 221[m]
Diesel production (current)[f]	198 060 858	248 809 785
% from biodiesel – Current[n]	0.81	3.64
% from all sources – Potential[o]	3.00	5.50

[a] National Renderers Association (2010); [b] Eurostat (2010); [c] Estimated from recovered vegetable oil data in Ireland (SEAI, 2003); [d] Includes rendered fats and estimated potential for yellow and trap grease; [e] Calculated from the production per capita estimations from NREL (Tyson, 2002, 2008) and metropolitan population in USA (US Census Bureau, 2010); [f] Energy Information Administration (2010); [g] European Biodiesel Board, 2010; [h] From Swisher (2010); [i] From Coelenbier (2010); [j] Considering 0.730 kg biodiesel/kg recovered vegetable oil (Fröhlich and Rice, 2005); [k] Considering 15 wt% lipid in dry sludge and 72 wt% saponifiable matter (Boockok et al, 1992); [l] From Dufreche *et al.* (2007); [m] Average production 2003–2009, Eurostat; [n] Current diesel substitution by biodiesel; [o] Potential diesel substitution including current production and potential production of biodiesel from wastes.

Regarding the use of animal fat for biodiesel production, this practice is more extensive in the USA, representing up to 23% of the total biodiesel raw material used there. In contrast, in the EU only 6% of the total biodiesel is produced from rendered animal fat (Swisher, 2010), despite the high production rate of biodiesel in this area. If all of the rendered animal fat was destined for biodiesel production, a trend which is being observed in recent years (USDA, 2009), these percentages would significantly vary, as the potential amount of biodiesel which can be obtained from these raw materials would increase the US production by 130%, whereas this percentage would be more moderate in EU – 32% – because of the already mentioned actual high production of biodiesel in this area. In this sense, it seems that the availability of animal fats for biodiesel production, while rather high, is not enough to maintain the increasing demand for raw materials for the biodiesel industry, especially in the EU. Therefore, other lipid wastes should be considered to ensure the supply of biodiesel feedstocks. For instance, several authors have considered the use of yellow and brown/trap grease as potential feedstocks for biodiesel production and thus several studies have been carried out in line with this (Wiltsee, 1998; Groschen, 2002).

According to Tyson (2002 and 2008), up to 4.02 kg/person/year of yellow grease – recycled cooking oil – and 6.06 kg/person/year of brown/trap grease – floating material from primary treatment in waste-water treatment plants – can be collected in metropolitan urban areas in the USA. These estimations have been used for the calculation of yellow and trap greases availability in the USA, whereas the availability of yellow grease in the EU was calculated from data published by SEAI for the actual and potential collection of WCO in Ireland in 2003, combined with the population of the island. This combination leads to an estimated collection of yellow grease of 2.56 kg/person/year and a potential collection of 3.58 kg/person/year, whereas there is no available data about calculations on brown/trap grease production. Using this data in combination with the US and EU population estimations for 2009 (US Census Bureau, 2009; Eurostat, 2010) the potential availability of yellow and brown/trap greases, as well as the potential production of biodiesel from the same feedstock, were calculated considering a conversion yield of 0.73 kg biodiesel/kg recovered oil (Fröhlich and Rice, 2005). Estimations based on these assumptions lead to a potential production of biodiesel from all the considered waste lipids of 3.6 million tonnes and 3.2 million tonnes of biodiesel in the USA and EU, respectively.

Considering the availability of these feedstocks together with the actual diesel production in refineries (Energy Information Administration, 2010), waste fats and oils, both rendered and recovered, together with the actual biodiesel production from animal fats and oils, would allow the substitution of up to 2.6% of the total conventional diesel produced from crude oil in the USA. This substitution is even higher in the EU, where the combination of the actual biodiesel production with the potential biodiesel production from waste oils and fats accounts for a conventional diesel substitution of 5.0%, even though the trap grease production in waste-water treatment plants has not been considered. Nevertheless, this figure is still far from matching the requirements established in the Commission Directive 2003/30/EC. In this sense, the use of sewage sludge produced in the waste-water treatment plants seems to be a rather attractive option, since this is a plentiful raw material; the estimated yearly production of sewage sludge in the USA is 6.2 million tonnes (Dufreche *et al.*, 2007), whereas in EU it reaches 12 million tonnes (Eurostat, 2010). Furthermore, there is already an existing system for its collection in a concentrated manner – the waste-water treatment plants themselves – and it contains 12–18 wt% lipids. Considering these features, the total amount of available grease to be extracted from sewage sludge in the USA and EU (15 wt% of lipid content) accounts for 930 000 tonnes and 1 801 565 tonnes, respectively, although it has to be considered that the unsaponifiable matter is rather abundant in this extract, up to 28 wt% (Boocock *et al.*, 1992), which significantly reduces these numbers. Nevertheless, the contribution of this latter waste as raw material for the production of biodiesel would allow the diesel substitution rate by biodiesel to increase by 3.00% and 5.50% in the USA and EU, respectively.

From the previous results the importance of the role of lipid wastes in the production of biodiesel is obvious. However, their contribution is not high enough to satisfy the requirements of European laws for the substitution of conventional fuels by renewable sources and, thus, these must be seen just as a further contribution in the pool of renewable lipid feedstock.

7.3 Technology for biodiesel production from waste lipids

7.3.1 Specific considerations for lipid waste raw materials

The use of waste oils and fats implies several features, like dealing with a large concentration of impurities. This may invalidate the use of a certain raw material in the production of biodiesel because the required transformation process would be too complex to be built at industrial scale. With regards to the composition, the closer a feedstock is to containing pure triglycerides, the easier is its conversion into biodiesel and the higher its cost and *vice versa* (van Gerpen *et al.*, 2003). Because of their nature and origin, waste lipids include several constituents that differ from triglycerides in their composition, such as FFAs, water, solids and other impurities like phosphorous, sulphur and nitrogen compounds or unsaponifiable matter. Other chemical properties, such as the degree of unsaturation of the alkyl chains in the fatty acids, are also relevant because of their influence on the final fuel properties. Also, physical parameters such as viscosity, density or titre (the temperature at which oil changes from solid to liquid) (Karmakar *et al.*, 2010) could also be important, not only because they may condition the properties of the resultant biodiesel but also because of the difficulties in processing and handling the raw materials. In the following, the parameters and properties of waste oils and fats that are considered more influential in the production of biodiesel are described.

Acid value/free fatty acid content

The acid value is a way to express the FFA content in a lipid raw material, usually expressed as mg of potassium hydroxide consumed to neutralize the acidity of a gram of raw material (Table 7.2). In the case of waste lipids, the acid values are substantially higher than for other conventional feedstocks for biodiesel production like vegetable oils, with certain exceptions such as Palanga, *Jatropha curcas L.* or Pongamia oils (Leung *et al.*, 2010; Azócar *et al.*, 2010). The reason for this high acidity might lie on the contact between water and the waste oils. For instance, heating cooking oil causes breakage of the glycerides by hydrolysis with water from food, leading to the formation of FFAs. In addition, FFAs attract water molecules owing to their relatively hygroscopic properties and, thus, the process is fed back.

Table 7.2 Properties of waste lipids evaluated as biodiesel feedstocks

Property	Beef tallow[a]	Waste lard	Category 1 fat	Waste cooking oil	Yellow grease[b]	Trap/brown grease[c]
Acid value (mg_{KOH} g^{-1})	–	0.47	33.45	4.06	<30	100–180
Density at 40°C (kg m^{-3})	920	894	829	918	915	880
Kinematic viscosity at 40°C (mm^2 s^{-1})	51.2	48.7	47.4	66.5	132.10	>150
Fatty acid profile (wt%)						
Myristic acid (14:0)	3–6	1.4	3.1	0.2	2.43	1–2
Palmitic acid (16:0)	24–32	26.6	28.1	9.3	23.24	18–30
Palmitoleic acid (16:1)	–	3.0	0.0	0.5	3.79	1–2
Stearic acid (18:0)	20–25	14.2	20.0	3.9	12.96	5–6
Oleic acid (18:1)	37–43	47.0	41.6	54.5	44.32	38–43
Linoleic acid (18:2)	2–3	6.2	6.0	29.7	6.97	11–19
Linolenic acid (18:3)	–	0.4	0.5	0.3	0.67	1–2
Unsaponifiable matter (wt%)	2.6	2.9	22.6	1.0	1–5	1–25
Water content (wt%)	0.1	–	0.1	0.4–1.5		>60
Solids (wt%)				2		3–36

[a] Ali *et al.* (1995).
[b] Canakci and van Gerpen (2001); Diaz-Felix *et al.* (2009).
[c] Wang *et al.* (2008); Garro *et al.* (2003).

The presence of FFAs in waste feedstocks makes their processing in conventional biodiesel production plants more complicated, especially because of the saponification reaction that takes place between FFA and the alkaline compounds used as catalysts. The use of an excess of catalyst can compensate for the consumption of the alkali (Canakci and van Gerpen, 2001), but this option is limited to feedstocks containing low amounts of FFA (the exact limit varies in the literature, but most studies coincide on a value below 2% FFA (Dorado *et al.*, 2002; Kee Lam *et al.*, 2010)) and it still leads to a reduction of the final product yield and to an increase in the viscosity of the reaction mixture. This makes the separation of esters from glycerol much more difficult (Predojević, 2008). In this way, the reduction of the FFA content is a compulsory step which must be accomplished for the treatment of waste lipid raw materials for the production of biodiesel if conventional alkali-driven processes are to be used.

Water

Moisture or water content in the feedstock used for the production of biodiesel is a parameter which must be carefully controlled since it can exert a dramatic influence on the transformation of the raw material into biodiesel. Although the organic hydrophobic nature of lipids facilitates the separation of water, low-grade fats and oils showing high FFAs content, or a high oxidation degree, tend to display high moisture levels. As already mentioned, the presence of water facilitates the formation of FFAs and consequently the production of soap during the transesterification process. Unlike yellow grease and usually brown grease, which contain less than 1 wt% of water (these greases come from previous dewatering processes), the rest of conventional lipid wastes usually display high water content. Thus, animal lipid offal used for the extraction of fats usually displays moisture levels over 50 wt%, but rendering plants typically reduce these values to below 1 wt% (van Gerpen *et al.*, 2003). In the same sense, trap grease consists of more than 90 wt% water and solids (with water representing almost 60 wt%) (Garro *et al.*, 2003). In this way, a drying step would be needed when processing these waste lipids.

A simple flash-evaporation step seems to be enough treatment to decrease the water content to very low levels (Karmakar *et al.*, 2010). A typical treatment is heating the lipid feedstock at 70–90°C under vacuum, although, if needed, distillation can also be employed (Garro *et al.*, 2003). Nevertheless, atmospheric pressure processes can also be used. For instance, Hammed *et al.* (2009) employed a heating pretreatment for waste oil at 120°C for 1 h under atmospheric pressure and determined that adhered water was completely removed. Likewise, such thermal pretreatments are also believed to decrease the acid value, the peroxide value, the content of unsaponifiable matter and the viscosity of the waste oil, owing to the evaporation of volatile compounds and impurities along with the water.

Impurities and unsaponifiable matter

Together with moisture, the content of this fraction that is distinct from fatty matter is usually described as MIU (moisture, impurities and unsaponifiables). Solids are one of these impurities, not so often found in processed oils and fats, but in the case of wastes they can represent a high amount of the overall residue. Thus, in the case of trap grease, the primary sludge coming from a first decantation can represent more than 35 vol% of the initial grease trap waste (Garro *et al.*, 2003) and its nature is very complex. On the other hand, in the case of WCO and animal fats, the amount of solids represents less than 2 wt% of the initial feedstock and its origin can be found both in the remains of food and bone as well as insoluble matter such as gums and steryl glucosides (Tang *et al.*, 2008), which make up a fraction of the unsaponifiable matter.

Other impurities found in these waste lipids are phosphorous, sulphur and nitrogen compounds which do not really affect the transesterification process itself (van Gerpen *et al.*, 2003), but their presence can be transferred to the resultant biodiesel, their content being limited (ASTM, 2010; CEN, 2008). Likewise, some of these compounds can have a deactivating effect on the catalyst used for the transesterification. These impurities are usually more abundant in fats and oils extracted from animal organisms (Dufreche *et al.*, 2007), owing to the presence of phospholipids in animal membrane cells and nitrogen in proteins (Nebel and Mittelbach, 2006). The presence of sulphur is not well understood but it is believed that its origin can be from food preservatives or proteins (for instance, hair and chicken feathers in poultry fat). These compounds should be removed to avoid technical and environmental problems (Lapuerta *et al.*, 2009; Wyatt *et al.*, 2005) and this is usually accomplished through a simple washing step or, if required, a degumming process for the removal of phosphorous. Finally, the unsaponifiable matter is another type of impurity which includes all those constituents which cannot be saponified and cannot thus be converted into biodiesel. Typically, unsaponifiable matter contains sterols, hydrocarbons, terpenes, tocopherols, pigments, waxes, higher molecular weight alcohols, and so on (Mittelbach and Remschmidt, 2006; Karmakar *et al.*, 2010), although the exact composition depends on the nature of the waste lipid and the processes which led to its origin. The removal of this impurity can be carried out through water washing either before (feedstock) or after (biodiesel) the transesterification reaction process (Seidel *et al.*, 2010), since this is not usually affected by its presence. Nevertheless, unsaponifiable matter can exert a dramatic deactivating effect on the catalyst used, especially by means of strong adsorption if heterogeneous catalysts are to be used.

Fatty acid profile

Although not a crucial property in determining whether a certain waste lipid raw material can be processed in a conventional biodiesel production plant or not, it has a major impact on the final biodiesel fuel properties. Most of the feedstocks used for biodiesel production display a fatty acid profile ranging from caprylic (C_8) to arachidic (C_{20}) acids, but typically, regardless of their origin, more than 90% of the total lipid fraction is formed by C_{16} and C_{18} acids (Karmarkar *et al.*, 2010; Azócar *et al.*, 2010). In this sense, the length of the alkyl chain of the fatty acids is not the main factor conditioning the final biodiesel properties (although important properties like the flashpoint are directly related to this factor), owing to the similarities in the different lipid wastes at this point. However, the degree of unsaturation of the alkyl chains, usually measured through the iodine value of the material, is a crucial feature conditioning important properties of the final fuel. Some of these properties are stability against oxidation, heat content, cetane number and cold behaviour. Fatty acid profiles corresponding to animal fats usually include a higher ratio of saturated fatty acids (Table 7.2) than vegetable oils, leading, on the one hand, to less oxidizable biodiesel with a higher cetane number but, on the other hand, showing poor cold properties (Ali *et al.*, 1995), such as high cloud and cold filter plugging points (Karmakar *et al.*, 2010; Lapuerta *et al.*, 2009). In fact, some of the biodiesels obtained from animal fats display gel consistency at room temperature and thus they must be blended with other biodiesels to match the required specifications for cold properties.

Physical properties

The physical properties of waste oils and fats, in a way as all the above-described characteristics, are very heterogeneous and vary largely from one type of waste to another, depending on their nature, origin and previous processing. One of the most important physical properties to be checked in a waste raw material used for biodiesel production is the titre, which is an indicator of the melting point. Since the transesterification process is basically a liquid process, oils and fats with high titre require heating, making the production cost and energy demand of the biodiesel plant higher. Usually, titres measured in animal fats are substantially higher than in vegetable-oil derived lipids, over 35°C, since this factor is usually conditioned by the fatty acid profile (the degree of unsaturation).

7.3.2 Process technologies

Alkali-driven processes

Although there is a huge variety of technically feasible procedures for carrying out the transformation of fatty acid glyceride-containing feedstocks into biodiesel, the current conventional procedure is based on the use of inexpensive homogeneous alkaline catalysts, NaOH, KOH, NaOMe and KOMe, together with methanol at mild temperatures (60–80°C) and pressures (approx. 1.4 atm) (Tyson *et al.*, 2004). Figure 7.1 displays the flow diagram including the typical unit operations in a

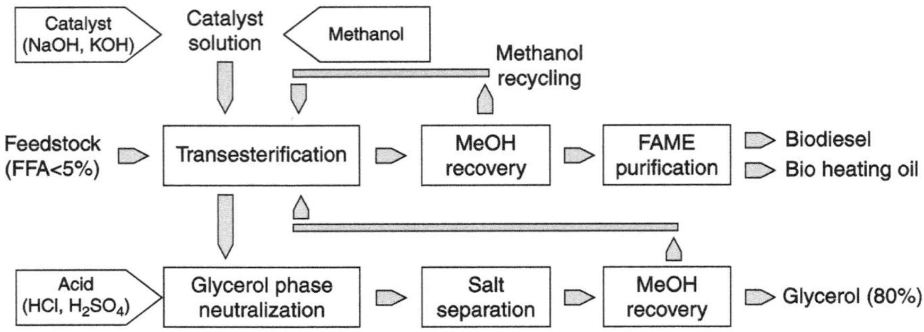

7.1 Conventional alkali-driven biodiesel production process.

conventional biodiesel production process. The process typically consists of unit operations of transesterification reaction, distillation for excess alcohol recovery, water washing for separation of biodiesel from glycerol, catalyst and alcohol, distillation for crude biodiesel purification, catalyst removal and glycerol purification. However, technical difficulties are found when treating complex feedstocks such as waste lipids in conventional plants because of the presence of impurities. As mentioned above, these can include a large variety of different compounds as a consequence of the feedstock origin – food ingredients, proteins, inorganic compounds, and so on – and, as a common feature, water, free fatty acids and unsaponifiables. Thus, when using lipid wastes in a conventional alkali-driven process for biodiesel production, the use of a conditioning step to treat such feedstocks is mandatory.

Phospholipids are one of the most undesirable impurities in raw materials for biodiesel production as they produce problems in the use of the fuel in engines (gum formation). The presence of these phosphorous-containing compounds, which are quite abundant in rendered animal fats, should therefore be avoided and typically a degumming preconditioning stage is employed. Most of the degumming procedures involve the treatment of the feedstock with an acid solution (mainly phosphoric acid, but citric or sulphuric acids are also used) followed by a precipitation or centrifugation step which allows the reduction of the concentration of phosphorous. In this conditioning step, usually proteins, carbohydrates and other inorganic residues are also removed (Verhé and Stevens, 2009).

As previously indicated, the presence of water also negatively affects the transesterification process, especially when base-catalyzed, because of soap formation and decrease in yield. Water is usually removed by heating the lipid feedstock, which is an energy-intensive separation operation. It is a quite expensive stage making the biodiesel production process less profitable. Thus, the use of the acid-catalysed route, which avoids the formation of soap, would significantly reduce the detrimental effect of the water presence in the raw material.

Considering the FFA content, the alkali-driven process is suitable when the FFA level is below approximately 2 wt% (Kee Lam *et al.*, 2010), otherwise the saponification of FFA with the base catalyst leads to a poor separation of the ester and glycerol phases, apart from the water formation, which can hydrolyze the triglycerides, contributing to more soap formation. If the FFA content is slightly over 2 wt% it is possible to add additional catalyst devoted to the compensation of the catalyst fraction spent in neutralizing the FFA, while still leaving enough alkali to drive the transesterification reaction. Nevertheless, this solution is limited in its application up to an FFA content of 5–6 wt% (van Gerpen *et al.*, 2003), and even lower if water is also present, since the resultant formed soap can produce the gelification of the reaction medium. Also, the profitability of the process is jeopardized when treating feedstocks with acid contents over 10 wt%, since a significant part of the feedstock is transformed into a waste instead of biodiesel. In this way, for feedstocks containing more than 5 wt% of free fatty acids, a different biodiesel production process has to be considered, either using the same alkali-driven process modified through the incorporation of a feedstock-conditioning step before the base-catalyzed transesterification step, or just selecting a completely different procedure such as an acid-driven alternative.

A large variety of procedures for the removal of FFAs from lipid feedstocks has been proposed, including adsorption (Lee *et al.*, 2002), the use of energy intense processes like microwave radiation (Porter and Jensen, 2006) or flow cavitation (Kozyuk and Brett, 2009), as well as the recycling of the produced glycerol to be used for FFA esterification in a process called glycerolysis (van Gerpen *et al.*, 2003; Jackam *et al.*, 2008). However, the most reported procedures for treating feedstocks with high FFA content are vacuum stripping with steam (Garro *et al.*, 2003; de Greyt *et al.*, 2008; Schols and Vila-Ayala, 2009) and the esterification of FFA by acid-driven catalysis (Canakci and van Gerpen, 2001, 2003). In the first option, the FFA and the triglyceride-containing fractions are separated by steam distillation. Thus, the FFA stream is treated with methanol to produce the FAME product, whereas the triglyceride containing stream, if desired, is processed through the conventional alkali-catalyzed pathway. However, although steam distillation leads to highly pure FFA, the efficiency of the separation of FFA from the rest of the feedstock is not very good and this procedure requires a high temperature (200–220°C), so it is not always profitable for the treatment of waste oils and fats for biodiesel production. In comparison, acid-driven FFA esterification seems to be a much better option (Leung *et al.*, 2010), since the FFA fraction is converted into biodiesel and the required process is a mature technology.

Figure 7.2 displays the modified conventional biodiesel production process combining acid catalysis followed by alkali catalysis. The use of homogeneous acid catalysts, mainly sulphuric acid, is one of the most extended options when treating highly acidic feedstocks (Ruhmer, 2010) in already deployed industrial biodiesel production plants operating with animal fats, WCO or trap

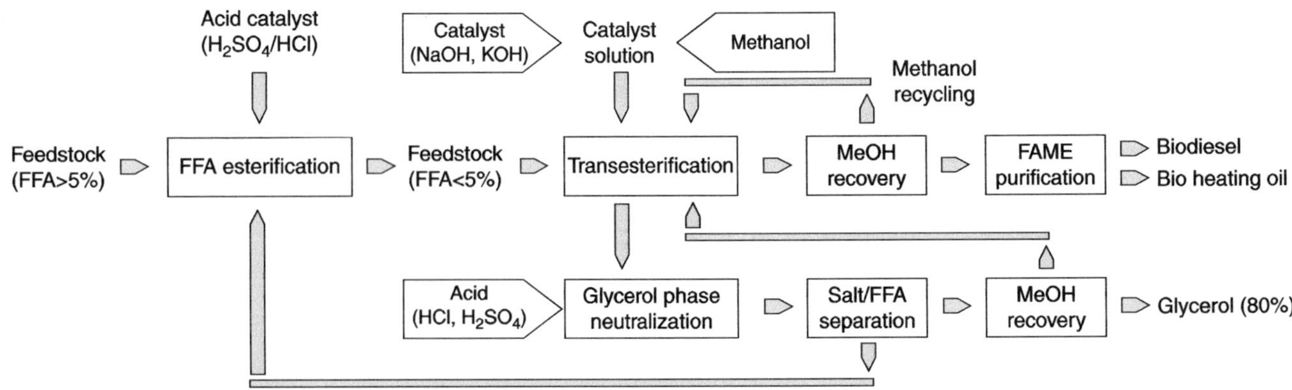

7.2 Combined acid and base-catalyzed biodiesel production process.

grease wastes (ASB Biodiesel, 2007). Nevertheless, it must also be pointed out that this two-step process (acid-catalyzed esterification followed by base-catalyzed transesterification) generates a huge amount of waste effluent from the neutralization of both catalysts and thus it is not environmentally benign.

On the other hand, economic analyses of this and other production processes indicate that the use of an acid esterification step leads to processes that are not as profitable as the conventional production process of biodiesel production from high-grade vegetable oils.

Acid-driven processes

As an alternative, calculations suggest that the use of an acid-driven esterification/transesterification-based process would be competitive for the transformation of waste lipids into biodiesel, since the lower reaction rate of the acid pathway can be largely compensated by the low cost of the waste feedstock and the simplifications introduced in the production process. In essence, an acid-catalyzed process is similar to the alkali-catalyzed process described above. However, owing to the lower reaction rates and the larger excess of alcohol needed, the transesterification reactor and alcohol distillation column should be larger than those of an equivalent alkali-driven process. Homogeneous acid catalysts such as H_2SO_4, HCl, BF_3 or H_3PO_4 have been proposed to promote simultaneously the esterification of FFAs and the transesterification of triglycerides in a single catalytic step, thus avoiding the above conditioning steps when using low cost feedstock with high FFA content (Wang *et al.*, 2008).

Furthermore, acid catalysts are more tolerant to the higher levels of water content in low-cost feedstocks than the base catalysts owing to a lack of saponification side reactions. However, the presence of water can still affect the acid-driven process, owing to hydrophilic nature of the acid sites which leads to an accumulation of water molecules in their vicinity (Morales *et al.*, 2008), hindering the progress of the transesterification reaction by reducing the contact of the catalyst with the hydrophobic oil molecules to a greater extent. For instance, Guan *et al.* (2009) reported that when water concentration in oil is above 0.5 wt%, the yield to biodiesel decreases linearly. They reported that hydrogen ion in the acid site can bind to water molecules more effectively than to methanol, resulting in a weaker acid. Furthermore, an increase in water concentration in the reaction medium leads to water-rich clusters around the catalytic proton. This produces an inhibitory effect on the acid catalyst, which would slow down the reaction rate. Also, owing to the lower activity of acid catalysts for transesterification than alkaline catalysts, reaction conditions need to be harsher to yield adequate transesterification reaction rates: higher temperature (and consequently higher pressure), higher excess of methanol and higher catalyst loading. This invariably leads to the use of more expensive reaction systems, especially due to the use of higher pressure.

Nevertheless, there is still a further approach to improve the performance of the acid-catalyzed biodiesel production from waste oils and fats, which is the use of a heterogeneous acid catalyst. Apart from the use of low-cost feedstock with none or few conditioning pretreatments, this option also minimizes the production costs owing to the use of a simplified process with a lower number of steps, as a benefit of using a heterogeneous catalyst. In addition, the resultant glycerol is not so contaminated with salts and inorganic compounds from the usual catalyst neutralization steps (in homogeneous alkali or acid-catalyzed processes), allowing easier commercialization and increasing the overall profitability of the process. As an additional consequence, this procedure also minimizes the generation of waste effluents, thus being a more environmentally friendly process. A possible scheme of this integrated process using acid heterogeneous catalysts is depicted in Fig. 7.3.

Several authors have reported the use of heterogeneous acid catalysts able to perform the direct conversion of highly acid raw materials into biodiesel in a single step (McNeff *et al.*, 2008a, 2008b), but the reaction conditions are still harsh in order to achieve an acceptable biodiesel yield. The use of sulphonic acid-functionalized polymer resins for the production of biodiesel from waste oils and fats (Bianchi *et al.*, 2010; Par *et al.*, 2010; Lu *et al.*, 2010) has also been reported. In this sense, the development of highly active heterogeneous acid catalysts would help to deploy this technology at industrial scale. Acid-functionalized mesoporous silicas, either with organosulphonic acid or zirconium moieties, have been revealed as interesting catalysts for the production of biodiesel from low-grade feedstock. These materials are characterized by high surface area, large uniform pores, high thermal stability, accessible Brönsted acid sites and the capability to control the surface hydrophilic/hydrophobic balance as well as the strength and concentration of acid sites. At a laboratory scale, these mesoporous acid catalysts have shown an excellent catalytic performance in the synthesis of biodiesel from acid crude vegetable oils, giving high yields of FAME at relatively moderate temperatures and methanol-to-oil molar ratios (Melero *et al.*, 2010; Iglesias *et al.*, 2011).

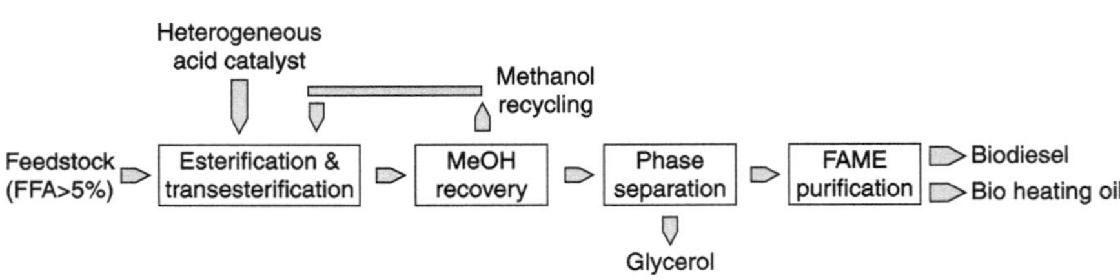

7.3 Biodiesel production process using heterogeneous acid catalysts.

Other approaches

Despite the clear advantages of heterogeneous systems over the traditional homogeneous processes for biodiesel production from waste oils and fats, the management of immiscible phases at a commercial scale is not an easy task. The scaling up of such systems would need mass and heat transfer limitations to be carefully addressed. Recent advances in chemical reaction engineering technologies have shown potential to overcome this problem. These technologies facilitate the transesterification reaction in terms of better mixing intensity and heat and mass transfer rates. Furthermore, they may also be more energy efficient than conventional heating systems. In the following, some prominent examples are described.

An oscillatory flow reactor (OFR) is a type of continuous flow reactor consisting of tubes containing equally spaced orifice plate baffles that impose an oscillatory motion upon the net flow of the process fluid. This creates flow patterns leading to efficient heat and mass transfer whilst maintaining an overall plug flow regime (Harvey *et al.*, 2003). The use of an OFR device allows, for a given reaction volume, longer residence times, as the mixing is not dependent on the net flow and the reactor length–diameter ratio can thus be reduced. Obviously, this is an important factor for the commercialization of the process at industrial scale. Harvey *et al.* (2003) applied this technology to the production of biodiesel from WCO using pure sodium hydroxide as base catalyst. The reactor provided nearly 99% oil conversion with negligible amounts of di- or triglycerides in less than 30 min at 50°C, evidencing the efficiency of such a reaction system. The advantages of using OFR with heterogeneous catalysts for biodiesel production should be emphasized as it is an ideal option for suspending solid catalysts or polymer supported catalysts, although no attempts have been reported so far.

Another interesting option is the application of microwave irradiation to the production of biodiesel. It has several advantages such as higher yields of usually cleaner products, minimum energy consumption and is environmentally friendly compared to conventional heating (Groisman and Gedanken, 2008). The heat is generated by molecular friction originating in the rapid rotation of molecules and ions subjected to microwave irradiation. Barnard *et al.* (2007) developed a continuous flow microwave reactor to produce biodiesel from WCO using 1 wt% KOH as catalyst. After only 10 min at 50°C the oil conversion reached 98.9%. This study also reported that a microwave irradiation process would be significantly more energy efficient than conventional heating in a continuous biodiesel production process. However, the major drawbacks of using this technology for biodiesel synthesis are the scaling up of the process to the industrial scale and safety aspects (Vyas *et al.*, 2010)

Ultrasound has proven to be a very useful tool in enhancing the reaction rates in a variety of reacting systems. It has successfully increased the conversion, improved the yield, changed the reaction pathway, and/or initiated the reaction in biological, chemical and electrochemical systems (Thompson and Doraiswamy, 1999). The use of this technology in transesterification has provided an efficient mixing intensity and sufficient activation energy to initiate the reaction (Singh *et al.*, 2007). Ultrasonic-assisted transesterification not only shortens reaction time, but also minimizes the required alcohol excess and reduces energy consumption compared to conventional mechanical stirring. It has been applied to the base-catalyzed transesterification of WCO (Refaat and El Sheltawy, 2008) and to the enzymatic transesterification of high acid value waste oil with Novozym 435 (Wang *et al.*, 2007), leading to an overall conversion of 94.86% in only 50 min. Furthermore, ultrasonic sound waves tend to reduce the adsorption of FAME and glycerol on the surface of the catalyst or biocatalyst and, as a result, the recyclability of the catalyst is enhanced.

The use of supercritical methanol in non-catalytic processes has also been described in the literature to reduce the multiple steps of the homogeneous base-catalyzed process (Saka *et al.*, 2006). In this technology the reaction takes place in a single homogeneous phase since the supercritical methanol is fully mixable with the oil or fat. Moreover, the reaction rate is very high and the subsequent purification is much simpler than in the conventional process. For instance, the yield of conversion rises to 50–95% within very few minutes of reaction without the use of catalyst (Demirbas, 2008). The supercritical route is also characterized by the simultaneous transesterification of triglycerides and esterification of fatty acids, leading to high yields. Additionally, far from being an impediment, the presence of water positively affects the formation of methyl esters in the supercritical methanol method (Demirbas, 2008). Therefore, this approach appears to be an adequate procedure for the production of biodiesel from highly acidic and water-containing low-grade feedstocks. On the other hand, the main drawback is that the non-catalytic supercritical process requires very high temperatures (350–400°C) and pressures (200–400 bar) and thus operational costs become a serious concern, especially at industrial scale.

7.4 Advantages and limitations

In order to reach the projected real potential of biodiesel in the short and medium terms, a great effort is required to supply a new generation of biodiesel based on, among other sources, the conversion of low-grade waste oils and fats, thereby avoiding the use of refined feedstocks employed in the food industry. The advantages of using waste oils and fats for the preparation of biodiesel are related to two different aspects. First, it is an environmentally friendly strategy that takes advantage of an abundant renewable biomass source and at the same time solves the problems associated with the management of otherwise problematic wastes. Second, the lower cost of the raw material is an economical driving force as it can lead to cost-effective biofuel that can compete with fossil fuels under better conditions.

On the other hand, the above-discussed poor properties of these feedstocks (FFA levels usually from 10–25%, the presence of large concentrations of water and other impurities, etc.) demand more complicated processes to transform these raw materials into biodiesel. In this sense, whereas the use of waste oils and fats herein disclosed is economically desirable because of their lower prices, their complicated composition could play an important role in the feasibility of their transformation into biodiesel. Likewise, economic concerns, such as the price of the raw material, are dependent on several factors like the availability of the feedstock in question, which is affected by the production rate and the interest of other sectors in using the same waste material. In addition, the collection costs, which are determined by the concentration in which the waste is produced, also have a significant impact on the final price and availability of the raw material.

In this sense, several studies have been performed on the profitability of the production of biodiesel from lipid wastes. For instance, Araujo *et al.* (2010) analyzed the economic assessment of biodiesel production from WCO. They concluded that plant capacity is an important factor affecting production processes. Thus, as the volume of collected WCO in a determined area was limited and would not allow for large plant capacities, a biodiesel plant was considered as a multi-oil plant, justifying higher capacities. Catalyst choice is also relevant for the final cost because waste oils and fats, as previously commented, have a free fatty acid percentage that makes the transesterification more difficult. Depending on this percentage, the process needs a pretreatment, an acid catalyst or a supercritical transesterification. Table 7.3 summarizes the estimated production costs for biodiesel from WCO for different technologies and plant capacities. From this analysis, production cost (per litre of biodiesel produced) is very competitive when using a non-catalytic supercritical process or an alkali-driven process at large scale, making the acid-catalyzed route less cost-effective. It must be noted, however, that the analysis by Araujo *et al.* (2010) assumes that the chosen fast food chains work with known standards, thereby guaranteeing the necessary

Table 7.3 Production costs of biodiesel from waste cooking oil

Process	Plant capacity (ton/year)	Production cost (US$ L^{-1})	Reference
Supercritical	125 000	0.09	van Kasteren and Nisworo (2007)
Supercritical	80 000	0.12	van Kasteren and Nisworo (2007)
Supercritical	8000	0.40	van Kasteren and Nisworo (2007)
Alkaline	115 000	0.10	Bender (1999)
Alkaline	12 000	0.15	Bender (1999)
Alkaline	8000	0.52	Zhang *et al.* (2003)
Acid	8000	0.48	Zhang *et al.* (2003)
Acid	36 036	0.31	Marchetti *et al.* (2008)

Source: adapted from Araujo *et al.*, 2010.

quality of WCO for biodiesel production, so the collected oil does not need a pretreatment; therefore, there are no additional costs. This premise would not be valid for the use of, for example, highly acidic waste fats or greases as the alkaline process would be clearly affected.

The economical assessment of different biodiesel production processes using homogeneous alkali and acid catalysts, as well as heterogeneous acid catalysts and supercritical methods, has also been recently discussed by West *et al.* (2008). They used HYSYS software to design and simulate four continuous processes to convert waste vegetable oil feedstock into biodiesel: two with traditional homogeneous alkali and acid catalysts, another one with a heterogeneous acid catalyst and a fourth process based on a supercritical method. While all four processes were capable of producing biodiesel of high purity, the heterogeneous and supercritical processes were the least complex and had the smallest number of unit operations. Material and energy flows, as well as sized unit operation blocks, were used to conduct an economic assessment of each process. The total capital investment, total manufacturing cost and after tax rate-of-return were calculated for each process. The heterogeneous acid catalyst process had the lowest total capital investment and manufacturing costs and had the only positive after tax rate-of-return. Therefore, the results indicate that the heterogeneous acid-catalyzed process is clearly advantageous over the other processes studied. At this point, the design of stable and non-expensive solid acid catalysts for the one-step production of biodiesel from waste oil and fats under moderate reaction conditions is so far the most limiting aspect for the commercial development of this promising technology. Heterogeneous catalysts, and in particular solid acid catalysts, which have been the backbone of the petrochemical and refining industry, will most likely play a key role in the transition to the so-called 'second generation' biodiesel. The main challenges in acid heterogeneous catalysts are related to the stability of acid sites and the control of surface properties to favour the diffusion of oily feedstock towards the acid sites and to promote expelling off of the polar compounds (glycerol and water, etc.) from the catalyst framework.

Additionally, feedstocks used for biodiesel production need to match several quality requirements regarding their properties, some being crucial in the production of biodiesel, since these exert a dramatic influence on the final properties of the produced biofuel (Canakci and Sanli, 2008). Thus, the nature of the lipid fraction and the quantity and nature of the impurities can determine whether a certain waste is useful or not for biodiesel production. The final biodiesel must meet the specifications designated by standards EN 14214 (CEN, 2008) in the EU and ASTM D-6751 (ASTM, 2010) in the USA, so that the selection of the starting feedstock must guarantee that the final product fulfils the required specifications (Canakci and Sanli, 2008; Canoira *et al.*, 2008; Karmakar *et al.*, 2010). For instance, a lipid raw material showing a low iodine index improves the oxidative stability of the final biodiesel, but deteriorates the cold flow properties of the same. Finally, legislation and regulation policies, such as those adopted through the Commission Directive 2003/30/EC, can also play an important role in markets trading fats and oils and thus these, together with other factors (Rosillo-Calle *et al.*, 2009), can either promote or depress the demand for raw materials for the production of biofuels, affecting the rest of the features mentioned.

7.5 Concluding remarks and future trends

Biodiesel is currently produced through transesterification of edible vegetable oils. However, the high prices of these oils in world markets have produced an important increase in the overall biodiesel production cost, making it not economically viable in many geographic regions without the implementation of government subsidies and mandates for use. In a positive outlook for the biodiesel industry, Pike Research, a cleantech analysis firm (Boulder, Colorado, USA), has predicted robust growth over the long-term for the global renewables market. They forecast that, despite significant challenges, the combined biodiesel and ethanol markets will reach US$247 billion in sales by 2020, up from just US$76 billion in 2010. Growth on the supply side is linked to game changing technologies related to biofuels processing, since in the biofuels world feedstock is currently the most important issue. In this sense, the use of recycled oils and fats has potential in the short and medium term. Nevertheless, commercial technology is not yet completely suitable for the transesterification of such wastes owing to their high FFA, water and impurities content. Conventional alkali-driven processes cannot be directly applied to waste lipid feedstocks without previous conditioning steps that increase the overall production cost in a significant manner. Nonetheless, the transesterification reaction can alternatively be carried out by using acid catalysts (homogeneous or heterogeneous), or by non-catalytic supercritical methods. These methods are much less affected by the FFA or impurities content of the waste lipids and therefore show important advantages relative to the alkali-driven process. However, they must face other challenges relative to low reaction rates, catalyst deactivation and regeneration, mass and energy transfer limitations, or high operation costs (especially in supercritical methods). Thus, apart from conventional techniques for biodiesel production, the use of more sophisticated technologies is being explored in order to improve the process by achieving better mixing intensities, higher heat and mass transfer rates and higher energy efficiencies.

7.6 References

Ali Y, Hanna M A and Cuppett S L (1995), 'Fuel properties of tallow and soybean oil esters', *J Am Oil Chem Soc*, **72**, 1557–64.

Araujo V K W S, Hamacher S and Scavarda L F (2010), 'Economic assessment of biodiesel production from waste frying oils', *Bioresour Technol*, **101**, 4415–22.

ASB Biodiesel Ltd (2007), 'Development of a biodiesel plant at Tseung Kwan O Industrial Estate. Project profile'. Available from: http://www.epd.gov.hk/eia/register/profile/latest/dir158/dir158.pdf [Accessed: 08/02/2011].

ASTM (2010), *Standard Specification for Biodiesel Fuel Blend Stock (B100) for Middle Distillate Fuels*, ASTM D6751-10, American Society for Testing Materials, West Conshohocken, USA.

Azócar L, Ciudad G, Heipieper H J and Navia R (2010), 'Biotechnological processes for biodiesel production using alternative oils', *Appl Microbiol Biotechnol*, **88**, 621–36.

Barnard T M, Leadbeater N E, Boucher M B, Stencel L M and Wilhite B A (2007), 'Continuous-flow preparation of biodiesel using microwave heating', *Energy Fuels*, **21**, 1777–81.

Berrios M, Gutiérrez M C, Martín M A and Martin A (2009), 'Application of the factorial design of experiments to biodiesel production from lard', *Fuel Process Technol*, **90**, 1447–51.

Bender M (1999), 'Economic feasibility review for community-scale farmer cooperatives for biodiesel', *Bioresour Technol*, **70**, 81–7.

Bianchi C L, Boffito D C, Piorola C and Ragaini V (2010), 'Low temperature de-acidification process of animal fat as a pre-step to biodiesel production', *Catal Lett*, **134**, 179–83.

Boocock D G B, Konar S K, Leung A and Ly L D (1992), 'Fuels and chemicals from sewage sludge. 1. The solvent extraction and composition of a lipid from a raw sewage sludge', *Fuel*, **71**, 1283–9.

Bruederle C E, Hnasko R M, Kraemer T, Garcia R A, Haas M J, *et al.* (2008), 'Prion infected meat-and-bone meal is still infectious after biodiesel production', *PLoS ONE*, **3**, e2969.

Canakci M (2007), 'The potential of restaurant waste lipids as biodiesel feedstocks', *Bioresour Technol*, **98**, 183–90.

Canakci M and Sanli H (2008), 'Biodiesel production from various feedstocks and their effects on the fuel properties', *J Ind Microbiol Biotechnol*, **35**, 431–41.

Canakci M and van Gerpen J (2001), 'Biodiesel production from oils and fats with high free fatty acids', *Trans ASAE*, **44**, 1429–36.

Canakci M and van Gerpen J (2003), 'A pilot plant to produce biodiesel from high free fatty acid feedstocks', *Trans ASAE*, **46**, 945–54.

Canoira L, Rodríguez-Gamero M, Querol E, Alcántara R, Lapuerta M and Oliva F (2008), 'Biodiesel from low-grade animal fat: production process assessment and biodiesel properties characterization', *Ind Eng Chem Res*, **47**, 7997–8004.

CEN (2008), *Automotive Fuels – Fatty Acid Methyl Esters (Fame) For Diesel Engines – Requirements and Test Methods*, EN 14214:2008, European Committee for Standardization, Brussels, Belgium.

Commission Directive 2003/30/EC of 8 May 2003 relating to the promotion of the use of biofuels or other renewable fuels for transport. *Official Journal of the European Union*, L123, 17.05.2003, pp 42–6.

Commission Regulation 2002R1774 of 3 October 2002 relating to the health rules concerning animal by-products non intended for human consumption. Corrigendum OJ L273, 10.10.2002, p 1.

Coelenbier P (2010), 'Overview of the European animal by products industry', *EFPRA Congress 2010*, Budapest, 2–5 June.

de Greyt W, Kellens M, Holm H C, Christensen M W and Nielsen P M (2008), *Method for Producing Biodiesel*, International patent application 125574A1. 23-October-2008.

Demirbas A (2008), 'Comparison of transesterification methods for production of biodiesel from vegetable oils and fats', *Energy Convers Manage*, **49**, 125–30.

Dias J M, Alvim-Ferraz M C M and Almeida M F (2009), 'Production of biodiesel from acid waste lard', *Bioresour Technol*, **100**, 6355–61.

Diaz-Felix W, Riley M R, Zimmt W and Kazz M (2009), 'Pretreatment of yellow grease for efficient production of fatty acid methyl esters', *Biomass Bioenerg*, **33**, 558–63.

Dorado M P, Ballesteros E, Almeida J A, Schellert C, Lohrlein H P and Krause R (2002), 'An alkali-catalized transesterification process for high free fatty acid waste oils', *Trans ASAE*, **45**, 525–9.

Dufreche S, Hernandez R, French T, Sparks D, Zappi M and Alley E (2007), 'Extraction of lipids from municipal waste water plant microorganisms for production of biodiesel', *J Am Oil Chem Soc*, **84**, 181–7.

Energy Information Administration (2010). *International Energy Statistics*. Available from: http://www.eia.gov/emeu/international/contents.html [Accessed: 28/01/2010].

Espinosa da Cunha M, Canielas Krause L, Aranda Moraes M S, Schmitt Faccini C, Assis Jacques R, *et al.* (2009), 'Beef tallow biodiesel produced in a pilot scale', *Fuel Process Technol*, **90**, 570–575.

European Biodiesel Board (2010). *Statistics of the EU Biodiesel Industry*. Available from: http://www.ebb-eu.org/stats.php [Accessed: 29/01/2010].

Eurostat, 2010. European Comission statistics database. Available from: http://epp.eurostat.ec.europa.eu/portal/page/portal/statistics/search_database [Accessed: 03/02/2011].

Fröhlich A, Rice B (2005), 'Evaluation of recovered vegetable oil as a biodiesel feedstock', *Irish Journal of Agricultural and Food Research*, **44**, 129–39.

Garro J M, Lemieux A, Jollez P and Cadoret N (Prolab Technologies Inc) (2003). *Method for Fractionating Grease Trap Waste and Uses of Fractions Therefrom*. International patent application 038020 A1. 2003-May-08.

Groisman Y and Gedanken A (2008), 'Continuous flow, circulating microwave system and its application in nanoparticles fabrication and biodiesel synthesis', *J Phys Chem C*, **112**, 8802–8.

Groschen R (2002), 'The feasibility of biodiesel from waste/recycled greases and animal fats', Minnesota Department of Agriculture. Available from: http://www.angelfire.com/ks3/go_diesel/files042803/wastefatsfeasability.pdf [Accessed: 02/02/2011].

Haas M J (2005), 'Improving the economics of biodiesel production through the use of low value lipids as feedstocks: vegetable oil soapstock', *Fuel Process Technol*, **56**, 1087–96.

Hammed B H, Goh C S and Chin L H (2009), 'Process optimization for methyl ester production from waste cooking oil using activated carbon supported potassium fluoride', *Fuel Process Technol*, **90**, 1532–3.

Harvey A P, Mackley M R and Seliger T (2003), 'Process intensification of biodiesel production using a continuous oscillatory flow reactor', *J Chem Technol Biotechnol*, **181**, 217–22.

Iglesias J, Melero J A, Bautista L F, Morales G, Sánchez-Vázquez R, *et al.* (2011), 'Zr-SBA-15 as an efficient acid catalyst for FAME production from crude palm oil', *Catal Today*, **167**, 46–55.

Jackam J P, Pierce J M, Jones J D and Talley R H (2008), *Production of Biodiesel and Glycerin from High Free Fatty Acid Feedstocks*, European Patent Application 1889899A1. 2008-February-20.

Karmakar A, Karmakar S and Mukherjee S (2010), 'Properties of various plants and animals feedstocks for biodiesel production', *Bioresour Technol*, **101**, 7201–10.

Kee Lam M, Teong Lee K and Rahman Mohamed A (2010), 'Homogeneous, heterogeneous and enzymatic catalysis for transesterification of high free fatty acid oil (waste cooking oil) to biodiesel: a review', *Biotechnol Adv*, **28**, 500–18.

Kondamudi N, Mohapatra S K and Misra M (2008), 'Spent coffee grounds as a versatile source of green energy', *J Agric Food Chem*, **56**, 11757–60.

Kondamudi N, Strull J, Misra M and Mohapatra S K (2009), 'A green process for producing biodiesel from feather meal', *J Agric Food Chem*, **57**, 6163–6.

Kozyuk O V and Brett M (2009), *Method for Reducing Free Fatty Acid Content of Biodiesel Feedstock*. International Patent Application 021248A2. 2009-February-12.

Lapuerta M, Rodríguez-Fernández J, Oliva F and Canoira L (2009), 'Biodiesel from low-grade animal fats: diesel engine performance and emissions', *Energy Fuels*, **23**, 121–9.

Lee K T, Foglia T A and Chang K S (2002), 'Production of alkyl ester as biodiesel from fractionated lard and restaurant grease', *J Am Oil Chem Soc*, **79**, 191–5.

Leung D Y C, Wu X and Leung M K H (2010), 'A review on biodiesel production using catalyzed transesterification', *Appl Energy*, **87**, 1083–95.

Lin C Y and Li R J (2009), 'Fuel properties of biodiesel produced from the crude fish oil from the soapstock of marine fish', *Fuel Process Technol*, **90**, 130–6.

Lu P, Yuan Z, Li L, Wang Z and Luo W (2010), 'Biodiesel from different oil using fixed-bed and plug-flow reactors', *Renewable Energy*, **35**, 283–7.

Ma F, Clements L D and Hanna M A (1998), 'Biodiesel fuel from animal fat. Ancillary studies on transesterification of beef tallow', *Ind Eng Chem Res*, **37**, 3768–71.

Marchetti J, Miguel V and Errazu A (2008), 'Techno-economic study of different alternatives for biodiesel production', *Fuel Process Technol*, **89**, 740–8.

Marulanda V F, Anitescu G and Tavlarides L L (2010), 'Investigations on supercritical transesterification of chicken fat for biodiesel production from low-cost lipid feedstocks', *J Supercrit Fluids*, **54**, 53–60.

McNeff C V, McNeff L C, Yan B, Nowlan D T, Rasmussen N, *et al.* (2008a), 'A continuous catalytic system for biodiesel production', *Appl Catal A General*, **343**, 39–48.

McNeff C V, McNeff L C, Gyberg A E, Yan B and Nowlan D T (2008b), *Methods and Apparatus for Producing Alkyl Esters from Lipid Feed Stocks and Systems Including the Same*, US Patent Application 51592A1. 2008-February-28.

Melero J A, Bautista L F, Morales G, Iglesias J and Sánchez-Vázquez R (2010), 'Biodiesel production from crude palm oil using sulfonic acid-modified mesostructured catalysts', *Chem Eng J*, **161**, 323–31.

Mittelbach M and Remschmidt C (2006), *Biodiesel – The Comprehensive Handbook*, Graz, Austria.

Mittelbach M, Pokits B, Müller H, Müller M and Riesner D (2007), 'Risk assessment for prion protein reduction under the conditions of the biodiesel production process', *Eur J Lipid Sci Technol*, **109**, 79–90.

Mondala A, Liang K, Toghiani H, Hernández R and French T (2009), 'Biodiesel production by in situ transesterification of municipal primary and secondary sludges', *Bioresour Technol*, **100**, 1203–10.

Morales G, Athens G, Chmelka B, van Grieken R and Melero J A (2008), 'Aqueous-sensitive reaction sites in sulfonic acid-functionalized mesoporous silicas', *J Catal*, **254**, 205–17.

Mustapic N and Stanisa B (2006), 'Possibility of biodiesel production from waste fats and oils and animal fats', *Energy and the environment, Vol II. 20th International Congress on Energy and the Environment*, Opatija, Croatia, pp. 215–24.

National Renderers Association (2010). *US Production, Consumption and Export of Rendered Products for 2004–2009*. Available from: http://nationalrenderers.org/economic/statistics/ [Accessed: 02/02/2010].

Nebel B A and Mittelbach M (2006), 'Biodiesel from extracted fat out of meat and bone meal', *Eur J Lipid Sci Technol*, **108**, 398–403.

Par J Y, Lee J S, Wang Z M and Kim D K (2010), 'Production and characterization of biodiesel from trap grease', *Korean J Chem Eng*, **27**, 1791–5.

Porter M J and Jensen S (2006), *Microwave-Enhanced Process to Maximize Biodiesel Production Capacity*. US Patent Application 0162245A1. 2006-July-27.

Predojević Z J (2008), 'The production of biodiesel from waste frying oils: A comparison of different purification steps', *Fuel*, **87**, 3522–8.

Refaat A A and El Sheltawy S T (2008), 'Comparing three options for biodiesel production from waste oil feedstocks by solid acid catalysis', *Process Saf Environ Prot*, **86**, 441–7.

Rice B, Frölich A and Leonard R (1999), *Cost Reduction in Bio-Diesel Production*, TEAGASC – The Irish Agriculture and Food Development Authority, Dublin, Ireland.

Rosillo-Calle F, Pelkmans L and Walter A (2009), *A Global Overview of Vegetable Oils, with Reference to Biodiesel*, Final report for the IEA Bioenergy Task 40. International Energy Agency Bioenergy.

Ruhmer K (2010) 'Turning trap grease waste into biodiesel', *Waste To Fuels Conference & Trade Show 2010*, Jacksonville, 18–20 April.

Saka S, Kusdiana D and Minami E (2006), 'Non-catalytic biodiesel fuel production with supercritical methanol technologies', *J Sci Ind Res*, **65**, 420–25.

Schols B and Vila Ayala J (2009), *Fatty Waste Material Purification Process*, UK Patent Application 2451581A, 2009-February-04.

SEAI (2003), *A Resource Study on Recovered Vegetable Oil and Animal Fats*, Sustainable Energy Authority of Ireland. Available from: http://www.seai.ie/Grants/Renewable_Energy_RD_D/Projects_funded_to_date/Biomass/A_resource_Study_on_Recovered_Vegetable_Oil_and_Animal_Fats/ [Accessed: 03/02/2011].

Seidel B, Alm M, Peters R, Kördel W and Schäffer A (2007), 'Safety evaluation for a biodiesel process using prion-contaminated animal fat as a source', *Environ Sci Pollut Res*, **13**, 125–30.

Seidel E, Bauer I and Bönsch R (2010), *Producing Fatty Acid Alkyl Ester Comprises e.g. Mixing Fats or Oils, Alkyl Alcohols and Catalyst, Separating Mixture into Low, Ester-Rich Phase and Heavy Glycerin-Rich Phase and Separating Fatty Acid Alkyl Ester from Low, Ester-Rich Phase*. EP Patent Application 1788065 A1. 2010-August-05.

Singh A K, Fernando S D and Hernández R (2007), 'Base-catalyzed fast transesterification of soybean oil using ultrasonication', *Energy Fuels*, **21**, 1161–4.

Swisher K (2010). 'Market report 2009', Available from: http://rendermagazine.com/assets/4d90dcf5dabe9d7aaa0000a0/marketreport2009.pdf [Accessed: 27/10/2011].

Tang H, de Guzman R C, Salley S O and Ng K Y S (2008), 'Formation of insolubles in palm oil-, yellow grease- and soybean oil-based biodiesel blends after cold soaking at 4°C', *J Am Oil Chem Soc*, **85**, 1173–82.

Thompson L H and Doraiswamy L K (1999), 'Sonochemistry: science and engineering', *Ind Eng Chem Res*, **38**, 1215–49.

Tyson K S (2002), 'Brown grease feedstocks for biodiesel'. Available from: http://www.nrbp.org/pdfs/pub32.pdf. National Renewable Energy Laboratory [Accessed: 16/01/2010].

Tyson K S (2008), 'Brown grease to biodiesel'. Available from: http://info.ornl.gov/sites/publications/Files/Pub10998.pdf. Oak Ridge National Laboratory [Accessed: 16/01/2010].

Tyson K S, Bozell J, Wallace R, Petersen E and Moens L (2004), *Biomass Oil Analysis: Research needs and recommendations*, Renewable Energy Laboratory, NREL/TP-510-34796.

US Census Bureau (2009), 'Metropolitan and micropolitan statistical area estimates. Annual estimates of the population – CBSA-EST2009-01'. Available from: http://www.census.gov/popest/metro/CBSA-est2009-annual.html [Accessed: 18/01/2010].

USDA (2009), 'Oil crops yearbook (89002)', Economic Research Service – U.S. Department of Agriculture. Available from: http://usda.mannlib.cornell.edu/usda/ers/89002/2009/index.html [Accessed: 16/01/2010].

van Gerpen J, Shanks B, Pruszko R, Clemens D and Knothe G (2003), *Biodiesel Production Technology*, Renewable Energy Laboratory, NREL/SR-510-36244.

van Kasteren J and Nisworo A (2007), 'A process model to estimate the cost of industrial scale biodiesel production from waste cooking oil by supercritical transesterification', *Resour Conserv Recycl*, **50**, 442–58.

Veljkovic V, Lakicevic S, Stamenkovic O, Todorovic Z and Lazic M (2006), 'Biodiesel production from tobacco (Nicotiana tabacum L.) seed oil with a high content of free fatty acids', *Fuel*, **85**, 2671–5.

Verhé R and Stevens C V (2009), 'Production of biodiesel from waste lipids', *Biofuels*, Soetaert W and Vandamme E J (eds), John Wiley & Sons, Chichester, 153–70.

Vyas A P, Verma J L and Subrahmanyam N (2010), 'A review on FAME production processes', *Fuel*, **89**, 1–9.

Wang J X, Huang Q D, Huang F H, Wang J W and Huang Q J (2007), 'Lipase-catalyzed production of biodiesel from high acid value waste oil using ultrasonic assistant', *Chin J Biotechnol*, **23**, 1121–8.

Wang Z M, Lee J S, Park J Y, Wu C Z and Yuan Z H (2008), 'Optimization of biodiesel production from trap grease via acid catalysis', *Korean J Chem. Eng*, **25**, 670–4.

West A H, Posarac D and Ellis N (2008), 'Assessment of four biodiesel production processes using HYSYS.Plant', *Bioresour Technol*, **99**, 6587–6601.

Wiltsee G (1998), *Urban Waste Grease Resource Assessment*. National Renewable Energy Laboratory, NREL/SR-570-26141. Available from: http://www.epa.gov/region9/waste/biodiesel/docs/NRELwaste-grease-assessment.pdf [Accessed: 10/01/2011].

Wyatt V T, Hess M A, Dunn R O, Foglia T A, Haas M J and Marmer W N (2005), 'Fuel properties and nitrogen oxide emission levels of biodiesel produced from animal fats', *J Am Oil Chem Soc*, 82, 585–91.

Zhang Y, Dubé M, McLean D and Kates M (2003), 'Biodiesel production from waste cooking oil: 2. Economic assessment and sensitivity analysis', *Bioresour Technol*, **90**, 229–40.

8
Biodiesel from microbial oil

L. F. BAUTISTA and G. VICENTE, Universidad Rey Juan Carlos, Spain and V. GARRE, Universidad de Murcia, Spain

Abstract: Fatty acid methyl esters (biodiesel) are usually obtained from plant oils. Oleaginous microorganisms present an alternative to plant oils because they can accumulate high levels of lipids and do not require arable land. In particular, heterotrophic microorganisms (bacteria, fungi and yeasts) can be grown on waste or low-grade biomass as a carbon and energy source. After biomass production, biodiesel can be produced by transformation of extracted microbial lipids or direct transformation of dry microbial biomass. Direct transformation of lipids into FAMEs gives cost savings and increases lipid extraction. Careful characterization of the lipid composition of each microbial candidate should be carried out before it can be adopted since biodiesel fuel as an alternative needs to comply with existing standards. Genetic engineering and biorefinery-based production strategies can reduce costs by increasing lipid accumulation of microorganisms.

Key words: biodiesel, single cell oil, oleaginous microorganisms, microbial biodiesel, biodiesel from biomass, biorefinery.

8.1 Introduction

Biodiesel is composed of esters of fatty acids and short-chain monoalcohols that can be used in diesel engines as fuel alone or in different mixtures with mineral diesel. Today, at industrial scale, most biodiesel is produced by esterification and transesterification of edible vegetable oils such as soybean oil, rapeseed oil, and palm oil. Wastes such as cooking oil and animal fats make a minor contribution. The rapid increase in the production of biodiesel in the last few years makes evident that traditional energy crops will deliver the future demand for biofuels at the expense of reducing the agricultural land available for food supply (Azócar *et al.*, 2010) altering prices and availability of edible oils.

In this context, new and alternative lipid sources capable of surpassing the above limitations must be found. Many studies on alternative non-edible oils from oleaginous crops grown on marginal or semi-arid areas, for example, *Jathropha curcas*, *Brassica carinata*, *Cynara cardunculus*, *Ricinus communis* (castor oil), have been reported, although, similarly to the use of edible oils, the cost of these alternative oils still accounts for 70–85% of the final biodiesel production cost (Meng *et al.*, 2009).

Microorganisms contain different types of lipids, commonly referred to as single cell oils (SCO), ranging from those with structural function, such as sphingolipids, saccharolipids and phospholipids, to those acting as carbon and energy reservoirs, such as triacylglicerides (TAG). Microbes capable of accumulating >20% lipids are named oleaginous microorganisms (Ratledge, 1991). Microalgae, bacteria, fungi and yeasts are the main oleaginous microbial species. Microbial lipids have been used for their application as food additives, pharmaceuticals and feed ingredients for aquaculture (Molina Grima *et al.*, 1993; Becker, 1994; Alvarez and Steinbuchel, 2002; Ratledge, 2004; Szczęsna-Antczak *et al.*, 2006). Recently, microbial lipids have been evaluated as feedstocks for biodiesel production since they have an appropriate fatty acid profile, do not compete with food supply and do not require agricultural land.

Some microalgae, which capture carbon dioxide by transformation into lipids using sunlight, have attracted attention for biofuel production because of their higher oil productivity and faster growth compared to conventional energy crops (Chisti, 2007; Li *et al.*, 2008a), although these photosynthetic microorganisms have problems associated with their growth in bioreactors owing to the necessity of light supply, large acreages and fouling. The above-mentioned problems can be overcome if appropriate heterotrophic microorganisms are used. These microorganisms can be grown on waste or low-grade biomass as carbon and energy sources in well-controlled large-scale fermenters, with high efficiency, improved biomass yield and reduced cost of biomass and oil.

8.2 Heterotrophic microorganisms for oil production

Little information has been reported on the use of SCO from heterotrophic microorganisms for biodiesel production. However, the accumulation of SCO in fungus, yeasts, bacteria and even in microalgae grown heterotrophically has been described extensively because of interest in the production of specific lipids, especially polyunsaturated fatty acids (PUFA) for the pharmaceutical and food industry.

8.2.1 Fungi

Oleaginous fungi have also been considered as potential oil sources for biodiesel production because they accumulate large amounts of lipids. Amongst these microorganisms, particular attention has been dedicated to various oleaginous Zygomycetes species, like *Mortierella isabelina* and *Cunninghamella echinulata*, which may accumulate up to 86% and 57% of lipids in dry biomass, respectively (Papanikolaou *et al.*, 2004; Fakas *et al.*, 2008, 2009). These fungi are able to grow and accumulate large amounts of lipids in cultures containing raw glycerol, derived from the commercial first-generation biodiesel process, as a carbon source.

However, microorganisms considered so far as a feedstock for biodiesel production lack appropriate genetic engineering techniques to improve fatty acid profiles that would produce high quality biodiesel (Liu and Zhao, 2007). Besides, their genomes have not been sequenced, which makes it even more difficult to improve strategies based on genetic manipulation. In this context, the genome of *Mucor circinelloides*, which was used for the first commercial production of microbial lipids (Ratledge, 2004), has already been sequenced at the Joint Genome Institute under a project financed by the Department of Energy (DoE) of the USA. In addition, a large number of genetic engineering techniques are available for this oleaginous fungus (Nicolás *et al.*, 2003; Navarro *et al.*, 2001). The regulation of lipid accumulation in this fungus has been extensively studied (Aggelis *et al.*, 1988; Aggelis *et al.*, 1990; Ratledge and Wynn, 2002; Ratledge, 2004) and key genes have been identified (Wynn *et al.*, 1999).

M. circinelloides shows many relevant features favouring its use for biodiesel production since it can accumulate around 25% of lipids, grow on a wide range of carbon sources (McIntyre *et al.*, 2002) and has been proved to grow in large industrial stirred-tank fermenters (220 m^3) to produce oil rich in γ-linolenic acid (Ratledge, 2004). In this respect, biodiesel produced from SCO extracted from *M. circinelloides* (Vicente *et al.*, 2009) and by direct methanolysis of biomass (Vicente *et al.*, 2010) met both the European Standard (EN 14214) and United States Standard (ASTM D6751).

8.2.2 Yeasts

Different genera of the group of yeasts are known to be oleaginous microorganisms. Since TAG are the main fraction of total lipids in yeasts (Ratledge, 2002), their assessment as biodiesel feedstock has been studied. In this context, the amount of accumulated lipids in several yeasts species grown on nitrogen-limited cultures with lignocellulosic-derived compounds as a carbon source have been reported (Chen *et al.*, 2009), that is, *Trichosporon cutaneum* (38.9%), *Lipomyces starkeyi* (37.2%), *Rhodotorula glutinis* (16.7%) and *Rhodosporidium toruloides* (39.3%). In that study, *T. cutaneum* was shown to be the most adopted strain by most of the inhibitors formed after lignocellulose degradation.

When cultivation was performed on more readily available carbon sources such as xylose and glycerol, *R. glutinis* increased its lipid content to 32% and other species of the genus *Rhodotorula* can also accumulate significant amount of lipids, such as *R. graminis* (21.26%) or *R. minuta* (24.62%) (Pan *et al.*, 2009). Higher lipid accumulation yields were also obtained on molasses media cultures of *R. mucilaginosa* (69.5%), *Candida tropicalis* (46.8%) and *Yarrowia lypolitica* (59.9%) (Karatay and Dönmez, 2010). In another work, Liu and Zhao (2007) reported the direct methanolysis of biomass from two yeasts, *L. starkeyi* and *R. toruloides*, obtaining FAME yields of 96.8% and 98.1%, respectively.

8.2.3 Bacteria

Bacteria usually have high growth rates compared to other microorganisms and so must be considered as a potential source of biofuels. Although many bacteria are not capable of producing significant amounts of SCO, they can often accumulate complex lipids (for example polyhydroxyalkanoates, fatty alcohols, glycolipids or phospholipids) mainly in their outer membrane so that must be difficult to extract (Meng *et al.*, 2009). However, some studies proved the intracellular accumulation of significant amounts of fatty acid triglycerides in some species of the Actynomycetes group of bacteria corresponding to the genera *Rhodococcus, Nocardia* and *Mycobacterium* (Subramaniam *et al.*, 2010; Azócar *et al.*, 2010).

The deep existing knowledge of the culture and physiology of *Escherichia coli*, as well as extensive developments in metabolic engineering and synthetic biology led to *E. coli* being engineered for the production of fatty acid ethyl esters (FAEE). For this purpose, the ethanol pathway from the ethanol-producing bacterium *Zymomonas mobilis* was expressed along with an acyltransferase gene from the Gram-negative proteobacteria *Acinetobacter baylyi*, resulting in the production of FAEE by esterification of endogenously produced ethanol from glucose with exogenously supplemented fatty acids (Kalscheuer *et al.*, 2006). Another work goes further (Steen *et al.*, 2010) since *E. coli* was engineered to overproduce free fatty acids (FFA) combined with the introduction of both an ethanol pathway and the *A. baylyi* acyltransferase gene to yield FAEE. In addition, the above modified *E. coli* was again engineered with the genes encoding endoxylanase catalytic domain from *Clostridium stercorarium* and a xylanase from *Bacteroidetes ovatus* allowing the final strain to use hemicellulose as a carbon source. However, biodiesel yields are still low and further improvements must be achieved for economic viability.

8.2.4 Heterotrophic microalgae

By changing culture conditions or using genetic engineering modifications, some photoautotrophic microalgae can be converted into heterotrophic microalgae, accumulating oils by using organic carbon instead of CO_2 (Li *et al.*, 2008b). Heterotrophic growth of *Chlorella protothecoides* results in higher biomass production and higher lipid accumulation in cells in comparison to the autotrophic growth of this microalga (Miao and Wu, 2006). These authors reported an integrated method for biodiesel production from heterotrophic *C. protothecoides* oil by acidic transesterification. In comparison to the autotrophic microalgae, the growth of these microorganisms can be carried out in conventional microbial bioreactors. A similar pattern has been reported for the heterotrophic *C. zofingienensis* which can accumulate up to 42.1% of fatty acids after 14 days cultivation on glucose as a carbon source (Liu *et al.*, 2011).

8.3 Carbon sources for microorganism production

Carbon follows a cycle on the Earth occupying different 'reservoirs'. During Earth's evolution, organic carbon was trapped and compressed on the seabed by billions of tonnes of silt and sand for several million years forming fossil fuels. The development of society and transportation is based on burning these fuels which release carbon in the form of carbon dioxide to the atmosphere. Atmospheric CO_2 level has been increasing at an accelerating rate since the start of the Industrial Revolution and is believed to be the main factor responsible for the rise in the Earth's temperature. Plants and photosynthetic microorganisms (microalgae and cianobacteria) can capture this atmospheric carbon through photosynthesis in organic molecules that may be used for the generation of biofuels, in this particular case, lipids that can be transformed into biodiesel. Although this is an ideal way to produce biodiesel, it generates large amounts of biomass that is not a direct feedstock for biodiesel production. This is particularly important in the case of plants, including soybean and oil palm, which are being widely used to produce biodiesel, because they produce oils in amounts that are miniscule, for example less than 5% of total biomass basis (Chisti, 2007). Photosynthetic microorganisms represent a better situation because oils levels are higher than those found in plants and, for example, oil levels of 20–50% are quite common in microalgae (Chisti, 2007). Therefore, in the best case, 50% of the biomass containing atmospheric carbon will be transformed into biodiesel. In addition, human activities generate a huge amount of residual biomass from agriculture, food-producing industry and municipal and industrial wastewaters that contain enough energy to meet a significant fraction of the world's entire energy demand, if it could be collected and converted efficiently to useful energy forms (Rittman, 2008), in our case biodiesel.

Heterotrophic microorganisms have their opportunity to contribute to the production of biodiesel, or other biofuel, by recovering the carbon captured in organic molecules, mainly carbohydrates, present in residual biomass derived from human activities, including energy generation. Oleaginous microorganisms divert an important part of the carbon from carbohydrates they use in their growth to lipids, which has structural, regulatory or energetic functions. Independently of its function, any saponifiable lipid may be transformed into biodiesel (Vicente *et al.*, 2009).

In this section, the possible carbon sources to support the growth of heterotrophic organisms are analysed, with particular attention to examples where accumulation of lipids has been described. However, few works have analysed the transformation of heterotrophic microorganism biomass into biodiesel in detail (Table 8.1), including optimization of the process and characterization of the biodiesel (Vicente *et al.*, 2009, 2010).

Most oleaginous microorganisms prefer simple sugar, like glucose, to grow, but the use of glucose or other pure sugars to obtain microbial biomass for the production of biodiesel is not economically viable and probably will never be. However, any feedstock commonly used in the biotechnology industry to grow microorganisms could be used, although ideal carbon sources should have zero or negative acquisition costs. The design of cost-effective processes for biodiesel

Table 8.1 Growth of oleaginous heterotrophic microorganisms using carbon in residual biomass

Type of residue	Residue	Microorganism	Lipid content (% dry mass)	FAME production	Reference
Rich in single carbohydrates	Molasses	*Candida lipolytica*	59.9	Yes	Karatay and Dönmez, 2010
	Molasses	*Candida tropicalis*	46.8	Yes	Karatay and Dönmez, 2010
	Molasses	*Rhodotorula mucilaginosa*	69.5	Yes	Karatay and Dönmez, 2010
	Pretreated cane molasses	*Trichosporon fermentans*	53.6	Yes	Zhu *et al.*, 2008
	Beet molasses	*Cryptococcus curvatus*	ND	Yes	Takakuwa and Saito, 2010
	Cheese whey	*Cryptococcus curvatus*	ND	Yes	Takakuwa and Saito, 2010
Starchy residues	Starch wastewater	*Rhodotorula glutinis*	35	No	Xue *et al.*, 2010
	Sweet potato starch	*Cunninghamella echinulata*	45.3	No	Du *et al.* 2007
Plant cell wall materials	Orange peel	*Cunninghamella echinulata*	2	No	Gema *et al.*, 2002
	Orange waste, tomato peel, etc.	*Rhodococcus opacus*	41–96	No	Gouda *et al.*, 2008
	Orange waste, tomato peel, etc.	*Gordonia sp*	50–75	No	Gouda *et al.*, 2008
	Apple pectin	*Mortierella isabellina*	24	No	Papanikolaou *et al.*, 2007
	Apple pectin	*Cunninghamella echinulata*	10	No	Papanikolaou *et al.*, 2007
	Hydrolysed corn stalk, tree leaves and rice straw	*Rhodotorula glutinis*	28.6	Yes	Dai *et al.*, 2007
Glycerol	Glycerol	*Cryptococcus curvatus*	52	No	Liang *et al.*, 2010
	Glycerol	*Cunninghamella echinulata*	27	No	Fakas *et al.*, 2008
Other	Sewage sludge	Various (unidentified)	27	Yes	Dufreche *et al.*, 2007
	Monosodium glutamate wastewater	*Rhodotorula glutinis*	11.6	Yes	Xue *et al.*, 2006

production implies the selection of the most appropriate feedstocks to provide the carbon molecules required to make the lipids, which will be transformed in biodiesel. In many cases, and particularly in the cheapest ones, they consist of complex mixtures of compounds, including those that are difficult to assimilate and/or toxic substances that may inhibit microbial growth. This complexity favours the use of filamentous fungi because they usually produce enzymes for hydrolysis of carbohydrate polymers, which are frequently absent in bacteria and yeasts. The main sources of carbon for growth of heterotrophic organisms can be grouped (Cardona and Sánchez, 2007) into carbon sources rich in simple carbohydrates (mono and disaccharides), starchy sources, plant cell wall materials and glycerol.

8.3.1 Carbon sources rich in simple carbohydrates

Raw materials that are basically rich in sucrose, sugar cane, either in the form of cane juice or cane molasses and beet molasses, have been extensively used for microbial growth. Molasses is a by-product of the sugar industry readily available at low cost, which consists of water, approximately 50% (w/w) total sugars (sucrose, glucose, fructose), suspended colloids, heavy metals, vitamins and nitrogenous compounds, and so on (Najafpour and Shan, 2003). Another by-product used for microbial growth that is rich in a monosaccharide, lactose, is cheese whey (Guimarães *et al.*, 2010).

Few studies have analysed the accumulation of lipids suitable for transformation into biodiesel in oleaginous microorganisms grown in molasses media (Table 8.1). One remarkable study analysed these parameters in 11 yeasts growing in molasses media (Karatay and Dönmez 2010). Several of these strains showed high levels of biomass and lipids, with promising results for *Candida lipolytica, Candida tropicalis* and *Rhodotorula mucilaginosa.* Although their lipids were transformed into FAMEs, the process was not optimized and would require a detailed study.

Although this type of carbon source is probably the best among the low cost raw materials, the presence of heavy metals in high concentrations may cause a critical problem during fermentation because they can inhibit the growth of microorganisms, influence the pH of the substrate and are involved in the inactivation of the enzymes associated with biosynthesis of products. Different pretreatments of molasses may reduce this problem (Roukas, 1998; Kotzamanidis *et al.*, 2002), but will add costs to biodiesel production. The yeast *Trichosporon fermentans* was grown using cane molasses pretreated with sulphuric acid and produced 36.4 g L^{-1} of biomass with a lipid content of 38.2%. The lipid content obtained with waste molasses was much lower than that with nitrogen-limited culture medium owing to a low carbon/nitrogen ratio (C/N) molar ratio (about 12.5). Thereby, addition of various sugars to the pretreated waste molasses increased the C/N molar ratio and consequently enhanced the lipid accumulation of *T. fermentans* up to 53.6% (Table 8.1). Similar to vegetable oils, the lipid mainly contained palmitic acid, stearic acid, oleic acid and linoleic acid and the unsaturated fatty acids amounted to about 64% of the total fatty acids. This microbial oil was transesterified to biodiesel by base catalysis after removal of free fatty acids and a high methyl ester yield of 92% was obtained (Zhu *et al.*, 2008).

Trends in biofuels production tend toward the isolation or generation of microorganisms able to produce biofuels themselves. The lipases present in many microorganisms can catalyse the transesterification process of lipids in the presence of methanol. A recent survey of 81 yeasts isolated from raw milk identified a *Cryptococcus curvatus* strain that produced FAMEs when lysates of the cell were incubated with methanol. The biomass of this strain grown in beet molasses and cheese whey media showed a transesterification rate of its triglycerides higher than 97% (Takakuwa and Saito, 2010). Although FAME levels were low (around 70 mg L^{-1}), the strain grew well in both carbon sources and further improvements could be made to increase FAME levels.

8.3.2 Starchy carbon source

Starch is a polysaccharide made by glucose units joined together by α-glycosidic bonds, which can be easily hydrolysed into the constituent sugars by amylases. Owing to the presence of these enzymes in many microorganisms, they are able to use starch as a carbon source. Corn and wheat starch are being extensively used for the production of bioethanol (Cardona and Sánchez, 2007) and also for the growth of oleaginous microorganisms, whose biomass is a potential feedstock for biodiesel production (Papanikolaou *et al.*, 2007). However, alternative starchy carbon sources, that is agro-industrial waste materials, should be used in order to avoid competition with human food (Table 8.1).

Promising results have been obtained with the oleaginous yeast *Rhodotorula glutinis* to reduce environmental pollution caused by starch wastewater from starch production plants. Pilot-scale production of the yeast in a 300-L fermenter using starch wastewater without sterilization and pH adjustment yielded 40 g L^{-1} biomass with a 35% lipid content and 80% chemical oxygen demand (COD) reduction after only 30–40 h of cultivation (Xue *et al.*, 2010). Similarly, Du *et al.* (2007) found that *Cunninghamella echinulata* reduced COD in sweet potato starch processing waste producing a biomass with 45.3% lipids. Other industrial wastes with more than 52% starch content, such as potato peel waste and thippi, have been used for fuel ethanol production and could be used for the growth of oleaginous microorganisms (Arapoglou *et al.*, 2010; Patle and Lal, 2008).

8.3.3 Plant cell wall materials

Although previous sources are important from the point of view of reduction of environmental contamination, plant cell wall is the most abundant biological material on the Earth. The main polymeric constituents of this structure are pectin, cellulose, hemicellulose and lignin (Sticklen, 2008). Plant primary cell walls contain cellulose, hemicellulose and pectin that can be hydrolysed to provide fermentable sugars. The growth of oleaginous microorganisms in waste material rich in primary cell walls, such as orange peel (Gema *et al.*, 2002; Gouda *et al.*, 2008), or their constituents, like pectin, has been described (Table 8.1) (Papanikolaou *et al.*, 2007).

Secondary plant cell walls are made from lignocellulose which contains lignin in addition to cellulose and hemicellulose. Lignin is highly resistant towards chemical and biological degradation and acts as cement between wood fibres. Therefore, degradation of lignin is a key step in the degradation of lignocellulose (Martínez *et al.*, 2005; Himmel *et al.*, 2007). Current technology using lignocellulose is based on harsh expensive pretreatments that release monosaccharides and oligosaccharides, which can support microbial growth. However, these treatments are expensive and produce compounds that inhibit microbial growth. Therefore, an ideal situation would be that heterotrophic microorganisms use this material without previous treatments. Filamentous Basidiomycetes are the main group of microorganisms able to grow on this material because of their ability to degrade or modify lignin (Martínez *et al.*, 2005) but, to our knowledge, there is a unique report describing their use as a feedstock for biodiesel production. In this study, selected strains of *Rhodotorula glutinis* showing high xylose assimilating capacity were capable of accumulating a significant amount of lipids cultured with chemically prehydrolysed corn stalk, tree leaves and rice straw as carbon sources (Table 8.1). Although these SCO were transformed into biodiesel, a full characterization was not reported (Dai *et al.*, 2007).

8.3.4 Glycerol

The current technology for biodiesel production generates about 10% (w/w) glycerol as the main by-product (Da Silva *et al.*, 2009). Crude glycerol is a mixture of glycerol (65–85%, w/w), methanol and soap (Mu *et al.*, 2006). The significant expansion of biodiesel production has resulted in the generation of large quantities of glycerol. In various European countries, crude glycerol is presently treated as a new kind of industrial wastewater owing to the obligatory requirement for biodiesel production and the huge excess that results (Mu *et al.*, 2008). The carbon present in this molecule can be captured in lipids by microorganisms that have the ability to use glycerol as the carbon source (Meesters *et al.*, 1996).

Crude glycerol has been used in the production of several products by microbial fermentation, but in a few cases it has been used to produce lipid-rich microbial biomass. The oleaginous yeast *Cryptococcus curvatus* grown on crude glycerol derived from yellow grease supplemented with nitrogen, using a two-stage fed-batch operation, produced 32.9 g L^{-1} of biomass with a lipid content of 52% (Liang *et al.*, 2010). The effect of growth on glycerol over the degree of saturation has been studied in the yeast *Rhodotorula glutinis*, although in this case pure glycerol was used instead of crude glycerol. Yeast cultured for 24 h on a medium containing (i) dextrose, (ii) xylose, (iii) glycerol, (iv) dextrose and xylose, (v) xylose and glycerol, or (vi) dextrose and glycerol accumulated 16, 12, 25, 10, 21 and 34% TAG on a dry weight basis, respectively. A difference in the degree of saturation for the carbon sources tested was observed. Thus, the cells cultivated on glycerol alone had the highest degree of unsaturated fatty acids, at 53%, while xylose had the lowest, at 25% (Easterling *et al.*, 2009).

In contrast to the theoretical calculations that showed that conversion of glycerol into lipid should be poor compared with glucose, they were similar when the fungus *Cunninghamella echinulata* was cultured using tomato waste hydrolysate as a nitrogen source (Table 8.1) (Fakas *et al.*, 2008). Therefore, the overall idea is that glycerol could be an efficient substrate for production microbial biomass rich in lipids.

8.4 Biodiesel production from heterotrophic microorganisms

8.4.1 Microbial biomass production

Lipid accumulation, in terms of lipid profile, amount, productivity and conversion yield, depends primarily on microorganism physiology, but it is also influenced by various operating conditions, such as nutrient limitation, pH, aeration and temperature conditions (revised by Beopoulos *et al.*, 2009). However, initiation of lipid accumulation is mainly triggered by a nutrient imbalance in the culture medium. When cells run out of a key nutrient, usually nitrogen, excess carbon substrate continues to be assimilated by the cells and converted into storage lipids (Ratledge and Wynn, 2002). The key to the initiation of lipid accumulation in response to nitrogen limitation is a severe limitation of isocitrate dehydrogenase activity caused by the decrease in adenosine monophosphate (AMP) concentration, which is provoked by an increase in AMP deaminase activity induced by low nitrogen levels (revised by Ratledge and Wynn, 2002). As a carbon source has to be available when nitrogen is low, the key parameter in lipid accumulation induced by nitrogen limitation is the carbon/nitrogen (C/N) ratio.

Three different modes of culture are commonly used to grow heterotrophic microorganisms: batch, fed-batch and continuous mode. Most of the processes analysing lipid accumulation refer to batch mode. In this culture mode, the microbial growth follows two phases: a first phase of growth when the C/N ratio is high and a second phase characterized by slow growth and lipid accumulation (Ratledge and Wynn, 2002). The global conversion yield of glucose into lipids depends on the duration of the growth phase and on the transition to the accumulation phase, which depends on the C/N ratio. The total substrate-to-lipid conversion yield therefore is a function of the initial C/N ratio of the batch culture.

A low initial C/N ratio gives a poor substrate-to-lipid conversion, whereas very high ratios create a severe nitrogen deficiency, leading to a rapid decrease in cell viability before the cells are able to enter the lipid accumulation stage (Beopoulos *et al.*, 2009). The most frequent C/N ratios used in batch culture favouring lipid accumulation and lipid productivity are in range of 50–75.

The fed-batch mode seems to be a good culture system, because carbon and nitrogen flows can be perfectly controlled, making it possible to dissociate growth and lipid accumulation (Beopoulos *et al.*, 2009). *Cryptococcus curvatus* was cultured in crude glycerol using fed-batch mode produced biomass with up to 52% lipid content at the end of 12 days of culture (Liang *et al.*, 2010).

The process of lipid accumulation can also be achieved in continuous culture, where it is necessary to grow the cells at a sufficiently low dilution rate to allow the cells to assimilate the carbon (Ratledge and Wynn, 2002). In this context, continuous growth of the yeast *Yarrowia lipolytica* in a medium with a C/N ratio of 66 produced 9.3 g dry biomass L^{-1} with a lipid content of 33% dry weight (Aggelis and Komaitis, 1999). Lipid production in continuous culture in this yeast has been modelled (Ykema *et al.*, 1986). Moreover, the best conversion of glucose into oil that has been achieved in practice (which was with an oleaginous yeast growing in continuous culture) was 22 g oil from 100 g glucose (Ratledge and Wynn, 2002).

Although nitrogen limitation is the easiest condition to control and is generally the most efficient type of limitation in inducing lipid accumulation, limitation of other elements, like magnesium, zinc, iron or phosphorus also triggers lipid accumulation (Gill and Ratledge, 1972; Granger 1992). Moreover, the type of nitrogen source seems to affect lipid accumulation because organic nitrogen from tomato waste hydrolysate enhances glucose uptake and lipid accumulation in the fungus *Cunninghamella echinulata* (Fakas *et al.*, 2008).

The lipid profile can be modified by adjusting the culture temperature because the degree of saturation generally decreases with decreasing temperature (Kendrick and Ratledge, 1992). This modulation of the lipid profile probably results from an increase in desaturase stability at low temperatures, with no such increase in stability observed for the other enzymes. However, low temperatures do not favour lipid production, because they also lead to large decreases in cellular activity and metabolism (Beopoulos *et al.*, 2009). There are few alternatives to genetic modification for the modulation of fatty acid profile. One of these alternatives is the use of growth inhibitors, such as cerulenin (Georgopapadakou *et al.*, 1987), or natural antimicrobial compounds, such as *Teucrium polium* extracts (Aggelis and Komaitis, 1999).

8.4.2 Oil extraction

A large number of methods are currently available to extract lipids from all kinds of biological materials, including microbial biomass. Most of them use organic solvents, which have to demonstrate a high solubility for all lipids compounds and must be sufficiently polar to remove the lipids from their association with cell membranes and lipoproteins (Smedes and Askland, 1999). Chloroform and methanol mixtures apply well, as was studied by Folch *et al.* (1957). This approach was adapted by Bligh and Dyer (1959) resulting in a method which has become the standard method for total lipid determination for over 40 years. This procedure consists of a three-step extraction using chloroform, methanol and water (wet extraction), the lipids being separated in the chloroform layer. However, the method of Bligh and Dyer is not totally appropriate for extraction of lipids from microbial biomass (Zhu *et al.*, 2002), since lipids in membranes do not extract effectively. In this sense, the lipid yield of dry extraction using chloroform and methanol is higher than that of wet extraction.

Restrictions on the use of chlorinated solvents under the Montreal Protocol made it necessary to develop an alternative method to the Folch and Blight and Dyer methods. Thus, several different solvent mixtures were tested as a replacement for chloroform (Smedes and Askland, 1999). Of the non-chlorinated solvents tested, the solvent pairs diisopropyl ester–methanol and cyclohexane–2 propanol were the best combinations for lipid extraction from marine tissues.

However, conventional solvent extractions have some drawbacks (Sahena *et al.*, 2009; Boutin and Badens, 2009). Thus, solvents commonly used present problems of toxicity and the corresponding extraction methods usually produce large amounts of hazardous solvent wastes. In general, they are carried out at high temperatures which can be responsible for the destruction of valuable products. Moreover, the use of organic solvents can lead to product contamination with solvent residues. For these reasons, new procedures for lipid extraction have recently been developed. Among them, the more outstanding ones are the use of supercritical fluids, microwave extraction and the combination of lipid extraction with FAME production in a single step (Sahena *et al.*, 2009).

The supercritical fluid extraction method has been demonstrated to be very advantageous and environmentally friendly over the conventional solvent extraction methods for recovering lipids (Randolph, 1990; Sahena *et al.*, 2009). This process consists of a separation technology that uses supercritical fluid as the solvent. Every fluid is characterized by a critical point, which is defined in terms of the critical temperature and critical pressure. For instance, CO_2 is supercritical above 31°C and 7.38 MPa, which makes it an ideal solvent for extracting thermally sensitive materials. Thus, many studies about supercritical CO_2 extraction of lipids in the food industry have been reported (List *et al.*, 1984, 1989; Arul *et al.*, 1987; Chao *et al.*, 1991; Froning *et al.* 1998). However, extraction using supercritical CO_2 mainly recovered non-polar lipids, whereas polar lipids do not extract effectively because of their lower solubility in supercritical CO_2 (Sahena *et al.*, 2009). Microorganisms contain a significant quantity of polar lipids in their membranes. Consequently, supercritical CO_2 extraction is not completely suitable for the total extraction of lipids from microbial biomass.

In addition, the use of microwaves for recovering lipids has recently been studied (Leray *et al.*, 1995; Paré *et al.* 1997). According to Paré *et al.* (1997), treatment with microwaves energy has allowed the development of rapid, safe and cheap methods for extracting lipids and does not require samples devoid of water. However, it can result in lipid oxidation and in quantitative modifications of the fatty acid composition.

Oil characterization

Microbial lipids (SCO) composition includes more lipids than triglyceride and free fatty acids, the main components of crude vegetable oils and animal fats. In fact, not all lipids obtained from microbial biomass are suitable for making biodiesel (Vicente *et al.*, 2009, 2010). Only lipids with fatty acid ester linkages (also referred to as saponifiable lipids) and free fatty acids can produce FAMEs, which can be used as biodiesel if they comply with the current existing standards (ASTM D 6751 in the United States or EN 14213 and 14214 in the European Union). These lipids include energy storage and structural lipids. The energy storage lipids are neutral lipids (mono-, di- and triglycerides) and free fatty acids. The structural lipids are the polar lipids, including phospholipids, sphingolipids, saccharolipids and sterol esters. As conventional vegetable oils, microbial saponifiable lipids and free fatty acids can be converted into FAMEs through a transesterification and esterification reaction with methanol, respectively, in the presence of a suitable catalyst. The lipids extracted from oleaginous microorganisms also include a small fraction of non-saponifiable lipids, which consists of carotenoids, sterols, tocopherols and retinoids. These lipids, as they do not have fatty acid ester linkages, cannot be transformed into FAMEs.

8.4.3 Obtaining FAMEs

Currently, biodiesel is produced in industry from vegetable oils, animal fats and waste cooking oils. The most common technology for producing biodiesel is by transesterification with methanol (also called methanolysis). In this reaction, triglycerides, as the main components of vegetable oils and animal fats, react with methanol to produce FAMEs and glycerol. Generally, this reaction is catalysed by a basic or an acid catalyst. However, the basic catalysts (e.g. sodium and potassium hydroxide or sodium methoxide) are the most commonly used in industry, because the process proves faster and the reaction conditions are moderated (Freedman *et al.*, 1984; Reid, 1911).

The microbial lipids can potentially be used as raw materials to produce biodiesel, but their high free fatty acid content, as is the case in waste cooking oils and low-value animal fats, makes it unsuitable for the alkaline catalysed transesterification technology currently in use in most biodiesel plants in Europe. The utilization of a basic catalyst in microbial oil methanolysis produces soaps by neutralizing the free fatty acids in this oil. This undesirable side reaction partially consumes the catalyst, decreases the biodiesel yield and complicates the separation and purification steps. In this case, the use of a basic catalyst is not appropriate for obtaining biodiesel using microorganisms. However, soap formation from free fatty acid neutralization can be avoided by using an acid catalyst such as sulphuric or hydrochloric acids. The acids catalyse the free fatty acid esterification with methanol producing FAMEs and increasing the biodiesel yield. Nevertheless, the acid-catalysed transesterification is much slower than the base-catalysed reaction and also needs more extreme temperatures and pressure conditions (Freedman *et al.*, 1984; Schwab *et al.*, 1987).

More recently, there has been an increase in the development of heterogeneous catalysts for producing FAMEs mainly from plant oils (Melero *et al.*, 2009), because their utilization in the transesterification reaction greatly simplifies and economizes the post-treatment of the products (separation and purification). Besides, the use of heterogeneous catalysts does not produce soaps through free fatty acid neutralization and triglyceride saponification. However, the heterogeneous catalysed reaction also requires extreme reaction conditions, while the methyl ester yield and the reaction time are still unfavourable compared to the alkali catalysts (Vicente *et al.*, 1998).

8.4.4 Direct transformation of microbial biomass

In addition to the biodiesel production from extracted microbial oil, a single-step method has been developed that transforms lipids present in dried microbial biomass into FAMEs, without previous lipid extraction (Fig. 8.1) (Vicente *et al.*, 2009, 2010). This method combines lipid extraction, acid-catalysed transesterification of

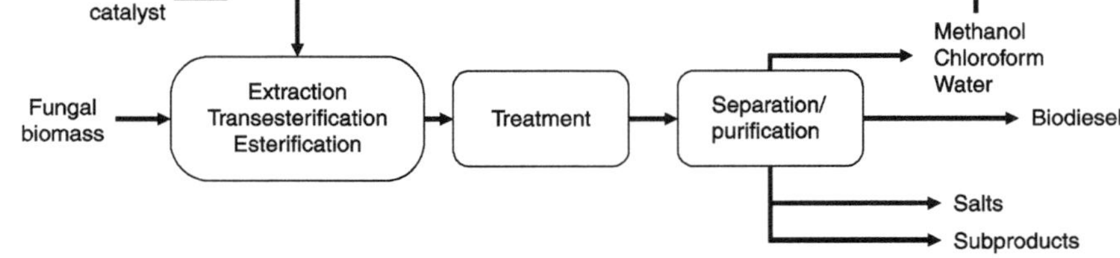

8.1 Schematic flow sheet for a single-step biodiesel production process from fungal biomass.

the extracted saponifiable lipids and acid-catalysed esterification of the extracted free fatty acids in one step and was initially proposed because of the substantial reduction in both time and solvents that this technique offers for analytical purposes (Lewis *et al.*, 2000). Similar procedures that avoid the lipid extraction step have already been developed from a variety of raw materials such as marine tissues (Meier *et al.*, 2006), yeast (Liu and Zhao, 2007), fungi (Liu and Zhao, 2007, Vicente *et al.*, 2009; Fakas *et al.*, 2009; Vicente *et al.*, 2010), bacteria (Dionisi *et al.*, 1999) and microalgae (Johnson and Wen, 2009; Ehimen *et al.*, 2010).

In this context, the use of this method for microorganisms may be especially advantageous, since the extraction of microbial lipids is usually accomplished via solvent extraction and not by using cheaper physical extraction methods as utilized for conventional oil crops (Ehimen *et al.*, 2010). The direct transformation of microbial oil in biomass has been shown to result in increased biodiesel purity and yields, compared to the conventional two-stage route (Lewis *et al.*, 2000; Vicente *et al.*, 2009). In addition, waste production is reduced using the method that combines lipid extraction and the acid-catalysed transesterification/esterification of the extracted lipids in one step.

8.5 Suitability of biodiesel

Biodiesel fuel, in the form of FAMEs, need to comply with existing standards to be considered as an alternative to diesel fuel. In the European Union, there are separate standards for biodiesel intended for vehicle use (Standard EN 14214) and for use as heating oil (Standard EN 14213). Conversely, the ASTM standard D6751 includes biodiesel specifications in the USA.

Biodiesel quality depends on the composition of raw materials and, consequently, not all microorganisms can be used as a feedstock for biodiesel production. Thus, careful characterization of the lipid composition of each microbial candidate should be carried out before its adoption by the industry.

Triglycerides are the main component in plant oils and animal fats, also being the most significant storage lipid in the lipid bodies from microbial biomass. Storage lipids together with structural lipids have to be transformed into FAMEs to obtain suitable biodiesel.

Microbial oils also differ from most vegetable oils and animal fats in being quite rich in unsaturated (two and three double bonds) and polyunsaturated (four or more double bonds) fatty acids (Meng *et al.*, 2009). In this sense, European Standard EN 14214 limits the content of linolenic acid methyl ester (three double bonds) and the polyunsaturated acid methyl esters in biodiesel for vehicle use to 12% and 1%, respectively. The high degree of unsaturation inherent in these FAMEs is evidence of lower oxidative stability during storage, causing the formation of various degradation products (polymerized products, acids, aldehydes, ketones and others), which can negatively affect engine operability (Schober and Mittelbalch, 2007). However, the oxidation stability of oil extracted from microorganisms and their corresponding FAMEs is also influenced by the content of natural antioxidants (tocopherols, carotenoids, etc.). Oils extracted from microorganisms have relatively high levels of these natural antioxidants. If these compounds are not separated during the purification process, microbial derived biodiesel will also have higher oxidation stabilities independent of their level of unsaturation. Conversely, the iodine value is a measure of the total unsaturation, the specified limits being 120 and 130 g iodine/100 g in the EN 14214 and EN 14213 European Standards, respectively.

8.6 Enhancement of microbial oil production

8.6.1 Genetic and metabolic engineering

Lipid accumulation is influenced by nutrient levels and culture conditions, but extensive use of microorganisms to produce an oleaginous biomass for biodiesel production would ideally require microorganisms that accumulate lipid independently of both factors. Genetic modifications to increase the lipid accumulation may increase the function of the lipid biosynthesis pathway, reduce the mobilization of storage lipid (β-oxidation) or increase carbon flux toward lipid biosynthesis. Few examples of these approaches can be found in literature, but recently genetic engineering has been used to improve TAG accumulation in both oleaginous and non-oleaginous yeast. Increases in lipid accumulation to levels three times those observed in wild-type strains have been obtained in the oleaginous yeast *Yarrovia lipolytica* by redirection of the carbon flux toward TAG assembly by deletion of the *GUT2* gene, which codes for a glycerol-3-phospate dehydrogenase situated at the crossroad of lipid metabolism and dihydroxyacetone phosphate production for glycolysis. Additional deletion of the *POX1* to *POX6* genes encoding the acyl-CoA oxidases, completely abolishes β-oxidation, preventing lipid mobilization. The resulting mutant strain had lipid levels four times higher than those of the wild-type strain (Beopoulos *et al.*, 2008).

Although *Saccharomyces cerevisiae* is a non-oleaginous microorganism, the lipid metabolism has also been extensively studied and has provided new insights into the enzymes and regulatory mechanisms of lipid metabolism in eukaryotic cells, including regulation of storage lipid biosynthesis. Storage lipids in this yeast accumulate by inositol deficiency, phosphate deficiency, low temperature and sporulation. Transposon insertion mutagenesis resulted in the identification of five genes negatively affecting lipid accumulation because disruption of these genes increased lipid content. Since these genes are not directly involved in storage lipid biosynthesis, a regulatory role was proposed (Kamisaka *et al.*, 2006). The role of one of these genes, *snf2*, has been studied in detail demonstrating that it encodes a DNA-dependent ATPase that forms part of the SWI/SNF (switching/sucrose non-fermenting) chromatin-remodelling complex (Peterson and Tamkun, 1995) which is involved in the transcriptional repression of phospholipid and

TAG biosynthesis enzymes (Chen *et al.*, 2007). Overexpression of *dga1* (diacylglycerol acyltransferase, which catalyses the formation of triglycerides from diacylglycerol and Acyl-CoA) and *faa3* (acyl-CoA synthetase, which coverts free FAs into acyl-CoA) enhanced lipid accumulation up to 30%, which makes a non-oleaginous species oleaginous (Kamisaka *et al.*, 2007).

The extent of lipid accumulation in microorganisms is considered to be controlled by the activity of malic enzyme (ME), which acts as the sole source of NADPH for fatty acid synthase (FAS) (Ratledge and Wynn, 2002). Consequently, overexpression of a gene coding for the ME isoform E of *Mucor circinelloides* resulted in a 2.5-fold increase in TAG accumulation (Zhang *et al.*, 2007).

Biodiesel characteristics depend upon the fatty acid composition of raw materials and, consequently, not all microorganisms can be used as a feedstock for biodiesel production (Pinzi *et al.*, 2009; Durrett *et al.*, 2008). Various studies suggest that biodiesel with high levels of methyl oleate will have excellent, if not optimal, characteristics with regard to ignition quality, NO*x* emissions and fuel stability (Durrett *et al.*, 2008). One way to generate microorganisms with ideal lipid composition for biodiesel production could be by genetic manipulation of key genes. This approach has been demonstrated to be feasible in the oleaginous fungus *Mortierella alpina* 1S-4, which is used for commercial production of polyunsaturated fatty acid. Mutants and strains generated by direct genetic manipulation of genes coding for elongases and desaturases showed different fatty acid profiles from the wild-type strain (reviewed by Sakuradani, 2010). Of particular interest is the gene coding for the Δ12-Desaturase, which desaturates oleic acid (18:1*n*-9) to linoleic acid (18:2*n*-6). Mutants in this gene or strains in which its expression has been silenced by RNA interference (RNAi) showed a fatty acid profile with 63% oleic acid content (Jareonkitmongkols *et al.*, 1992; Takeno *et al.*, 2005).

Recent results on synthesis, turnover and control of FAs, lipids and lipid particle assembly as well as events that lead to oleagenecity have provided a collection of promising access points for genetic and metabolic engineering. These tools have enabled alterations to be made in substrate specificities towards the use of a wider range of compounds, lipid accumulation in non-oleaginous species; FA overproduction; heterologous gene expression and direct biodiesel production (Kosa and Ragauskas, 2011).

8.6.2 Biorefinery approach

Future (second generation) biorefineries must integrate the production of a large variety of products such as fuels, commodities, fine chemicals, materials, and so on. To develop the concept of a second-generation refinery, the use of SCO must include not only the production of biodiesel as a commodity product but also other added-value fine chemicals such as the omega-6 fatty acids γ-linolenic and arachidonic for the pharmaceutical and nutraceutical industries.

The use of refined edible oils or cereal grains as feedstocks competing directly with the food industry must be replaced by wastes and low-grade non-edible renewable materials with high diversity and availability. In this respect, lignocellulosic materials are a promising feedstock for the production of biofuels. However, the main developments of lignocellulose use are aimed at fermentative microorganisms producing ethanol or other low-chain alcohols. The fungus *Microsphaeropsis* sp. is capable of producing SCO in solid state fermentation on a mixture of wheat straw and wheat bran, reaching 10.2% total lipids based on dry fermented substrate (Peng and Chen, 2008)

Glycerol is a by-product of the biodiesel process by transesterification of triglycerides. The glycerol market tends to be saturated and the price drops as biodiesel production increases so that new applications of glycerol need to be developed to improve the overall economy of the process (Yazdani and Gonzalez, 2007). Ethanol, butanol, propionic acid, 1,3-propane-diol or oxalic acid production along with SCO accumulation have been reported by fermentation processes using glycerol as a carbon source with significant yields (Demain, 2009; André *et al.*, 2010).

8.7 Conclusions and future trends

The rapid increase in the production of biodiesel in recent years makes evident that alternative lipid sources to traditional energy crops are required. The use of SCO derived from oleaginous microorganisms has attracted great attention in attempts to attain the above goal. While photosynthetic microorganisms are probably the best candidates because they capture carbon dioxide in lipids using sun energy (Chisti, 2007), human activities generate a huge amount of residual biomass that produces serious pollution problems but contains enough energy to meet a significant fraction of the world's entire energy demand (Rittman, 2008). Heterotrophic oleaginous microorganisms may contribute to the production of biodiesel, recovering the carbon captured in organic molecules, mainly carbohydrates, present in that residual biomass.

Organism selection for lipid production is a crucial aspect in the adoption of SCO for biodiesel production. Many of the cellular traits and biosynthetic characteristics that are sought after in an ideal lipid-producing microorganism can be found in isolated native organisms, although probably none of them combine them all. The required qualities include the ability to use a large number of molecules present in residual biomass, including lignocellulosic materials, resist inhibition by toxic substances present in the residual biomass, and biosynthesise specific fatty acids with high yield. The ideal microorganism that consolidates all these traits does not exist, because the combination of features that are required for SCO production at the industrial scale probably is distant from that required for living in nature. Therefore, metabolic engineering approaches are expected to be necessary to create such strains (Alper and Stephanopoulos, 2009). The genetic manipulation needed to improve lipid biosynthesis will require a previous deep knowledge of the lipid pathways and their regulation, in addition to the effect of the whole cellular metabolism in lipid accumulation. Although information about promising access points for genetic and metabolic engineering is available (Kosa and Ragauskas, 2011), future systems biology studies in model oleaginous microorganisms will identify additional cellular elements or pathways suitable for being modified to increase lipid accumulation.

Genetic manipulation has gone a step forward by the generation of bacterial strains that directly synthesize FAEE (Kalscheuer *et al.*, 2006; Steen *et al.*, 2010). This is an incipient trend in biodiesel biotechnology that pursues the creation of a process analogous to those in engineered fermentative systems that use microorganisms whose phases of growth and production are separated and whose carbon partitioning is controlled to achieve very high productivities (Stephanopoulos *et al.*, 1998). In these processes, cells partition carbon and free energy almost exclusively to produce and secrete a desired product, in this case biodiesel, while minimizing energy conversion losses due to growth-associated metabolism (Robertson *et al.*, 2011). With the advent of modern genetic tools and synthetic biology approaches, the day of having a set of heterotrophic microorganisms able to transform metabolically the carbon present in residual biomass into biodiesel is approaching. When that day arrives, technical improvements in chemical and physical transformation of SCO into biodiesel will be required to make it cost-effective.

8.8 References

Aggelis G, Ratomahenina R, Arnaud A, Galzy P, Martin-Privat P, *et al.* (1988). Etude de l'influence des conditions de culture sur la teneur en acide gamma linolenique de souches de *Mucor. Oléagineux*, **43**, 311–17.

Aggelis G, Pina M and Graille J (1990). Localisation de l'acide gamma linolenique dans les mycelium et dans les spores chez deux mucorales. *Oléagineux*, **45**, 229–32.

Aggelis G and Komaitis M (1999). Enhancement of single cell oil production by *Yarrowia lipolytica* growing in the presence of *Teucrium polium L.* aqueous extract. *Biotechnol Lett*, **21**, 747–9.

Alper H and Stephanopoulos G (2009). Engineering for biofuels: exploiting innate microbial capacity or importing biosynthetic potential? *Nature Rev Microbiol*, **7**, 715–23.

Alvarez HM and Steinbüchel A (2002). Triacylglycerols in prokaryotic microorganisms. *Appl Microbiol Biotechnol*, **60**, 367–76.

André A, Diamantopoulou P, Philippoussis A, Sarris D, Komaitis M and Papanikolaou S (2010). Biotechnological conversions of bio-diesel derived waste glycerol into added-value compounds by higher fungi: production of biomass, single cell oil and oxalic acid. *Ind Crop Prod*, **31**, 407–16.

Arapoglou D, Varzakas Th, Vlyssides A and Israilides C (2010) Ethanol production from potato peel waste (PPW). *Waste Manage*, **30**, 1898–1902.

Arul J, Boudreau A, Makhlouf J, Tardif R and Sahasrabudhe MR (1987). Fractionation of anhydrous milk fat by supercritical carbon dioxide. *J Food Sci*, **52**, 1231–6.

Azócar L, Ciudad G, Heipieper HJ and Navia R (2010). Biotechnological processes for biodiesel production using alternative oils. *Appl Microbiol Biotechnol*, **88**, 621–36.

Becker EW (1994). *Microalgae: Biotechnology and Microbiology*. Cambridge University Press, New York.

Beopoulos A, Cescut J, Haddouche R, Uribelarrea JL, Molina-Jouve C and Nicaud JM (2009). *Yarrowia lipolytica* as a model for bio-oil production. *Progr Lipid Res*, **48**, 375–87.

Beopoulos A, Mrozova Z, Thevenieau F, Le Dall MT, Hapala I, *et al.* (2008). Control of lipid accumulation in the yeast *Yarrowia lipolytica. Appl Environ Microbiol*, **74**, 7779–89.

Bligh EG and Dyer WJ (1959). A rapid method for total lipid extraction and purification. *Can J Biochem Physiol*, **8**, 911–17.

Boutin O and Badens E (2009). Extraction from oleaginous seeds using supercritical CO_2: experimental design and products quality. *J Food Eng*, **92**, 396–402.

Cardona CA and Sánchez OJ (2007). Fuel ethanol production: process design trends and integration opportunities. *Biores Technol*, **98**, 2415–57.

Chao RR, Mulvaney SJ, Baley ME and Fernando LN (1991). Supercritical CO_2 conditions affecting extraction of lipid and cholesterol from ground beef. *J Food Sci*, **56**, 183–7.

Chen M, Hancock LC and Lopes JM (2007). Transcriptional regulation of yeast phospholipid biosynthetic genes. *Biochim Biophys Acta*, **1771**, 310–21.

Chen X, Li Z, Zhang X, Hu F, Ryu D and Bao J (2009). Screening of oleaginous yeast strains tolerant to lignocellulose degradation compounds. *Appl Biochem Biotechnol*, **159**, 591–604.

Chisti Y (2007). Biodiesel from microalgae. *Biotechnol Adv*, **25**, 294–306.

Da Silva GP, Mack M and Contiero J (2009). Glycerol: a promising and abundant carbon source for industrial microbiology. *Biotechnol Adv*, **27**, 30–39.

Dai CC, Tao J, Xie F, Dai YJ and Zhao M (2007). Biodiesel generation from oleaginous yeast *Rhodotorula glutinis* with xylose assimilating capacity. *African J Biotechnol*, **6**, 2130–4.

Demain AL (2009). Biosolutions to the energy problem. *J Ind Microbiol Biotechnol*, **36**, 319–32.

Dionisi F, Golay PA, Elli M and Fay, LB (1999). Stability of cyclopropane and conjugated linoleic acids during fatty acid quantification in lactic acid bacteria. *Lipids*, **34**, 1107–15.

Du J, Wang HX, Jin HL, Yang KL and Zhang XY (2007). Fatty acids production by fungi growing in sweet potato starch processing waste water. *Chin J Bioproc Eng*, **5**, 33–6.

Dufreche S, Hernandez R, French T, Sparks D, Zappi M and Alley E (2007). Extraction of lipids from municipal wastewater plant microorganisms for production of biodiesel. *J Am Oil Chem Soc*, **84**, 181–7.

Durrett TP, Benning C and Ohlrogge J (2008). Plant triacylglycerols as feedstocks for the production of biofuels. *Plant J*, **54**, 593–607.

Easterling ER, French WT, Hernandez R and Licha M (2009). The effect of glycerol as a sole and secondary substrate on the growth and fatty acid composition of *Rhodotorula glutini. Biores Technol*, **100**, 356–61.

Ehimen EA, Sun ZF and Carrington CG (2010). Variables affecting the *in situ* transesterification of microalgae lipids. *Fuel*, **89**, 677–84.

Fakas S, Papanikolaou S, Galiotou-Panayotou M, Komaitis M and Aggelis G (2008). Organic nitrogen of tomato waste hydrolysate enhances glucose uptake and lipid accumulation in *Cunninghamella echinulata. J Appl Microbiol*, **105**, 1062–70.

Fakas S, Papanikolaou S, Batsos A, Galiotou-Panayotou M, Mallouchosa A and Aggelis G (2009). Evaluating renewable carbon sources as substrates for single cell oil production by *Cunninghamella echinulata* and *Mortierella isabellina. Biomass Bioenergy*, **33**, 573–80.

Folch J, Lees M and Stanley GHS (1957). A simple method for the isolation and purification of total lipids from animal tissues. *J Biol Chem*, **226**, 497–509.

Freedman B, Pryde EH and Mounts TL (1984). Variables affecting the yields of fatty esters from transesterified vegetable oils. *J Am Oil Chem Soc*, **61**, 1638–43.

Froning GW, Wehling RL, Cuppett SL and Niemann L (1998). Moisture content and particle size of dehydrated egg yolk affect lipid and cholesterol extraction using supercritical carbon dioxide. *Poultry Sci*, **77**, 1718–22.

Gema H, Kavadia A, Dimou D, Tsagou V, Komaitis M and Aggelis G (2002). Production of gamma-linolenic acid by *Cunninghamella echinulata* cultivated on glucose and orange peel. *Appl Microbiol Biotechnol*, **58**, 303–7.

Georgopapadakou NH, Dix BA, Smith SA, Freudenberger J and Funke PT (1987). Effect of antifungal agents on lipid biosynthesis and membrane integrity in *Candida albicans*. *Antimicrob Agents Chemother*, **31**, 46–51.

Gill CO and Ratledge C (1972). Effect of n-alkanes on the transport of glucose in *Candida* sp. strain 107. *Biochem J*, **127**, 59–60.

Gouda MK, Omar SH and Aouad, LM (2008). Single cell oil production by *Gordonia* sp. DG using agro-industrial wastes. *World J Microbiol Biotechnol*, **24**, 1703–11.

Granger L (1992). *Caractérisation Cinétique et Stoechiometrique de la Synthèse D'acide Gras Chez* Rhodotorula Glutinis. Institut National des Sciences Appliquées de Toulouse, Toulouse, p 247.

Guimarães PMR, Teixeira JA and Domingues L (2010). Fermentation of lactose to bio-ethanol by yeasts as part of integrated solutions for the valorisation of cheese whey. *Biotechnol Adv*, **28**, 375–84.

Himmel ME, Ding SY, Johnson DK, Adney WS, Nimlos MR, *et al.* (2007). Biomass recalcitrance: engineering plants and enzymes for biofuels production. *Science*, **315**, 804–7.

Jareonkitmongkols S, Imizua A and Amada H (1992). Fatty acid desaturation-defective mutants of an arachidonic-acidproducing fungus, *Mortierella alpina* 1s-4. *J General Microbiol*, **138**, 997–1002.

Johnson MB and Wen Z (2009). Production of biodiesel fuel from the microalga *Schizochytrium limacinum* by direct transesterification of algal biomass. *Energy Fuels*, **23**, 5179–83.

Kalscheuer R, Stolting T and Steinbuchel A. (2006). Microdiesel: *Escherichia coli* engineered for fuel production. *Microbiol*, **152**, 2529–36.

Kamisaka Y, Noda N, Tomita N, Kimura K, Kodaki T and Hosaka K (2006). Identification of genes affecting lipid content using transposon mutagenesis in *Saccharomyces cerevisiae*. *Biosci Biotechnol Biochem*, **70**, 646–53.

Kamisaka Y, Tomita N, Kimura K, Kainou K and Uemura H (2007). *DGA1* (diacylglycerol acyltransferase gene) overexpression and leucine biosynthesis significantly increase lipid accumulation in the Δ*snf2* disruptant of *Saccharomyces cerevisiae*. *Biochem J*, **408**, 61–8.

Karatay SE and Dönmez G (2010). Improving the lipid accumulation properties of the yeast cells for biodiesel production using molasses. *Biores Technol*, **101**, 7988–90.

Kendrick A and Ratledge C (1992). Lipid formation in the oleaginous mould *Entomophthora exitalis* grown in continuous culture: effects of growth rate, temperature and dissolved oxygen tension on polyunsaturated fatty acids. *Appl Microbiol Biotechnol*, **37**, 18–22.

Kosa M and Ragauskas AJ (2011). Lipids from heterotrophic microbes: advances in metabolism research. *Trends Biotechnol*, **29**, 53–61.

Kotzamanidis C, Roukas T and Skaracis G (2002). Optimization of lactic acid production from beet molasses by *Lactobacillus delbrueckii* NCIMB8130. *World J Microbiol Biotechnol*, **18**, 441–8.

Leray C, Gric T, Gutbier G and Bnouham M (1995). Microwave oven extraction procedure for lipid analysis in biological samples. *Analysis*, **23**, 65–7.

Lewis T, Nichols P and McMeekin T (2000). Evaluation of extraction methods for recovery of fatty acids from lipid-producing microheterotrophs. *J Microbiol Methods*, **43**, 107–16.

Li Y, Horsman M, Wu N, Lan CQ and Dubois-Calero N (2008a). Biofuels from microalgae. *Biotechnol Prog*, **24**, 815–20.

Li Q, Du W and Liu D (2008b). Perspectives of microbial oils for biodiesel production. *Appl Microbiol Biotechnol*, **80**, 749–56.

Liang Y, Cui Y, Trushenski J and Blackburn JW (2010). Converting crude glycerol derived from yellow grease to lipids through yeast fermentation. *Biores Technol*, **101**, 7581–6.

List GR, Friedrich JP and Christianson DD (1984). Properties and processing of corn oils obtained by extraction with supercritical carbon dioxide. *J Am Oil Chem Soc*, **61**, 1849–51.

List GR, Friedrich JP and King JW (1989). Supercritical CO_2 extraction and processing of oilseeds. *Oil Mill Gaz*, **95**, 28–34.

Liu B and Zhao Z (2007). Biodiesel production by direct methanolysis of oleaginous microbial biomass. *J Chem Technol Biotechnol*, **82**, 775–80.

Liu J, Huang J, Sun Z, Zhong Y, Jiang Y and Chen F (2011). Differential lipid and fatty acid profiles of photoautotrophic and heterotrophic *Chlorella zofingiensis*: assessment of algal oils for biodiesel production. *Biores Technol*, **102**, 106–10.

Martínez ÁT, Speranza M, Ruiz-Dueñas FJ. Ferreira P, Camarero S, *et al.* (2005). Biodegradation of lignocellulosics: microbial, chemical and enzymatic aspects of the fungal attack of lignin. *Int Microbiol*, **8**, 195–204.

McIntyre M, Breum J, Arnau J and Nielsen J (2002). Growth physiology and dimorphism of *Mucor circinelloides* (syn. racemosus) during submerged batch cultivation. *Appl Microbiol Biotechnol*, **58**, 495–502.

Meesters P, Huijberts GNM and Eggink G (1996). High-cell-density cultivation of the lipid accumulating yeast *Cryptococcus curvatus* using glycerol as a carbon source. *Appl Microbiol Biotechnol*, **45**, 575–9.

Meier S, Mjos SA, Joensen H and Grahl-Nielsen O (2006). Validation of a one-step extraction/methylation method for determination of fatty acids and cholesterol in marine tissues. *J Chromatogr A*, **1104**, 291–8.

Melero JA, Iglesias J and Morales G (2009). Heterogeneous acid catalysts for biodiesel production: current status and future challenges. *Green Chem*, **11**, 1285–1308.

Meng X, Yang J, Xu X, Zhang L, Nie Q and Xiang M (2009). Biodiesel production from oleaginous microorganisms. *Renew Energy*, **34**, 1–5.

Miao X and Wu Q (2006). Biodiesel production from heterotrophic microalgal oil. *Biores Technol*, **97**, 841–6.

Molina Grima E, Sánchez Pérez JA, García Camacho F, García Sánchez JL and López Alonso D (1993). n-3 PUFA productivity in chemostat cultures of microalgae. *Appl Microbiol Biotechnol*, **38**, 599–605.

Mu Y, Teng H, Zhang DJ, Wang W and Xiu ZL (2006). Microbial production of 1,3-propanediol by *Klebsiella pneumoniae* using crude glycerol biodiesel preparations. *Biotechnol Lett*, **28**, 1755–9.

Mu Y, Xiu Z and Zhang D (2008). A combined bioprocess of biodiesel production by lipase with microbial production of 1,3-propanediol by *Klebsiella pneumoniae*. *Biochem Eng J*, **40**, 537–41.

Najafpour GD and Shan CP (2003). Enzymatic hydrolysis of molasses. *Biores Technol*, **86**, 91–94.

Navarro E, Lorca-Pascual JM, Quiles-Rosillo MD, Nicolás FE, Garre V, *et al.* (2001). A negative regulator of light-inducible carotenogenesis in *Mucor circinelloides*. *Mol Genet Genomics*, **266**, 463–70.

Nicolás FE, Torres-Martínez S and Ruiz-Vázquez RM (2003). Two classes of small antisense RNAs in fungal RNA silencing triggered by non-integrative transgenes. *EMBO J.*, **22**, 3983–91.

Pan L, Yang D, Shao L, Li W, Chen G and Liang Z (2009). Isolation of the oleaginous yeasts from the soil and studies of their lipid-producing capacities. *Food Technol Biotechnol*, **47**, 215–20.

Papanikolaou S, Komaitis M and Aggelis G (2004). Single cell oil (SCO) production by *Mortierella isabellina* grown on high sugar content media. *Biores Technol*, **95**, 287–91.

Papanikolaou S, Galiotou-Panayotou M, Fakas S, Komaitis M and Aggelis G (2007). Lipid production by oleaginous *Mucorales* cultivated on renewable carbon sources. *Eur J Lipid Sci Technol*, **109**, 1060–70.

Paré JRJ, Matni G, Bélanger JMR, Li K, Rule C and Thibert B (1997). Use of the microwave-assisted process in extraction of fat from meat, dairy and egg products under atmospheric pressure conditions. *J AOAC Int*, **80**, 928–33.

Patle S and Lal B (2008). Investigation of the potential of agro-industrial material as low cost substrate for ethanol production by using *Candida tropicalis* and *Zymomonas mobilis*. *Biomass Bioenergy*, **32**, 596–602.

Peng X and Chen H (2008). Single cell oil production in solid-state fermentation by *Microsphaeropsis* sp. from steam-exploded wheat straw mixed with wheat bran. *Biores Technol*, **99**, 3885–9.

Peterson CL and Tamkun JW (1995). The SWI–SNF complex: a chromatin remodeling machine? *Trends Biochem Sci*, **20**, 143–6.

Pinzi S, Garcia IL, Lopez-Gimenez FJ, Luque de Castro MD, Dorado G and Dorado MP (2009). The ideal vegetable oil-based biodiesel composition: a review of social, economical and technical implications. *Energy Fuels*, **23**, 2325–41.

Randolph TW (1990). Supercritical fluid extractions in biotechnology. *Trends Biotechnol*, **8**, 78–82.

Ratledge C (1991). Microorganisms for lipids. *Acta Biotechnol*, **11**, 429–238.

Ratledge C (2002). Regulation of lipid accumulation in oleaginous microorganisms. *Biochem Soc Trans*, **30**, 1047–50.

Ratledge C (2004). Fatty acid biosynthesis in microorganisms being used for single cell oil production. *Biochimie*, **86**, 807–15.

Ratledge C and Wynn JP (2002). The biochemistry and molecular biology of lipid accumulation in oleaginous microorganisms. *Adv Appl Microbiol*, **51**, 1–51.

Reid EE (1911). Studies in esterification. IV. The interdependence of limits as exemplified in the transformation of esters. *Am Chem J*, **45**, 479–516.

Rittman BE (2008). Opportunities for renewable bioenergy using microorganisms. *Biotechnol Bioeng*, **100**, 203–12.

Robertson DE, Jacobson SA, Morgan F, Berry D, Church GM and Afeyan NB (2011). A new dawn for industrial photosynthesis. *Photosynth Res*, **107**, 269–77.

Roukas T (1998). Pretreatment of beet molasses to increase pullulan production. *Proc Biochem*, **33**, 805–10.

Sahena F, Zaidul ISM, Jinap S, Karim AA, Abbas KA, *et al.* (2009). Application of supercritical CO_2 in lipid extraction – A review. *J Food Eng*, **95**, 240–53.

Sakuradani E (2010). Advances in the production of various polyunsaturated fatty acids through oleaginous fungus *Mortierella alpina* breeding. *Biosc Biotechnol Biochem*, **74**, 908–17.

Schober S and Mittelbach M (2007). Iodine value and biodiesel: is limitation still appropriate? *Lipid Technol*, **19**, 281–4.

Schwab AW, Bagby MO and Freedman B (1987). Preparation and properties of diesel fuels from vegetables oils. *Fuel*, **66**, 1372–8.

Smedes F and Askland T (1999). Revisiting the development of the Bligh and Dyer total lipid determination method. *Marine Pollut Bull*, **38**, 193–201.

Song YD, Wynn JP, Li YH, Grantham D and Ratledge C (2001). A pre-genetic study of the isoforms of malic enzyme associated with lipid accumulation in *Mucor circinelloides. Microbiol*, **147**, 1507–15.

Steen EJ, Kang Y, Bokinsky G, Hu Z, Schirmer A, *et al.* (2010). Microbial production of fatty-acid-derived fuels and chemicals from plant biomass. *Nature*, **463**, 559–62.

Stephanopoulos GN, Aristidou AA and Nielsen J (1998). *Metabolic Engineering: Principles and Methodologies*, Chapter 6: Examples of pathway manipulations. Academic Press, San Diego.

Sticklen MB (2008). Plant genetic engineering for biofuel production: towards affordable cellulosic ethanol. *Nature Rev Genet*, **9**, 433–43.

Subramaniam R, Dufreche S, Zappi M and Bajpai R (2010). Microbial lipids from renewable resources: production and characterization. *J Ind Microbiol Biotechnol*, **37**, 1271–87.

Szczęsna-Antczak M, Antczak T, Piotrowicz-Wasiak M, Rzyska M, Binkowska N and Bielecki S (2006). Relationships between lipases and lipids in mycelia of two *Mucor* strains. *Enzyme Microb Technol*, **39**, 1214–22.

Takakuwa N and Saito K (2010). Conversion of beet molasses and cheese whey into fatty acid methyl esters by the yeast *Cryptococcus curvatus*. *J Oleo Sci*, **59**, 255–60.

Takeno S, Sakuradani E, Tomi A, Inohara-Ochiai M, Kawashima H, *et al.* (2005). Improvement of the fatty acid composition of an oil-producing filamentous fungus, *Mortierella alpina* 1S-4, through RNA interference with Δ12-desaturase gene expression. *Appl Environ Microbiol*, **71**, 5124–8.

Vicente G, Coteron A, Martínez M and Aracil J (1998). Application of the factorial design of experiments and response surface methodology to optimize biodiesel production. *Ind Crops Prod*, **8**, 29–35.

Vicente G, Bautista LF, Rodríguez R, Gutiérrez FJ, Sadaba I, *et al.* (2009). Biodiesel production from biomass of an oleaginous fungus. *Biochem Eng J*, **48**, 22–7.

Vicente G, Bautista LF, Rodríguez R, Gutiérrez FJ, Martínez V, *et al.* (2010). Direct transformation of fungal biomass from submerged cultures into biodiesel. *Energy Fuels*, **24**, 3173–8.

Wynn JP, Hamid AA and Ratledge C (1999). The role of malic enzyme in the regulation of lipid accumulation in filamentous fungi. *Microbiol*, **145**, 1911–17.

Xue F, Zhang X, Luo H and Tan T (2006). A new method for preparing raw material for biodiesel production. *Proc Biochem*, **41**, 1699–1702.

Xue FY, Gao B, Zhu Y, Zhang X, Feng W and Tan T (2010). Pilot-scale production of microbial lipid using starch wastewater as raw material. *Biores Technol*, **101**, 6092–5.

Yazdani SS and Gonzalez R (2007). Anaerobic fermentation of glycerol: a path to economic viability for the biofuels industry. *Curr Opin Biotechnol*, **18**, 213–19.

Ykema A, Verbree EC, van Verseveld HW and Smit H (1986). Mathematical modelling of lipid production by oleaginous yeasts in continuous cultures. *Anton Leeuw*, **52**, 491–506.

Zhang Y, Adams IP and Ratledge C (2007). Malic enzyme: the controlling activity for lipid production? Overexpression of malic enzyme in *Mucor circinelloides* leads to a 2.5-fold increase in lipid accumulation. *Microbiol*, **153**, 2013–25.

Zhu LY, Zong MH and Wu H (2008). Efficient lipid production with *Trichosporon fermentans* and its use for biodiesel preparation. *Biores Technol*, **99**, 7881–5.

Zhu M, Zhou PP and Yu LJ (2002). Extraction of lipids from *Mortirella alpina* and enrichment of arachidonic acid from fungal lipids. *Biores Technol*, **84**, 93–5.

9

Microalgae processing for biodiesel production

T. M. MATA and A. A. MARTINS, Faculty of Engineering of the University of Porto (FEUP), Portugal and N. S. CAETANO, School of Engineering of the Polytechnic Institute of Porto (IPP), Portugal

Abstract: Microalgae are increasingly seen as a potential alternative to traditional feedstocks for biodiesel, which are limited and may have economic, social and environmental impacts. However, processing microalgae for biofuels is remarkably different and poses significant challenges in ensuring that they are competitive when compared with the other feedstocks. This article describes and critically analyses the main aspects and methods that can be used for the downstream processing of microalgae for biodiesel production. A brief analysis is made of the current and potential biodiesel production processes from microalgae, focusing on their main advantages and problems. The biorefinery concept is also discussed, that is, how microalgae biomass can be used to produce not just biodiesel but also other biofuels and/or high value bio-products and what will enable the economic sustainability of the whole process to be increased.

Key words: biodiesel, biofuels, biorefinery, downstream processing, microalgae.

9.1 Introduction

Currently used feedstocks for biodiesel production face some serious problems that limit their capacity to become an effective substitute for fossil fuels in transportation. For instance the utilization of edible vegetable oils (e.g. palm oil or soybean oil), also used for human consumption, may lead to food price increases and even to food shortages. Moreover, the utilization of arable land will also increase pressure on the food supply and may lead to biodiversity loss, even for non-edible vegetable oils such as jatropha or hemp. Moreover, as the traditional feedstocks are based on annual crops, problems in storage and potential shortages of raw materials in bad agricultural years pose additional threats to current biodiesel production systems.

Therefore, there is increasing interest in the development and exploration of new feedstocks without the problems stated above, particularly not directly linked to the human food chain, and that may contribute to a lower environmental impact. Various possibilities are being devised right now, such as the processing of residual fats from the meat and/or fish industries, the direct processing of residual lignocellulosic materials for bioethanol production or extraction of microalgae oil for biodiesel production, among other possibilities.

The idea of producing biodiesel from microalgae took shape in the NREL project (Sheehan *et al.*, 1998), following the observation that under certain conditions they accumulate high amounts of lipids. Since then, a lot of research has focused on developing viable cultivation systems in an economic and environmental way, in particular in the identification of the optimal growth conditions that will ensure high oil content and the best methods for microalgae biomass processing. Owing to current concerns about global warming, the increase in fossil fuels consumption and the rising price of petroleum, the same idea continues to attract a lot of attention. In fact, microalgae are currently seen as a viable alternative feedstock for the production of biofuels and other bio-products.

There are already a modest number of companies exploring microalgae but investigation is mainly for the food markets (Torrey, 2008), where the final product is the algae itself, formulated into powders or pellets for final consumption (e.g. in functional foods, nutraceuticals, cosmetics, pharmaceuticals). Exploitation of microalgae for biofuels or combined applications for biofuels production and CO_2 mitigation are still under research and development (R&D) (Carlsson *et al.*, 2007).

When compared to currently dominant feedstocks for biodiesel production, microalgae can grow and be harvested almost continuously, reducing the problems linked with the seasonality of raw materials supply for the biofuels industry. Other significant aspects include the potential high productivity of microalgae biomass and lipids and the possibility of using waste materials as nutrients, such as wastewater and/or CO_2 flue gas emissions, which will contribute to CO_2 abatement (Mata *et al.*, 2010).

Besides lipids, microalgae biomass is also a source of proteins and carbohydrates (Satyanarayana *et al.*, 2011). Additionally, they contain a variety of other bio-compounds with high commercial value, some of them extensively used in the cosmetics and pharmaceutical industries (e.g. astaxanthin and long-chain polyunsaturated fatty acids, especially of ω-3 and ω-6 series). However, extensive investment in R&D is still needed in order to implement on an industrial scale the most effective, environmentally benign and economically sound production technologies, in particular for the extraction of these bio-compounds (Mata *et al.*, 2010).

Hence, this chapter looks at the current and future aspects relevant to the downstream processing of microalgae biomass for biodiesel production but also considering other potential products, with a brief overview of how microalgae processing can be integrated and will be valuable for future bio-refineries. A particular focus is given to lipids extraction, as it is a key step in biomass downstream processing. The topics of microalgae cultivation and growth, biodiesel production and its final usage and also of other potential chemicals are not discussed.

9.2 Microalgae biomass downstream processing

In the last few years R&D concerning microalgae paid much attention to strain development and bioreactor design, but their biomass downstream processing

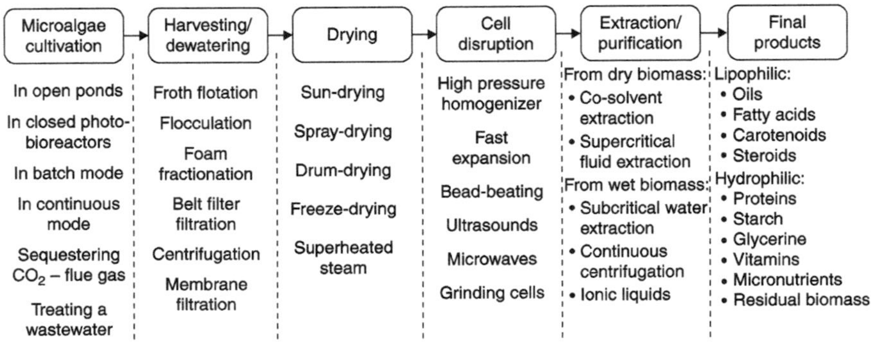

9.1 Microalgae biomass downstream processing.

into biofuels has been mostly neglected. In fact, the subjects of harvesting, dewatering and lipids extraction from microalgae are quite new and immature in terms of their application on an industrial scale, in sharp contrast to their practical relevance.

When aiming to produce biodiesel on a commercial and/or industrial scale, a very large volume of both biomass and culture medium must be processed. Also, the algal biomass separation and concentration should ensure an efficient lipid extraction process. The overall biomass downstream processing sequence is represented in Fig. 9.1, showing the main steps needed to obtain lipids from microalgae. The cultivation system is included in the figure for the sake of completeness, as some of the processing steps strongly depend on the physical characteristics of the microalgae. In the following sub-sections some of these downstream processing steps are analyzed in detail. Particular attention is given to the various methods proposed and/or available for lipids extraction.

9.2.1 Harvesting and dewatering

Microalgae cells are typically small, in the range of micrometers to tens of micrometers, making their separation from the culture media more challenging. Generally, harvesting is difficult and expensive since large volumes of water are handled (the cell density may be less than 0.5 g dry cell/L), posing a huge operation cost for dewatering, thereby rendering microalgae-based fuels usually less economically attractive. Harvesting of microalgae biomass may contribute 20–30% to the total cost of producing it (Grima *et al.*, 2003). Also, the dilute nature of the harvested microalgal cultures (with typically more than 99.6% water on a weight basis) makes dewatering a very important step.

Harvesting normally involves two processes: bulk harvesting and thickening. Bulk harvesting is a large-scale operation whose purpose is to separate biomass from the bulk culture (e.g. through flocculation and flotation). The thickening process follows and may use either centrifugation or filtration.

Several options are available for harvesting microalgae, including concentration through flocculation with adequate chemicals (e.g. aluminium sulphate and ferric chloride) or polymers to aggregate the microalgae cells (Knuckey *et al.*, 2006), electro-coagulation/flocculation (Poelman *et al.*, 1997), followed by centrifugation of the resulting mass (Heasman *et al.*, 2000), foam fractionation (Csordas and Wang, 2004), membrane filtration (Rossignol *et al.*, 2000), ultrasonic separation (Bosma *et al.*, 2003), froth flotation (Levin *et al.*, 1962; Koopman and Lincoln, 1983), three-phase continuous centrifugation (Benemann and Oswald, 1996), or a combination of these and other methods (Danquah *et al.*, 2009; Guelcher and Kanel, 1999; Howard *et al.*, 2008; Koopman and Lincoln, 1983; Lee *et al.*, 1998; Levin *et al.*, 1962; Oh *et al.*, 2001; Shelef *et al.*, 1984; Sim *et al.*, 1988). The interruption of the carbon dioxide supply to an algal system can also cause algae to flocculate on their own, which is called 'auto-flocculation'.

For large scale operations, harvesting by chemical flocculation is too expensive and if the water is brackish or salty it requires additional chemical flocculants to induce it. Other possibilities were proposed in the literature with a mixed degree of success. For example, Levin *et al.* (1962) developed an efficient froth flotation procedure for harvesting algae from dilute suspensions without addition of flotants. Harvesting is carried out in a long column containing the feed solution which is aerated from below. A stable column of foam is produced and harvested from a side arm near the top of the column.

Sim *et al.* (1988) investigated three different techniques for harvesting microalgae in pilot scale ponds, namely centrifugation, chemical flocculation followed by flotation and continuous filtration with a fine-weave belt filter. These authors concluded that none of these processes was completely satisfactory. Centrifugation gives good recovery and thickened slurry, but it is energy intensive and the capital investment is high. Chemical flocculation is more economical but the use of toxic flocculants hinders the incorporation of the final products/residuals into animal feed. Continuous filtration is more energy efficient, economical and chemical-free, but the size and algae morphology may be a problem.

Koopman and Lincoln (1983) proposed the auto-flotation of algae by photosynthetically produced dissolved oxygen as a rapid and effective harvesting technique, especially when used in conjunction with chemical flocculation. Danquah *et al.* (2009) studied the influence of microalgal growth phases and intercellular interactions during cultivation on dewatering efficiency of microalgae cultures, concluding that microalgal cultures harvested during a low growth rate phase allowed a higher rate of settling than those harvested during a high growth rate phase. Lee *et al.* (2010b) studied microbial flocculation, by incorporating a baffled hydraulic flocculator into a large scale system, as a technique for microalgae harvesting and made an estimation of the costs and energy required to perform it.

After harvesting and dewatering, the microalgae biomass still has about 80–85% water (on a mass basis) (Cooney *et al.*, 2009). Thus, an additional drying step may be required to drive off the remaining water and leave the biomass at a minimum of 99% (w/w) suspended solids.

9.2.2 Drying of biomass

Drying biomass can increase its shelf life of the final product and may be accomplished via several methods such as spray-drying (Leach *et al.*, 1998), convective drying (Desmorieux and Decaen, 2005), drum-drying, fluidized bed drying, freeze-drying (Cordero and Voltolina, 1997), refractance window dehydration technology (Nindo and Tang, 2007), low-pressure shelf drying and sun-drying (Prakash *et al.*, 1997).

Although cheaper, sun-drying is not a very effective method, since it requires a large drying surface, takes a long drying time and risks the loss of some bioreactive products (Li *et al.*, 2008). The other drying techniques can be more efficient but are not economically feasible for low value products, such as biodiesel (Mata *et al.*, 2010). Also, when microalgae are used to produce other biofuels, such as the fermentative biogas, complete drying at high temperatures for the substrates concentration should be avoided since the biogas production potential decreases significantly (Mussgnug *et al.*, 2010). Moreover, not all the applications require drying of microalgae biomass, as shown by Xu *et al.* (2011) who assess a dry and a wet route for biofuels production from microalgae, concluding that both the drying process in the dry route and the oil extraction process in the wet route consume a significant amount of energy. Xu *et al.* (2011) also show that the wet route has more potential to produce high valuable biofuels and the dry route has a higher fossil energy ratio (FER) which is defined as the ratio between the amount of energy that goes into the final fuel product (fuel energy output) and the amount of fossil energy input (non-renewable energy) required for fuel production. Therefore, there is a trade-off one should consider between drying efficiency and cost-effectiveness to maximize the net energy output of the microalgae biofuels (Mata *et al.*, 2011). The dried biomass can then be subjected to various methods to extract the lipids as preparation for its conversion into biodiesel.

9.3 Cell disruption methods

After drying there is a need to promote the disruption of the microalgae cell to release the metabolites of interest. This normally precedes extraction and consists of a prior step to disrupt the microalgae cell membrane physically in order to increase the rate at which the extracting agent contacts the target components.

Several methods can be used for cell disruption depending on the characteristics of the microalgae wall and on the nature of the product to be obtained either based on mechanical action (e.g. cell homogenizers, bead mills, ultrasounds, autoclave and spray drying) or non-mechanical action (e.g. freezing, organic solvents and osmotic shock and acid, base and enzyme reactions) (Middleberg, 1995).

Cell disruption processes include ultrasound and microwave assisted extractions (Cravotto *et al.*, 2008), autoclaving (Mendes-Pinto *et al.*, 2001), bead-beating (Lee *et al.*, 2010a), grinding cells while frozen in liquid nitrogen (Pernet and Tremblay, 2003), high-pressure homogenization (Kleinig and Middelberg, 1998), enzymatic hydrolysis (You *et al.*, 2011), electroporation or electropermeabilization (Sommerfeld *et al.*, 2010), three-phase continuous centrifugation (Benemann and Oswald, 1996), osmotic shock (Shepherd and Beilby, 1999) among other methods or combinations of these. Some of these methods are briefly discussed in the next sub-sections.

9.3.1 Ultrasound-assisted extraction (UAE) and microwave-assisted extraction (MAE)

MAE is a simple technique used for isolation of essential oils from plants that is based on a basic physical principle, namely the different ability of the species being subjected to the microwave irradiation to absorb microwave energy (measured as dielectric constant) depending on its chemical nature (Luque de Castro *et al.*, 1999).

UAE is a technique that uses ultrasonic waves to create cavitation bubbles in a solvent material. The sound waves propagate into the liquid media resulting in alternating high-pressure (compression) and low-pressure (rarefaction) cycles, with rates depending on the frequency (Luque de Castro *et al.*, 1999). During the low-pressure cycle, high-intensity ultrasonic waves create small vacuum bubbles in the liquid. When the bubbles collapse violently near the cell walls, during a high-pressure cycle (phenomenon known as cavitation), the resulting shear forces break the cell envelope causing the release of their contents into the solvent.

Cravotto *et al.* (2008) employed both UAE and MAE as complementary techniques to disrupt the *Crypthecodinium cohnii* microalgae cell wall prior to the application of solvent extraction techniques. Their results showed that ultrasounds worked better than microwaves to disrupt the tough algal cell wall, considerably improving the extraction yield from 4.8% (in Soxhlet) to 25.9%. Moreover, the use of ultrasound and microwaves, either alone or combined, can greatly improve the extraction of bioactive substances from microalgae at shorter reaction times and with much lower (up to 10 times) solvent consumption. As advantages, these techniques are more efficient, economical and less toxic than the co-solvents extraction approaches.

9.3.2 Ultrasonication and grinding

Pernet and Tremblay (2003) studied the use of ultrasonication, grinding and a combination of these two methods to facilitate the access of extracting solvent to the algal lipids of the diatom *Chaetoceros gracilis*.

Widjaja *et al.* (2009) studied the effect of CO_2 concentration, nitrogen depletion, harvesting time and extraction method on lipid production from freshwater microalgae *Chlorella vulgaris*. The best results were obtained with drying at very low temperature under vacuum, while drying at higher temperature decreased the lipid content. Ultrasonication did not affect the lipid content and the extraction time, as long as enough pulverization was applied to dried algae. Also, the highest lipid productivity was obtained during nitrogen depletion and at low to moderate CO_2 concentration.

9.3.3 Ultrasounds and solvents extraction

Burja *et al.* (2007) assessed various methods for their capacity to extract fatty acids from dried biomass of *Thraustochytrium sp.* ONC-T18, concluding that the combination of ultrasound with solvents (methanol:chloroform) is the most effective for breaking down the cell walls.

9.3.4 Mechanical homogenizer

Kleinig and Middelberg (1998) studied the tensions produced in the rigid wall of a yeast cell passing through a typical high-pressure homogenizer and determined the force required to break the yeast cells under compressive micromanipulation. Results show that inertial forces are an important mechanism for microbial cell disruption and can be used to improve cell wall breakage processes.

9.3.5 Enzymatic hydrolysis

Some red or green seaweeds are potential sources of proteins for commercialization, containing up to 20% protein (dry weight) (Fleurence, 1999). However, the microalgae cell wall is made up of anionic or neutral polysaccharides in large quantities which strongly hinders the solubilization of proteins during the application of classical extraction procedures. Fleurence (1999) discussed the use of enzymes for microalgae cell wall degradation in order to facilitate the extraction and solubilization of algal proteins.

Sander and Murthy (2009) investigated the use of enzymes to convert the cell wall cellulose to glucose and the membrane phospholipids to fatty acids and glycerol. They used cellulase and lipase enzymes for the degradation of the *Chlorella vulgaris* cell and membrane. The fermentation experiment resulted in glycerol formation when incubated with the cellulases and phospholipases. Glucose was formed when the pretreated mixed culture was incubated with cellulases but the yield in ethanol of the glucose after fermentation with yeasts was very low.

Wang *et al.* (2010), studied the effect of various protease and carbohydrase treatments on the extraction of polyphenols and other antioxidant ingredients from the red algae *Palmaria palmate* and concluded that all of the proteases tested significantly enhanced the extraction of polyphenols and other active components, when compared to carbohydrase, showing great potential for application in pulse extraction as antioxidants in functional foods and the nutraceuticals industry.

You *et al.* (2011) investigated the use of six different enzymes (papain, pectinase, snailase, neutrase, alcalase and cellulase) to extract lipids from *Mortierella alpina*. Their results showed that the lipid composition is not affected by the enzyme treatments according to the fatty acids profile.

9.3.6 Autoclaving, enzymatic hydrolysis, mechanical homogenizer

Mendes-Pinto *et al.* (2001) tested a range of physical and chemical processes to promote the disruption of the encysted cells of *Haematococcus pluvialis* in order to extract the carotenoid astaxanthin as a pigment used in the aquaculture industry. The processes tested in this study include autoclaving for 30 min at 121°C and 1 atm; acidification with HCl 0.1 M for 15 min and 30 min; alkalinization with NaOH 0.1 M for 15 min and 30 min; enzymatic treatment with a mixture of 0.1% protease K and 0.5% driselase in a phosphate buffer at pH 5.8 and 30°C, for one hour; spray drying, inlet 180°C, outlet 115°C; and mechanical disruption, with a cell homogenizer developed for this purpose. The results demonstrated that mechanical (homogenization) and autoclave treatments are the most effective in terms of their efficacy for astaxanthin recovery.

In a study of *Spirulina platensis* composition in phycocyanins, Boussiba and Richmond (1979) reported that total cell disruption was achieved after 3 min of the mechanical homogenizer operation, while the enzymatic procedure using lysozyme needed 24 h to accomplish disintegration of the cell wall.

9.3.7 Autoclaving, bead-beating, microwaves, sonication

Lee *et al.* (2010a) tested various cell disruption methods to identify the most effective for lipids extraction from *Botryococcus* sp., *Chlorella vulgaris* and *Scenedesmus* sp. Therefore these authors tested autoclaving, bead-beating, microwaves and sonication concluding that the microwave oven method for cells disruption is the most simple, easy and effective for lipid extraction from microalgae. Also, the highest oleic acid productivity was obtained for *Botryococcus* sp.

9.3.8 Quartz sand and solvents addition and ultrasound extraction

Wiltshire *et al.* (2000) examined the efficiency of pigment and fatty acid extraction from resistant algae such as *Scenedesmus obliquus*, concluding that by adding quartz sand and solvent to freeze-dried algal material with subsequent extraction in an ultrasound bath for 90 min at 4°C, they could extract over 90% of the different substances in just one step. Furthermore, these authors compared this extraction method with the addition of solvents to algal material with subsequent incubation, concluding that the first method, using ultrasound and sand, has about twice the extraction efficiency of the second one. Also, it is a simpler method and conserves the pigments and fatty acids present in the biomass, without alteration or breakdown of products. Moreover, this method is effective for a wide spectrum of algae, such as, *Cryptomonaserosa* (*Cryptophyceae*), *Cyclotella meneghiniana* (*Bacillariophyceae*), *Microcystis aeruginosa* (*Cyanophyceae*) and *Staurastrum paradoxum* (*Chlorophyceae, Desmidiaceae*).

9.3.9 Electroporation or electropermeabilization

Sommerfeld (2010) found that electroporation or electropermeabilization altered the cellular membranes and algal cell walls, thus improving the lipid extraction efficiency in terms of time and solvent use without affecting the composition of the extracted fatty acids.

9.3.10 Mechanical homogenizer and three-phase continuous centrifugation

Benemann and Oswald (1996) proposed the use of three-phase continuous centrifugation not only to concentrate the biomass but also to extract the lipids from microalgae in one single path. In this process three fractions are continuously separated: the lipid stream, the organic and protein rich water stream and the biomass cake. Before centrifugation, biomass is heated to about 60–70°C (by adding some hot oil recycled from the oil extraction step) and then passed directly through a homogenizer (or pressure expansion valve) which disrupts the cell walls and the internal lipid tissues forming a thick emulsion. The high temperatures, the solubility of the algal lipids in the same oil added for the emulsification and the shear forces required for homogenization, all favour the extraction.

Since centrifugation is too costly for producing lower-value products, such as extracting algal oils for biofuels applications, the optimal lipid composition for this process to be applied and to compensate for the operation costs must be about 35–40% by dry weight (Benemann and Oswald, 1996). This is higher than the lipid content found in many microalgae grown with nitrogen sufficiency (i.e. normally having less than 30% lipid). Therefore, the lipid productivity of microalgae necessarily has to be maximized, for example by changing factors such as temperature, light intensity and nitrogen supply (Courchesne *et al.*, 2009). Even so, the volume of lipids extracted using this method is relatively small when compared to the water volume being processed, which makes it difficult for reasonable recoveries to be achieved.

The next section revises and discusses the existing methods for lipids extraction from microalgae. Also, it identifies gaps and the missing information regarding potential pathways leading to process efficient extraction methodologies, having in mind their potential application on an industrial scale.

9.4 Lipids extraction from microalgae

In summary, microalgae cells are composed of lipid molecules together with proteins and carbohydrates. In particular, lipids are defined as any cellular compounds extractable in an organic solvent (e.g. chloroform, hexane).There are both polar and non-polar lipids. Polar lipids are the phospholipids and glycolipids that are functionally associated with the membrane structure and fluidity. Non-polar lipids are the mono, di and triglycerides, but also include certain pigments, vitamins, cyclic and acyclic hydrocarbons, wax esters and sterols. The saponifiable lipids represent about 30–40% of the total lipids and the free fatty acids (FFA) are also present at about 1–5% of the total lipid fraction (Benemann and Oswald, 1996).

In order to extract the desired bio-components (e.g. lipids) from microalgae, traditional extraction methods are of little use. For example, some solvent extraction or *in situ* biodiesel production is hindered by the presence of high amounts of bulk water and significant cell wall barriers in microalgae (Ehimen *et al.*, 2010). Although simpler, mechanical pressing cannot be applied effectively to microalgae because of their water content and small cell size, so that instead of being crushed, would flow with the water through the gaps that are always present in the pressing equipment. Moreover, mechanical crushing requires microalgae to have been dried previously and then pressed out. The associated energy costs would make this process not viable economically.

Depending on the cultivated species (or combination of species) it may be desirable to adapt the extraction technologies to the microalgae characteristics, that is, to develop a species-specific extraction process. A significant aspect to consider here is the presence, or not, of a cell wall and its composition, strength and structure when it exists. If, for example, a microalgae lacks a cell wall (as in the case of *Dunaliella salina*) it is more susceptible to shear breakage than other microalgae that contain a rigid cell wall, which has to be broken down in order to allow access to the lipids. These differences will have an impact on the applicability of the extraction method, its efficiency and the rate at which the target components can be extracted. Also, depending on the reactor type used to cultivate microalgae (e.g. if it is an open pond or a closed photo-bioreactor), one has to consider if there is dominance of the desired microalgae species or if any contamination exists that may preclude the possibility of using species-specific extraction processes. Therefore, the identification of an appropriate extraction technology heavily depends on the microalgal species and type of cultivation system.

Furthermore, when designing a solvent extraction system, the choice of solvents must take into consideration the components to be extracted. For example, non-polar solvents are excellent for non-polar organics, while for polar organics their ability to extract is often poor (Schafer, 1998). The amount and sequence in which solvents are added to the biomass can also affect the extraction efficiency (Lewis *et al.*, 2000). Moreover, depending on the microalgae species present in the medium, or the existence of contamination, it may be desirable to develop species-specific extraction processes adapted to their characteristics. Additionally, the toxicity, environmental impact and cost of the solvent(s) are also other aspects that need to be carefully considered.

9.4.1 Co-solvents lipid extraction systems

Studies have shown that the conventional Soxhlet extraction (e.g. using n-hexane as solvent) is not adequate for the microalgae lipid extraction (Richter *et al.*, 1996; Schafer, 1998) as they are binary co-solvent systems such as chloroform/methanol/water (Bligh and Dyer, 1959), hexane/isopropanol (Hara and Radin, 1978), hexane/ethanol (Grima *et al.*, 1994), hexane/2-propanol (Nagle and Lemke, 1990) and isopropanol/chloroform (Sobczuk and Chisti, 2010).

The co-solvent extraction systems are based on the concept 'like dissolves like', where the membrane-associated polar lipids require polar solvents (e.g. methanol) to disrupt the bonds between lipids and proteins, while lipids that are largely hydrophobic will favourably interact with the non-polar solvent molecules (e.g. chloroform). Also, these extraction systems have been promoted because an organic solvent (e.g. hexane, chloroform) and an alcohol (e.g. methanol, ethanol, isopropanol) will readily separate into two separate phases when water is added, thereby improving downstream separations (Cooney *et al.*, 2009). Some possibilities are presented below.

The Bligh and Dyer co-solvent system

The most well-known and currently the best method for total lipid quantification in marine biological tissues is the Bligh and Dyer co-solvent system (Bligh and Dyer, 1959; Grima *et al.*, 1994). In applying this method, the lipid-containing tissues (analyte) are exposed to a miscible co-solvent mixture comprising an alcohol (methanol) and an organic solvent (chloroform). More specifically, a mixture of chloroform/methanol/water in the ratios of 1:2:0.8 and 2:2:1.8 is used before and after dilution, respectively.

After extraction the homogenate obtained is diluted with water to produce a biphasic system, wherein water and chloroform separate into two immiscible phases. At this point, methanol self-associates with water to form the water/methanol phase containing the non-lipids. In terms of polarity methanol is more 'like' water and their molecular interactions (i.e. electrostatic, London forces, hydrogen bonds, hydrophobic bonding) are stronger than they are between methanol and chloroform molecules. By contrast, the lipid molecules are more 'like' chloroform (i.e. their interactions are stronger than between lipids and the water/methanol molecules) and they accumulate in the chloroform phase. After isolating the chloroform layer, the chloroform is evaporated and what remains is the purified lipid extract. This is done in a pre-weighed container to determine the exact total lipid concentration by gravimetry.

Although effective, the method selectivity is low as the organic solvent (chloroform) can extract more than just lipids (e.g. pigments and other contaminants) and for some types of lipid tissues (e.g. in fish tissues) it may underestimate the true lipid content (Iverson *et al.*, 2001). Furthermore, this method is difficult to scale up and has the problem of chloroform and methanol toxicities. Consequently, less volatile and less toxic solvents such as ethanol, 1-butanol, or isopropanol have been used by other authors in place of methanol (Nagle and Lemke, 1990; Grima *et al.*, 1994; Sobczuk and Chisti, 2010).

Other co-solvent approaches

Hara and Radin (1978) proposed the use of hexane/isopropanol co-solvent combination for extracting lipids from tissues, concluding this has advantages over the Bligh and Dyer method as these solvents are less toxic and cheaper, there is no interference in processing or contamination by proteolipid proteins, the two phases separate rapidly during the washing step, the solvent density is low enough to permit centrifugation of the homogenate as an alternative to filtration and the washed extract can be applied to a chromatographic column.

Nagle and Lemke (1990) evaluated the efficiency of three solvents for extracting lipids from microalgae *Monoraphidium minutum* and *Chaetoceros muelleri*, concluding that the most efficient solvent is 1-butanol (90% efficiency), followed by hexane/2-propanol and then by ethanol.

Grima *et al.* (1994) compared seven solvent mixtures for extracting the lipid fraction of lyophilized biomass of *Isochrysis galbana* and obtained the highest lipids yield (of 93.8%) with the Bligh and Dyer method, followed by ethanol 96% and a hexane/ethanol 96% mixture, in the ratio 1/2.5 (v/v) that resulted in 84.4% and 79.6% lipids yield, respectively.

Sobczuk and Chisti (2010) extracted lipids from microalgae *Choricystis minor* comparing two methods: (1) hot isopropanol (50–60°C) and a 1:1 v/v mixture of isopropanol and chloroform; and (2) the Bligh and Dyer solvent system, concluding that both methods performed equally and in view of this, the hot isopropanol method was preferred since it requires less chloroform and produces a monophasic solvent system that is easier to separate from the residual biomass.

Starting from freeze-dried biomass, Lewis *et al.* (2000) showed that the total amount of lipids and fatty acids extraction increased significantly (by about 30%) when solvents were added to biomass in order of increasing polarity (i.e., chloroform, methanol and then water) as opposed to decreasing polarity. This could be explained in terms of cell disruption (lysis) or disintegration prior to lipid extraction, where initial contact with the less polar co-solvents (e.g. chloroform or a mixture of chloroform and methanol) can weaken the cell structures and their association with lipids (e.g. through hydrogen bonding or electrostatic forces between membrane-bound lipids and proteins in the cell membrane), making them sufficiently permeable and easier for the second more polar solvent (e.g. water) to separate them. However, if a more polar co-solvent is added prior to the less polar co-solvents, it would form a solvent 'shell' around lipids making it more difficult to contact, solubilize and extract them using a second less polar co-solvent.

The problem of current solvent extraction techniques is that they usually require extraction conditions that are difficult to scale-up and the required solvent to biomass ratios are normally very high (Benemann and Oswald, 1996). Another significant problem is the degree to which microalgae cell walls are degraded or damaged, leading to problems in performing the separation, for example, if based upon density differences (Cooney *et al.*, 2009), or in the biodiesel production process itself. Other options were proposed in the literature, for example, ionic liquids (Young *et al.*, 2010) or the direct transesterification of glycerides (Lepage and Roy, 1984), but none has been implemented on an industrial scale.

9.4.2 Extraction at higher temperatures and pressures

Any successful extracting solvent must be able to penetrate the biomass structure and physically contact the target components. Therefore, extraction methods employing elevated temperatures and pressures can be used, such as accelerated solvent-extraction (ASE) (Richter *et al.*, 1996) or pressurized fluid extraction (PFE) (Denery *et al.*, 2004), supercritical fluid extraction (SFE) (Cheung, 1999) and subcritical water extraction (SCWE) (Eikani *et al.*, 2007). In these methods the key factors are solvent access to the material being extracted, the miscibility of the target components in the solvent and the solvent's polarity which must match as much as possible that of the components to be extracted.

The higher temperature and pressure increase their solvation power and the capacity of a particular solvent to solubilize the target components; in particular they will be more liquid-like as it is the case of SFE and reduce its surface tensions. In this way the solvent comes more rapidly into contact with the entire sample matrix, facilitating the extraction (Richter *et al.*, 1996).

Although more effective when compared to solvent methods, the effectiveness of these methods on the large scale remains unclear, in particular how they cope with large quantities of biomass and the associated energy and operating costs (Cooney *et al.*, 2009). Also, since they are generally energy intensive and the organic solvents used are expensive, these are critical issues when the final product is a commodity like biofuels that needs to compete in price with fossil fuels.

Pressurized fluid extraction (PFE) or accelerated solvent-extraction (ASE)

PFE also known as ASE (Richter *et al.*, 1996) is a potential technique for extracting particularly labile and light-sensitive chemicals such as nutraceuticals. By definition ASE uses non-aqueous solvents and is applied to dried biomass, where the solid or semi-solid sample is enclosed in a cartridge filled with an extraction solvent that statically performs the extraction at elevated temperature and pressure. Compressed inert gas is then used to purge the extract from the cell and into a collection vessel.

Pressurized fluid extraction is similar to Soxhlet extraction, except that the solvents are used near their supercritical region where they have high extraction properties. In this physical region the high temperature enables high solubility and high diffusion rates of lipid solutes in the solvent, while the high pressure keeps the solvent below its boiling point, enabling a high penetration of the solvent into the sample (Pascal and Segal, 2006). Organic solvents that are normally proposed in typical distillation based methods (e.g. Soxhlet extraction) can be used in PFE at temperatures above their boiling point (50–200°C) and high pressures (34–204 atm) for short periods of time (5–10 min) (Richter *et al.*, 1996).

For example, Richter *et al.* (1996) used ASE for lipid-extraction, reporting an extraction time of less than 15 min, without any evidence of thermal degradation of temperature-sensitive compounds. Also, Schafer (1998) applied ASE in an entire range of plant and animal tissues and compared it with the modified Folch procedure (Folch *et al.*, 1957), concluding that the fatty acid contents of cereal and yolk lipids extracted by ASE were highest when using isopropanol/hexane (2:3, v/v). However, the extraction of muscle lipids resulted in higher fatty acid contents when chloroform/methanol (2:1, v/v) was used.

Herrero *et al.* (2005) and Ibañez *et al.* (2006) used ASE for the extraction of antioxidants from the microalga *Spirulina platensis*. They tested four different solvents (hexane, petroleum ether, ethanol and water) and decided on ethanol as the extracting solvent owing to its known safety status and because it provides higher yields with medium antioxidant activities. Their work showed the possibility of using ASE as a fast and easy process to recover natural antioxidants from natural sources such as microalgae.

Mulbry *et al.* (2009) have also used ASE for the extraction of fatty acids from dried freshwater algae, mostly the filamentous green algae *Rhizoclonium hieroglyphicum*. They evaluated the efficiency of total oil and fatty acids extraction with three solvents (chloroform/methanol, isopropanol/hexane and hexane) with the ASE method and compared it to the Folch method. These authors concluded that the ASE method yielded higher values for total oil but similar values for fatty acids content and composition after four extraction cycles with chloroform/methanol, with a much higher yield in fatty acids extraction in the first cycle than using the Folch method. The extraction efficiency depends on the solvent used. Thus, fatty acids content determined using ASE with chloroform/methanol was lower than with isopropanol/hexane and the ASE procedure with hexane was the more efficient. Also, fatty acids content determined using the Folch method or ASE with chloroform/methanol did not show significant influence of sample particle size within the size range of 0.1–1 mm.

Denery *et al.* (2004) studied PFE for carotenoids extraction from *Haematococcus pluvialis* and *Dunaliella salina*, showing that under controlled temperature and pressure and in an oxygen and light-free environment, less solvent and shorter extraction times could be used to achieve higher or equal extraction efficiencies, while maintaining the integrity of chemical components, compared with traditional solvent extractions. Recent advances using ASE systems include procedures for selective removal of interferences during extraction, thus combining extraction and purification into a single step (Dionex, 2007).

Supercritical fluid extraction (SFE)

SFE is another technique, more recently applied to the extraction of essential oils from plants and lipids and other functional ingredients from microalgae (Cheung, 1999; Canela *et al.*, 2002; León, 2008). It is based on the enhanced solvating power of fluids (e.g. CO_2, water, nitrous oxide, ethane, n-butane, pentane, methanol and sulphur hexafluoride) above their critical point and on the combination of gas-like mass transfer properties and liquid-like solvating properties with diffusion coefficients greater than those of a liquid (Luque de Castro *et al.*, 1999).

However, because of supercritical CO_2 low polarity it is less effective in extracting more polar compounds from natural matrices. To overcome this, modifiers (also called co-solvents) are commonly used. Modifiers are highly polar compounds that when added in small amounts, can produce substantial changes in the solvent properties of pure supercritical CO_2 (Herrero *et al.*, 2006).

Normally, SFE is employed in batch mode but the continuous operation is also possible, though it may be difficult to scale up. Carbon dioxide is generally used as a means of extraction because of its preferred critical properties, low toxicity and chemical inertness (Luque de Castro *et al.*, 1999).

One of the main advantages of this method, after the extraction is completed, is that the product can be easily separated from the solvent just by lowering the temperature and pressure to atmospheric conditions, where the fluid returns to its original gaseous state and the extracted product remains as a liquid or solid. Also, by using SFE one avoids the use of organic solvents and increases the extraction efficiency by considerably shortening the extraction time. The main disadvantage is that SFE of lipids from microalgae requires high pressure equipment that is both expensive and energy intensive. Thus, this may increase the operating costs significantly and limit its applicability in biofuels production from microalgae for economic reasons.

Cheung (1999) concluded that the combined effect of pressure and temperature on the solubility of individual n-3 fatty acids in the supercritical carbon dioxide extraction process (SC-CO_2) varies with the carbon chain length. The concentrations of C18, C20 and C22 n-3 fatty acids, extracted under different pressure and temperature conditions, were significantly different. Proportions of total polyunsaturated fatty acids increased significantly and of total saturated fatty acids decreased significantly with increasing pressure. Nevertheless, Cheung concluded that SC-CO_2 extraction could be a potential process useful for food applications of algal lipids and *Hypneacharoides* could be an alternative non-conventional source of n-3 fatty acids.

Canela *et al.* (2002) used the supercritical fluid extraction with carbon dioxide to recover fatty acids and carotenoids from the microalgae *Spirulina maxima*. They conducted their work at temperatures of 20–70°C and pressures of 15–180 bar and concluded that neither the temperature nor the pressure significantly affected the total yield, but both the temperature and the pressure affected the extraction rate and the effect of the temperature prevailed over that of the pressure. The extracts were rich in essential fatty acids and carotenes but the latter would degrade at extraction temperatures higher than 50°C.

León (2008) used SFE with CO_2 (SC-CO_2) and 10% ethanol to recover antioxidants from *Spirulina platensis* in a pilot plant. He concluded that the higher extraction yield was obtained under low pressure (78 atm) and intermediate temperatures (55°C). But if only CO_2 was used, optimal vitamin E extraction conditions would be 361 atm and 83.3°C and the concentration of tocopherol in the extract would be of 29.4 mg g^{-1} (Mendiola *et al.*, 2008). Working with *Dunaliella salina*, León (2008) concluded that the optimal conditions to obtain a high yield in extraction of antioxidants consisted of high pressure (400 atm) and low temperature (15°C), that is, using CO_2under subcritical conditions he was able to attain high recovery yields in zeaxanthin, β-carotene, α-carotene, 13-*cis*-β-carotene, all *trans*-β-carotene, 15-*cis*-β-carotene and 9-*cis*-β-carotene. León (2008) also used SC-CO_2 to recover extracts with antimicrobial activity from *Chaetoceros muelleri* microalgae but with low yield owing to the presence of silica in the cell wall of diatoms.

Subcritical water extraction (SCWE)

The SCWE technique is based on the use of water in a dynamic mode, at temperatures just below the critical temperature (between 100 and 374°C) and a pressure high enough to maintain the liquid state. Luque de Castro *et al.* (1999) discuss the advantages and drawbacks of the use of supercritical CO_2 (SC-CO_2) extraction and especially of SCWE, as effective methods for the isolation of high-quality essential oils. More recently, SCWE has been applied to the extraction of antioxidant compounds from algae (Herrero *et al.*, 2006) and essential oils from oil seeds (Ayala and Castro, 2001; Eikani *et al.*, 2007).

As discussed above, elevated temperature and pressure facilitate solvent access to the target analytes. Also, after the extraction is completed and the temperature and pressure return to atmospheric, the water is no longer miscible with the extracted lipids and thus is easily separated from them by phase separation. Moreover, the use of SCWE can shorten the extraction time and avoid the use of toxic solvents. This method can also be directly applied to harvested microalgae eliminating the need for the dewatering step since the already present water can be used as the extracting agent. However, the major constraints of SCWE are that it is difficult to operate on the large scale and is an energy intensive process that poses difficulties in its applicability to biofuels production.

Nevertheless, Rodríguez-Meizoso *et al.* (2010) studied the influence of the extraction temperature (50, 100, 150 and 200°C) and solvent polarity on the extraction yield of compounds with antioxidant and antimicrobial activity from *Haematococcus pluvialis* microalgae in the red phase. The extraction yield achieved with this process was higher than 30% dry weight at a 200°C extraction temperature.

9.5 Advanced biodiesel production from microalgae

There are several processes for obtaining biodiesel from lipid feedstocks. The most widely used is the liquid phase transesterification of triglycerides with low molecular weight alcohols using a homogeneous acid or alkali catalyst, operated in batch mode in stirred reactors. This process has the advantages of being easy to implement, operate and control and the reactions occur under mild conditions of temperature and pressure. However, this process has several drawbacks, in particular the large reaction times needed for a complete conversion of oil, the batch operation mode and the complex separation stages after the reaction. Moreover, the transesterification reaction is very sensitive to feedstock purity, especially to water content (greater than 0.5% w/w) and to FFA content (greater than about 1% w/w), usually requiring pretreatment operations in order to avoid saponification and lower reaction yields (Canakci and Van Gerpen, 1999). Microalgae biomass has a high water content that has to be removed if the homogeneous catalyzed liquid phase transesterification process is used.

Additionally, the presence of water favours the hydrolysis of triglycerides and FFA, lowers the esters yield and renders the separation of the esters and glycerol and consequent esters purification difficult (Canakci, 2007; Aranda *et al.*, 2008). Also, the presence of water and FFA promotes the formation of soap in the presence of the alkalis catalysts, increasing catalyst consumption and diminishing its efficiency (Rice *et al.*, 1997).

Therefore, there is a need for more efficient production processes, suitable for microalgae-based biodiesel production, in order to respond to the expected increase in global demand for biodiesel, market competitiveness and need to fulfil the goals defined at governmental and regional levels. Some possibilities are being considered and studied in detail, some of them already implemented and used in industrial settings with varying degrees of success. Examples include, but are not limited to: enzymatic catalysis, reaction in supercritical fluids, reactor designs that promote improved mixing, such as static mixers, ultrasounds or cavitation, among others.

9.5.1 New and emerging processes for biodiesel production

The main goals of the new and emerging processes for biodiesel production are reduction in the reaction time, the possibility of operating a continuous model, which is essential to increase the production capacity to meet the expected demand for biodiesel, and reduced operating costs, in particular the processing of biomass for microalgae biodiesel production. Examples of processes being studied for the production of microalgae biodiesel include:

- single step extraction and conversion of microalgae lipids to biodiesel (Lepage and Roy, 1984; Wahlen *et al.*, 2011; Abou-Shanab *et al.*, 2011);
- *in situ* transesterification of microalgae lipids using co-solvents (Xu, 2010);
- homogeneous acid-catalyzed *in situ* transesterification of microalgae lipids (Ehimen *et al.* (2010);
- heterogeneous catalyzed continuous process using supercritical methanol (Krohn *et al.*, 2011; Patil *et al.*, 2011a);
- ultrasounds and microwave-assisted transesterification of microalgal biomass (Patil *et al.*, 2011b; Koberg *et al.*, 2011).

Lepage and Roy (1984) were the first to propose the direct transesterification method for the quantitative measurement of fatty acids, showing in their studies recoveries of fatty acid and triglyceride standards better than 96%. The main advantages of the direct transesterification process is that it is a one-step reaction (including extraction and conversion of microalgae lipids to biodiesel) carried out in the reactor vessel, without the extraction and purification steps. It makes possible effective production of fatty acid methyl esters from polar phospholipids (a key component of cellular walls), free fatty acids and triglycerides. Also, this process can be performed under milder conditions than, for example, those required for the supercritical methanol process, and it prevents the formation of by-products.

In their work, Whalen *et al.* (2011) applied the single step extraction and conversion method to produce biodiesel from microalgae and cyanobacteria lipids, showing that higher conversions of fatty acids and lipids can be obtained, even from phospholipids. These authors also studied the effect of varying the concentration of sulphuric acid catalyst on the FAME yield using the *in situ* transesterification method, showing that it has a modest effect on the biodiesel production.

Abou-Shanab *et al.* (2011) extracted the total lipids from the fresh microalgal biomass using a modified method of Bligh and Dyer (1959) and then applied a modified method of Lepage and Roy (1984) to the crude lipid extracted to perform direct transesterification. These authors also studied the effect of catalyst concentration on fatty acid methyl ester yield. The results of their study showed that the FAME yield increased from a minimum of 19.4% cellular dry weight for a 5 min reaction time to a maximum of 28.2% of cellular dry weight with 20 min reaction time. Increasing the reaction time beyond 20 min did not affect the FAME total yield.

Furthermore, when adding a co-solvent to the transesterification reaction system a single-phase is obtained, while without it the oil and methanol would be immiscible and form two phases, complicating the advancement of the reaction. In this regard, Xu (2010) studied the effect of various co-solvents (toluene, diethyl ether and dichloromethane) on the *in situ* transesterification of microalgal lipids. The results show that toluene has the best performance, resulting in higher biodiesel yield (42.60% w/w) than using dichloromethane (30.28% w/w) and diethyl ether (27.87% w/w).

The *in situ* transesterification of microalgae lipids can be homogeneously acid-catalyzed as demonstrated by Ehimen *et al.* (2010). In particular, these authors have shown that significant inhibition occurs only when the water content of the algal biomass is greater than 115% w/w. Also, stirring the reaction vessel and increasing the reacting alcohol volume and temperature lead to improved biodiesel conversion.

In contrast, it is also possible to use heterogeneous catalysts for microalgae biodiesel production, as was demonstrated by Krohn *et al.* (2011). These authors have shown the applicability and effectiveness of a heterogeneous catalyst (porous titania microspheres) in a fixed bed reactor operating in continuous mode, to catalyze the simultaneous transesterification and esterification of triglycerides and FFA, respectively, showing 85% efficiency.

Patil *et al.* (2011a) demonstrated the feasibility of the one-step process for direct liquefaction and conversion of wet algal biomass (containing about 90% of water) to biodiesel under supercritical methanol conditions. Various authors investigated the effect of microwave irradiation on the simultaneous extraction and transesterification (*in situ* transesterification) of dry algal biomass to biodiesel (Patil *et al.*, 2011b; Koberg *et al.*, 2011). Others applied microwave and ultrasound radiation together with the aid of a SrO catalyst for the direct transesterification of the *Nannochloropsis* biomass to biodiesel, concluding that the microwave oven method appears to be the most simple and efficient method for this purpose (Koberg *et al.*, 2011).

Although studies seem to indicate the feasibility of the new production processes, at least from a laboratory and pilot project point of view, work on the implementation on the industrial level is still lacking. Aspects that need to be further studied include: better processes to purify biodiesel in order to meet the regulations, better ways to incorporate biomass processing steps with the new biodiesel production processes and a full economic analysis of the new processes.

9.6 Process integration: the biorefinery concept

As already pointed out above, downstream processing of microalgae biomass may result not just in lipids suitable for the production of biodiesel but also in a wide variety of other chemicals, from simple base chemicals to other high value chemicals used for example in the pharmaceutical or nutraceutical industries. On the other hand, most of the described processing methods result in the biomass cake from lipids extraction that can be used for example for the production of bioethanol through hydrolysis and fermentation, or to generate power that can be used in the biodiesel production process itself, among other possibilities.

This situation shares similarities with the petrochemical industry, where crude oil is the basis for the production of not only fuels, but also other basic chemicals that are used in a multitude of processes and industries. This is one of the main reasons why crude oil is such a fundamental resource for modern society's needs and its development in the social, economic and even environmental spheres. In a similar fashion some authors have proposed the idea of biorefineries (Kamm *et al.*, 2006; Mussgnugetal., 2010; Demirbas and Demirbas, 2010; Gouveia, 2011), that is, a multi-product system adapted to the available raw materials of biological origin, for example, from agricultural origin, either crops specifically grown for it or residue such as straw (Annets and Audsley, 2003).The production of high value bio-products from microalgae is also a way of improving the profitability of the biodiesel production process (Li *et al.*, 2008).

The implementation of systems of this type strongly depends on their economic feasibility. Thus, a full analysis must contain all the stages associated with the production of the requested chemicals, starting from cultivation, harvesting, transportation, storage, biomass processing and chemicals production. Even though the price of raw materials is still considered the main factor affecting the profitability of existing and future biorefineries, their viability is also strongly dependent on how efficiently the production process can be operated.

Concerning microalgae, there is potential to obtain biomass on an almost continuous basis with high biomass productivities per unit area. Also, the residual materials contain many interesting chemicals for a wide array of practical applications (Cardozo *et al.*, 2007; Satyanarayana *et al.*, 2011). Figures 9.2 and 9.3 show, respectively, the large variety of potential compounds and several of the biofuel options that can be obtained from microalgae.

Regardless of the desired final products, biomass processing is a key step as shown in Fig. 9.4. Only the right selection of processing/separation processes will ensure efficient production of the required chemicals, either biodiesel, proteins or pigments, among other possibilities (Cardozo *et al.*, 2007).

Many of the processes described in Sections 9.2 and 9.3 can be applied in biorefineries, either alone or combined. However, as already stated, there is still a need to optimize their performance, particularly in terms of energy consumption and overall efficiency. There is still a lot of room for improvement, and even new

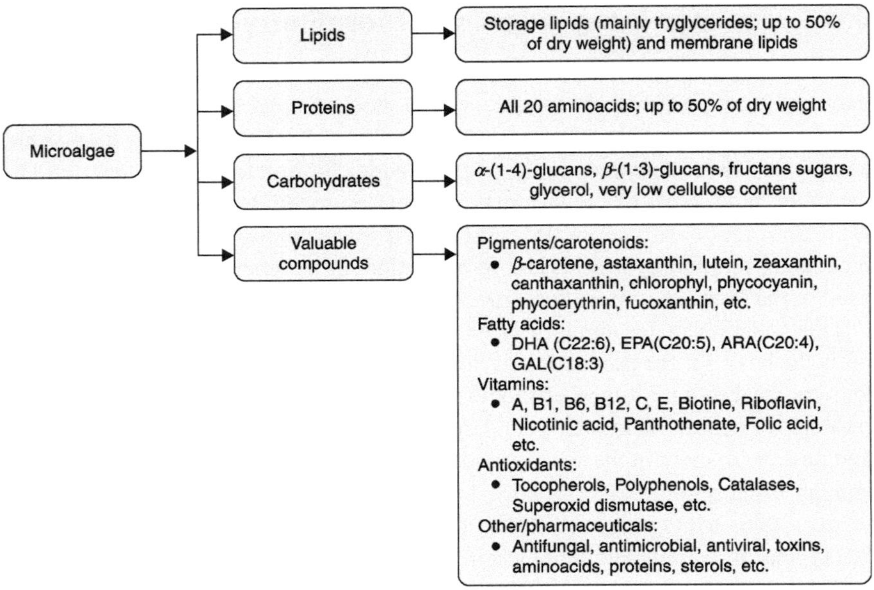

9.2 Main bio-components from microalgae.

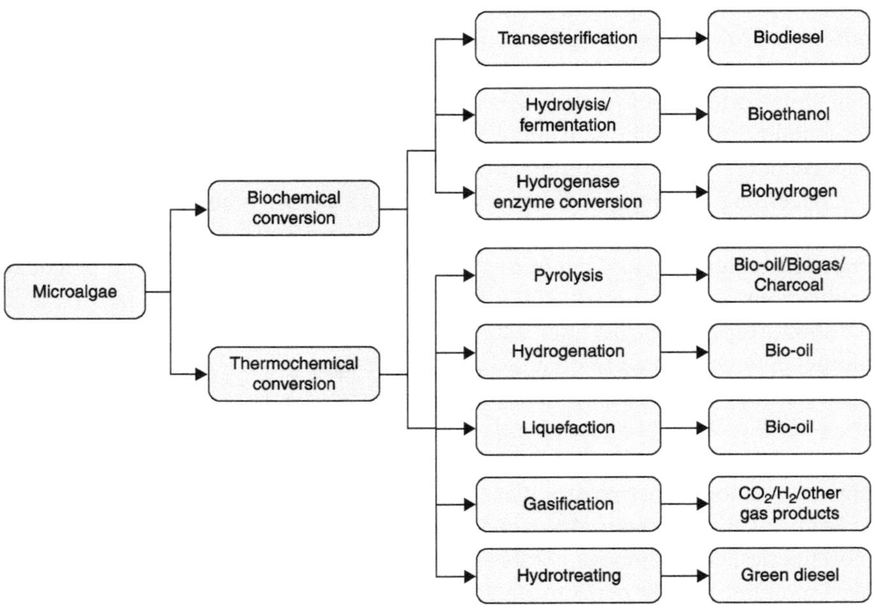

9.3 Microalgae biomass processes for biofuels.

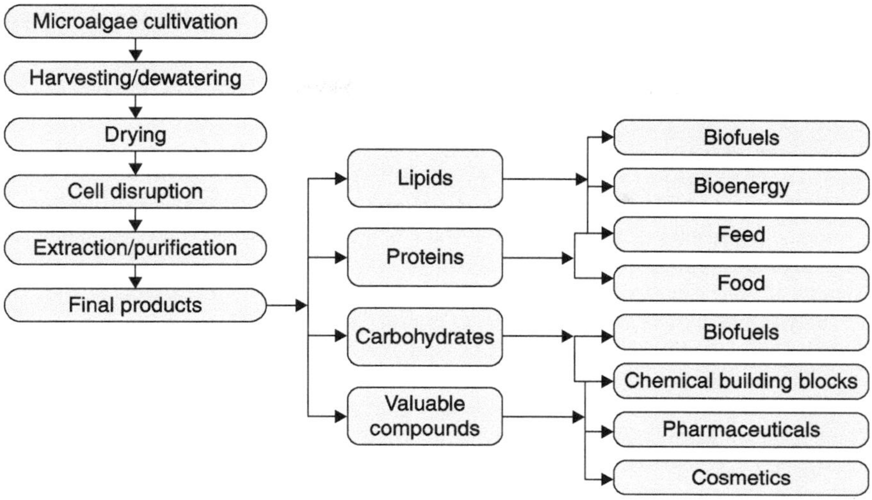

9.4 Utilization of microalgae biomass.

approaches to deal with this problem, representing today a very active area for research.

9.7 Conclusions

In this article the various possibilities for processing microalgae biomass are presented and critically assessed, focusing on lipid extraction for biodiesel production. The key aspects that have to be considered for their full scale development, implementation and operation are considered in detail. It can be concluded that energy consumption is a critical issue when producing biodiesel from microalgae, both from an economic and environmental point of view. Generally speaking, the large energy requirements for dewatering, drying, extraction and further biomass processing pose tremendous hurdles for any microalgae-based biofuel production technology. Moreover, the choice of the microalgae harvesting and dewatering techniques will determine the subsequent downstream unit operations, including the methods to be used for lipid extraction and possibly the biodiesel production process itself. Hence, to enable biofuels production or other high value bio-products from microalgae there are still many critical issues to be solved such as the choice of downstream biomass processing techniques to be used, which will have a direct impact on the design of the production process. Production of biodiesel from microalgae biomass using the current or future production processes is also briefly analysed, showing that new and improved processes are needed, in particular continuous operation, faster reaction times and lower dependence on the water content and other impurities.

Also discussed are the potential for using microalgae biomass in the production of other chemicals and even other types of biofuels, following the concept of biorefinery.

9.8 References

Abou-Shanab RAI, Hwang JH, Cho Y, Min B and Jeon BH (2011) Characterization of microalgal species isolated from fresh water bodies as a potential source for biodiesel production. *Applied Energy*, **88**(10), 3300–6.

Annetts JE and Audsley E (2003) Modelling the value of a rural biorefinery–part II: analysis and implications. *Agricultural Systems*, **76**(11), 61–76.

Aranda DAG, Santos RTP, Tapanes NCO, Ramos ALD and Antunes OAC (2008) Acid-Catalyzed Homogeneous Esterification Reaction for Biodiesel Production from Palm Fatty Acids. *Catalysis Letters*, **122**, 20–5.

Ayala RS and Castro L (2001) Continuous subcritical water extraction as a useful tool for isolation of edible essential oils. *Food Chemistry*, **75**(1), 109–13.

Benemann JR and Oswald WJ (1996) *Systems and Economic Analysis of Microalgae Ponds for Conversion of CO_2 to Biomass*. Final Report to the Pittsburgh Energy Technology Center. DOE/PE/93204-T5, Department of Energy, March 21.

Bligh E and Dyer W (1959) A rapid method of total lipid extraction and purification. *Canadian Journal of Physiology and Pharmacology*, **37**, 911–17.

Bosma R, van Spronsen WA, Tramper J and Wijffels RH (2003) Ultrasound, a new separation technique to harvest microalgae. *Journal of Applied Phycology*, **15**(2–3), 143–53.

Boussiba S and Richmond AE (1979) Isolation and characterization of phycocyanins from the blue-green alga *Spirulina platensis*. *Archives of Microbiology*, **120**(2), 155–9.

Burja AM, Armenta RE, Radianingtyas H and Barrow CJ (2007) Evaluation of fatty acid extraction methods for *Thraustochytrium sp.*ONC-T18. *Journal of Agricultural Food Chemistry*, **55**(12), 4795–4801.

Canakci M (2007) The potential of restaurant waste lipids as biodiesel feedstocks. *Bioresource Technology*, **98**(1), 183–90.

Canakci M and Van Gerpen J (1999) Biodiesel production via acid catalysis. *Transactions of the American Society of Agricultural Engineers*, **42**(5), 1203–10.

Canela APRF, Rosa PTV, Marques MOM and Meireles MAA (2002) Supercritical fluid extraction of fatty acids and carotenoids from the microalgae *Spirulina maxima*. *Industrial and Engineering Chemistry Research*, **41**(12), 3012–18.

Cardozo KHM, Guaratini T, Barros MP, Falcao VR, Tonon AP, *et al.* (2007) Metabolites from algae with economical impact. *Comparative Biochemistry and Physiology Part C*, **146**(1–2), 60–78.

Carlsson AS, van Beilen JB, Möller R and Clayton D (2007) Micro- and macro-algae: utility for industrial applications. In: *Outputs from the EPOBIO: Realising the Economic Potential of Sustainable Resources – Bioproducts from Non-food Crops Project*, Bowles, D. (ed). CNAP, CPL press, University of York, UK.

Cheung PCK (1999) Temperature and pressure effects on supercritical carbon dioxide extraction of n-3 fatty acids from red seaweed. *Food Chemistry*, **65**(3), 399–403.

Cooney M, Young G and Nagle N (2009) Extraction of bio-oils from microalgae. *Separation and Purification Reviews*, **38**(4), 291–325.

Cordero B and Voltolina D (1997) Viability of mass algal cultures preserved by freezing and freeze-drying. *Aquacultural Engineering*, **16**(4), 205–11.

Courchesne NMD, Parisien A, Wang B and Lan CQ (2009) Enhancement of lipid production using biochemical, genetic and transcription factor engineering approaches. *J. Biotechnology*, **141**(1–2), 31–41.

Cravotto G, Boffa L, Mantegna S, Perego P, Avogadro M and Cintas P (2008) Improved extraction of vegetable oils under high-intensity ultrasound and/or microwaves. *Ultrasonics Sonochemistry*, **15**(5), 898–902.

Csordas, A and Wang, JK (2004) An integrated photobioreactor and foam fractionation unit for the growth and harvest of *Chaetoceros spp.* in open systems. *Aquacultural Engineering*, **30**(1–2), 15–30.

Danquah MK, Gladman B, Moheimani N and Forde GM (2009) Microalgal growth characteristics and subsequent influence on dewatering efficiency. *Chemical Engineering Journal*, **151**(1–3), 73–8.

Demirbas A and Demirbas MF (2010) *Algae Energy: Algae as a New Source of Biodiesel*. Springer-Verlag, London.

Denery JR, Dragull K, Tang CS and Li QX (2004) Pressurized fluid extraction of carotenoids from *Haematococcus pluvialis* and *Dunaliella salina* and kavalactones from *Piper methysticum*. *Analytica Chimica Acta*, **501**(2), 175–81.

Desmorieux H and Decaen N (2005) Convective drying of *Spirulina* in thin layer. *Journal of Food Engineering*, **66**(4), 497–503.

Dionex (2007) *Accelerated Solvent Extraction Techniques for In-Line Selective Removal of Interferences*. Technical Note 210, Dionex, Sunnyvale, CA.

Ehimen EA, Sun ZF and Carrington CG (2010) Variables affecting the in situ transesterification of microalgae lipids. *Fuel*, **89**, 677–84.

Eikani MH, Golmohammad F and Rowshanzamir S (2007) Subcritical water extraction of essential oils from coriander seeds (*Coriandrumsativium L.*). *Journal of Food Engineering*, **80**(2), 735–40.

Fleurence J (1999) The enzymatic degradation of algal cell walls: a useful approach for improving protein accessibility? *Journal of Applied Phycology*, **11**(3), 313–14.

Folch J, Lees M and Sloane-Stanley GH (1957) A simple method for the isolation and purification of total lipids from animal tissues. *Journal of Biological Chemistry*, **226**(1), 497–509.

Gouveia L (2011) *Microalgae as a Feedstock for Biofuels*. Springer Heidelberg Dordrecht, London.

Grima EM, Medina AR, Giménez AG, Perez JAS, Camacho FG and Sanchez JLG (1994) Comparison between extraction of lipids and fatty acids from microalgal biomass. *JAOCS*, **71**(9), 955–9.

Grima EM, Belarbi EH, Fernández FGA, Medina AR and Chisti Y. (2003) Recovery of microalgal biomass and metabolites: Process options and economics. *Biotechnology Advances*, **20**(7–8), 491–515.

Guelcher SA and Kanel JS (1999) *Method for Dewatering Microalgae with a Bubble Column*. US Patent 5910254. Filed 20 Dec., 1996.

Haesman M, Diemar J, O'Connor W, Soushames T and Foulkes L (2000) Development of extended shelf-life microalgae concentrate diets harvested by centrifugation for bivalve molluscs – a summary. *Aquaculture Research*, **31**(8–9), 637–59.

Hara A and Radin NS (1978) Lipid extraction of tissues with a low toxicity solvent. *Analytical Biochemistry*, **90**(1), 420–6.

Herrero M, Martín-Álvarez PJ, Señoráns FJ, Cifuentes A and Ibáñez E (2005) Optimization of accelerated solvent extraction of antioxidants from *Spirulina platensis* microalga. *Food Chemistry*, **93**(3), 417–23.

Herrero M, Cifuentes A and Ibáñez E (2006) Sub- and supercritical fluid extraction of functional ingredients from different natural sources: Plants, food-by-products, algae and microalgae – A review. *Food Chemistry*, **98**(1), 136–48.

Howard EE, Alianell GA, Riding TJ, Barile PJ and Foster TR (2008) *Systems and Methods for Large Scale Production and Harvesting of Oil-Rich Algae*. US Patent No. US 2008/0096267A1.

Ibañez E, Herrero M, Martín-Álvarez PJ, Señoráns FJ, Reglero G and Cifuentes A (2006) Accelerated solvent extraction: A new procedure to obtain functional ingredients from natural sources. In: Turner C (ed) *Modern Extraction Technology Food and Agricultural Samples*. ACS Symposium Series 926, American Chemical Society, Washington DC, Chapter 5, 65–78.

Iverson SJ, Lang SLC and Cooper MH (2001) Comparison of the Bligh and Dyer and Folch methods for total lipid determination in a broad range of marine tissue. *Lipids*, **36**(11), 1283–7.

Kamm B, Kamm M, Gruber PR and Kromus S (2006) Biorefinery systems, an overview. In *Biorefineries – Industrial processes and products. Status quo and future directions*. Practical approach book. Kamm B, Gruber PR and Kamm M. (ed), Vol. 1, Wiley-VCH, Weinheim, 3–40.

Kleinig AR and Middleberg APJ (1998) On the mechanism of microbial cell disruption in high-pressure homogenisation. *Chemical Engineering Science*, **53**(5), 891–8.

Knuckey RM, Brown MR, Robert R and Frampton DM F (2006) Production of microalgal concentrates by flocculation and their assessment as aquaculture feeds. *Aquacultural Engineering*, **35**(3), 300–13.

Koberg M, Cohen M, Ben-Amotz A and Gedanken A (2011) Bio-diesel production directly from the microalgae biomass of Nannochloropsis by microwave and ultrasound radiation. *Bioresource Technology*, **102**, 4265–9.

Koopman B and Lincoln EP (1983) Autoflotation harvesting of algae from high-rate pond effluents. *Agricultural Wastes*, **5**(4), 231–46.

Krohn BJ, McNeff CV, Yan B and Nowlan D (2011). Production of algae-based biodiesel using the continuous catalytic Mcgyan process. *Bioresource Technology*, **102**, 94–100.

Leach G, Oliveira G and Morais R (1998) Spray-drying of *Dunaliella salina* to produce a ß-carotene rich powder. *Journal of Industrial Microbiology and Biotechnology*, **20**, 82–5.

Lee AK, Lewis DM and Ashman PJ (2010b) Energy requirements and economic analysis of a full-scale microbial flocculation system for microalgal harvesting. *Chemical Engineering Research and Design*, **88**(8), 988–96.

Lee J-Y, Yoo C, Jun, CY, Ahn CY and Oh HM (2010a).Comparison of several methods for effective lipid extraction from microalgae. *Bioresource Technology*, **101**(1), S75–S77.

Lee SJ, Kin SB, Kim JE, Kwon GS, Yoon BD and Oh HM (1998) Effects of harvesting method and growth stage on the flocculation of the green alga *Botryococcus braunii*. *Letters in Applied Microbiology*, **27**(1), 14–18.

León JAM (2008) *Extracción de compuestos bioactivos de microalgas mediante fluidos supercríticos*. Doctoral thesis. Departamento de Química-Física Aplicada, Facultad de Ciencias, Universidad Autónoma de Madrid.

Lepage G and Roy CC (1984) Improved recovery of fatty acid through direct transesterification without prior extraction or purification. *Journal of Lipid Research*, **25**(12), 1391–6.

Levin GV, Clendenning JR, Gibor A and Bogar FD (1962) Harvesting of algae by froth flotation. *Applied Microbiology*, **10**(2), 169–75.

Lewis T, Nichols PD and McMeekin TA (2000) Evaluation of extraction methods for recovery of fatty acids from lipid-producing microheterotrophs. *Journal of Microbiological Methods*, **43**(2), 107–16.

Li Y, Horsman M, Wu N, Lan CQ and Dubois-Calero N (2008) Biofuels from microalgae. *Biotechnology Progress*, **24**(4), 815–20.

Luque de Castro MD, Jiménez-Carmona MM and Fernández-Pérez V (1999) Towards more rational techniques for the isolation of valuable essential oils from plants. *Trends in Analytical Chemistry*, **18**(11), 708–15.

Mata TM, Martins AA and Caetano NS (2010) Microalgae for biodiesel production and other applications. *Renewable and Sustainable Energy Reviews*, **14**(1), 217–32.

Mata TM, Martins AA, Sikdar S and Costa CAV (2011) Sustainability considerations of biodiesel based on supply chain analysis. *Clean Technologies and Environmental Policy*, **13**(5), 655–71.

Mendes-Pinto MM, Raposo MFJ, Bowen J, Young AJ and Morais R (2001) Evaluation of different cell disruption processes on encysted cells of *Haematococus pluvialis*: effects on astaxanthin recovery and implications for bio-availability. *Journal of Applied Phycology*, **13**(1), 19–24.

Mendiola JA, García-Martínez D, Rupérez FJ, Martín-Álvarez PJ, Reglero G, *et al.* (2008) Enrichment of vitamin E from *Spirulina platensis* microalga by SFE. *Journal of Supercritical Fluids*, **43**(3), 484–9.

Middleberg APJ (1995) Process-scale disruption of microorganisms. *Biotechnology Advances*, **13**(3), 491–551.

Mulbry W, Kondrad S, Buyer J and Luthria D (2009) Optimization of an oil extraction process for algae from the treatment of manure effluent. *Journal of the American Oil Chemists Society*, **86**(9), 909–15.

Mussgnug JH, Klassen V, Schlüter A and Kruse O (2010) Microalgae as substrates for fermentative biogas production in a combined biorefinery concept. *Journal of Biotechnology*, **150**(1), 51–6.

Nagle N and Lemke P (1990) Production of methyl ester fuel from microalgae. *Applied Biochemistry and Biotechnology*, **24–25**(1), 355–61.

Nindo CI and Tang J (2007) Refractance window dehydration technology: a novel contact drying method. *Drying Technology*, **25**(1), 37–48.

Oh HM, Lee SJ, Park MH, Kim HS, Kim HC, *et al.* (2001) Harvesting of *Chlorella vulgaris* using a bioflocculant from *Paenibacillus sp.* AM49. *Biotechnology Letters*, **23**(15), 1229–34.

Pascal SL and Segal R (2006) Phytosterols – biological active compounds in food. *Journal of Agroalimentary Processes and Technologies*, **12**(1), 149–58.

Patil PD, Gud VG, Mannarswamy A, Deng S, Cooke P, *et al.* (2011a) Optimization of direct conversion of wet algae to biodiesel under supercritical methanol conditions. *Bioresource Technology*, **102**, 118–22.

Patil PD, Gud VG, Mannarswamy A, Cooke P, Munson-McGee S, *et al.* (2011b) Optimization of microwave-assisted transesterification of dry algal biomass using response surface methodology. *Bioresource Technology*, **102**, 1399–1405.

Pernet F and Tremblay R (2003) Effect of ultrasonication and grinding on the determination of lipid class content of microalgae harvested on filters. *Lipids*, **38**(11), 1191–5.

Poelman E, DePauw N and Jeurissen B (1997) Potential of electrolytic flocculation for recovery of microalgae. *Resources Conservation and Recycling*, **19**(1), 1–10.

Prakash J, Pushparaj B, Carlozzi P, Torzillo G, Montaini E and Materassi R (1997) Microalgal biomass drying by a simple solar device. *International Journal of Sustainable Energy*, **18**(4), 303–11.

Rice B, Frohlich A, Leonard R and Korbitz W (1997) *Bio-diesel Production Based on Waste Cooking Oil: Promotion of the Establishment of an Industry in Ireland*. Final Report, Altener Contract No. XVII/4.1030/AL/77/95/IRL.

Richter BE, Jones BA, Ezzell JL, Porter NL, Avdalovic N and Pohl C (1996) Accelerated solvent extraction: a technique for sample preparation. *Analytical Chemistry*, **68**(6), 1033–9.

Rodríguez-Meizoso I, Jaime L, Santoyo S, Señoráns FJ, Cifuentes A and E Ibáñez (2010) Subcritical water extraction and characterization of bioactive compounds from *Haematococcus pluvialis* microalga. *Journal of Pharmaceutical and Biomedical Analysis*, **51**(2), 456–63.

Rossignol N, Lebeau T, Jaouen P and Robert JM (2000) Comparison of two membrane-photobioreactors, with free or immobilized cells, for the production of pigments by a marine diatom. *Bioprocess and Biosystems Engineering*, **23**(5), 495–501.

Sander KB and Murthy GS (2009) Enzymatic degradation of microalgal cell walls. *ASABE Annual Meeting*, paper #096054, Reno, Nevada, 21–24 June.

Satyanarayana K G, Mariano AB and Vargas JVC (2011) A review on microalgae, a versatile source for sustainable energy and materials. *International Journal of Energy Research*, **35**(4), 291–311.

Schafer K (1998) Accelerated solvent extraction of lipids for determining the fatty acid composition of biological material. *Analytica Chimica Acta*, **358**(1), 69–77.

Sheehan J, Dunahay T, Benemann J and Roessler P (1998) *A Look Back at the U.S. Department of Energy's Aquatic Species Program: Biodiesel from Algae*. NREL/TP–580–24190, National Renewable Energy Laboratory, USA.

Shelef G, Sukenik A and Green M (1984) *Microalgae Harvesting and Processing: A Literature Review*. SERI/STR–231–2396.

Shepherd VA and Beilby MJ (1999). The effect of an extracellular mucilage on the response to osmotic shock in the Charophyte alga *Lamprothamnium papulosum*. *Journal of Membrane Biology*, **170**(3), 229–42.

Sim TS, Goh A and Becker EW (1988) Comparison of centrifugation, dissolved air flotation and drum filtration techniques for harvesting sewage-grown algae. *Biomass*, **16**(1), 51–62.

Sobczuk TM and Chisti Y (2010) Potential fuel oils from the microalga *Choricystis minor*. *Journal of Chemical Technology and Biotechnology*, **85**(1), 100–8.

Sommerfeld M, Chen W, Hu Q, Giorgi D, Navapanich T, *et al.* (2010) *Application of Electroporation for Lipid Extraction from Microalgae*. Poster available on-line. Assessed 2/3/2011 at http://www.optiswitch.com/products/pdfs/Application%20of%20Electroporation%20for%20Lipid%20Extraction%20from%20Microalgae%20Poster%20Session.pdf.

Torrey M (2008) Algae in the tank. *Tribology and Lubrication Technology*, **64**(12), 26–32; **19**(7), 432–7.

Wahlen BD, Willis RM and Seefeldt LC (2011). Biodiesel production by simultaneous extraction and conversion of total lipids from microalgae, cyanobacteria and wild mixed-cultures. *Bioresource Technology*, **102**, 2724–30.

Wang T, Jonsdottir R, Kristinsson HG, Hreggvidsson GO, Jonsson JO, *et al.* (2010) Enzyme-enhanced extraction of antioxidant ingredients from red algae *Palmaria palmate*. *LWT – Food Science and Technology*, **43**(9), 1387–93.

Widjaja A, Chien C-C and Ju Y-H (2009) Study of increasing lipid production from fresh water microalgae *Chlorella vulgaris*. *Journal of the Taiwan Institute of Chemical Engineers*, **40**(1), 13–20.

Wiltshire KH, Boersma M, Möller A and Buhtz H (2000) Extraction of pigments and fatty acids from the green alga *Scenedesmus obliquus* (Chlorophyceae). *Aquatic Ecology*, **34**(2), 119–26.

Xu L, (Wim) Brilman DWF, Withag JAM, Brem G and Kersten S (2011). Assessment of a dry and a wet route for the production of biofuels from microalgae: energy balance analysis. *Bioresource Technology*, **102**(8), 5113–22.

Xu R (2010) *Downstream Processing of Microalgae Biodiesel Production*. MSc. Thesis in Chemical Engineering, Hong Kong University of Science and Technology.

You J-Y, Peng C, Liu X, Ji X-J, Lu J, *et al.* (2011) Enzymatic hydrolysis and extraction of arachidonic acid rich lipids from *Mortierella alpina*. *Bioresource Technology*, **102**(10), 6088–94.

Young G, Nippgen F, Titterbrandt S and Cooney MJ (2010) Lipid extraction from biomass using co-solvent mixtures of ionic liquids and polar covalent molecules. *Separation and Purification Technology*, **72**(1), 118–21.

10

Valorization of the glycerol by-product of biodiesel production

C. J. A. MOTA, Universidade Federal do Rio de Janeiro, Brazil

Abstract: This chapter describes some uses of glycerol, a by-product of biodiesel production, as a valuable feedstock for the synthesis of several chemicals. Glycerol can be converted into polymer precursors, such as propanediols, epichloridrin, acrylic acid and propene. In addition, functionalization of glycerol in ethers, acetal/ketals and esters may provide potential fuel additives for gasoline and the biodiesel itself.

Key words: catalysis, ethers, fuel additives, glycerol, polymers.

10.1 Introduction

The use of biofuels is spreading all over the world. The scenario for the next decades indicates that bioethanol and biodiesel will share a significant part of the fuel market, contributing to the control of global warming. If on the one hand these actions bring sustainability to the fuel sector, on the other hand one must be aware of the by-products of the biofuel industry, in order to find economical and environmentally acceptable solutions.

Biodiesel is mainly produced by the transesterification of vegetable oils or animal fat, the triglycerides. In this process, methanol reacts with the triglyceride in the presence of a basic or acidic catalyst to afford fatty acid methyl esters, the biodiesel itself and glycerol (Fig. 10.1).

The source of the triglyceride may vary, from soybean and canola to tallow and algae, but roughly, for each 100 m^3 of oil processed about 10 m^3 of glycerol is produced. In recent years, the surplus of glycerol from biodiesel fabrication has increased enormously and today represents about 65% of the world's glycerol production. There is still one plant in operation using the petrochemical route.

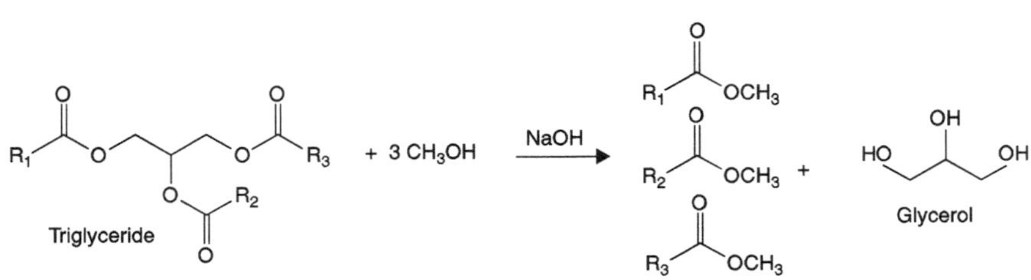

10.1 Transesterification of triglycerides to produce fatty acid methyl esters (biodiesel) and glycerol.

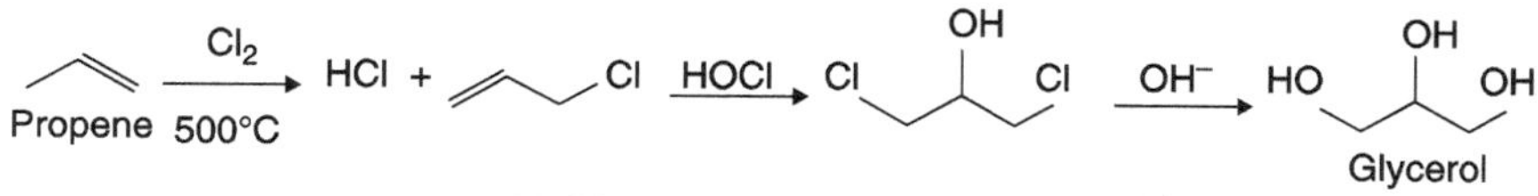

10.2 Petrochemical route of glycerol production.

It is based on the chlorination of propene to allyl chloride, followed by electrophilic addition of hydrochlorous acid and further hydrolysis in alkaline conditions (Fig. 10.2).

Glycerol is traditionally used in cosmetics, soaps and pharmaceuticals. However, these sectors cannot drain the enormous amount of glycerol that comes from biodiesel production. In the context of green chemistry (Anastas and Eghbali, 2009), which applies 12 principles in the development of environmentally friendly chemical processes, wastes and residues must be minimized. Thus, the glycerol by-product of biodiesel production must find new applications, capable of draining the increasing supply of this chemical and adding value to the biodiesel chain. The purpose of this chapter is to show some potential applications of glycerol as feedstock for producing polymers and fuel additives.

10.2 Composition and purification of glycerol produced from biodiesel

Glycerol is 1,2,3-propanetriol. It was first identified by Carl Scheele in 1779, upon heating olive oil with litharge (PbO). The viscous, transparent liquid that separated from the oil phase was named glycerol owing to its sweet taste (from the Greek *glykos* = sweet). The term glycerine applies to commercial products, which are rich (e.g. 95%) in glycerol. However, with the increasing production of biodiesel, there are many commercial glycerine products with different glycerol contents and other impurities, such as water, salts and organic compounds. Table 10.1 shows the typical composition of the glycerine phase obtained in a Brazilian biodiesel plant. The minimum glycerol content is 80 wt% and there is about

Table 10.1 Average composition of crude glycerine from a Brazilian biodiesel plant

Composition	wt%
Glycerol	80.0 min
Water	10.0 max
Methanol	1.0 max
NaCl	10.0 max

10 wt% water. The remaining 10 wt% is mainly methanol and dissolved salts, such as NaCl, formed upon acid neutralization of the homogenous basic catalyst.

The traditional transesterification process, using homogeneous basic catalysts, produces a glycerine phase with many impurities. The polarity of the glycerol favours the dissolution of NaOH, used as catalyst or formed upon hydrolysis of sodium methoxide (the most used catalyst nowadays), soap (sodium salts of fatty acid), which is formed upon neutralization of free fatty acids present in the oil or by hydrolysis of the triglyceride, as well as mono and diglycerides, formed by incomplete transesterification. In addition, methanol and water may also be found in the glycerine phase.

Glycerol has a boiling point of 290°C and a viscosity of 1.5 Pa s at 20°C. Therefore, purification procedures are normally time-consuming and costly. The most traditional process involves distillation at reduced pressure. A product with a content of glycerol of at least 99.5 wt% can be obtained by thin film distillation and meets the requirements of the United State Pharmacopeia (USP). Other purification processes such as extraction, ion-exchange, adsorption, crystallization and dialysis can also be applied to the glycerine phase. The costs and purity of the purified product will have an impact on the commercial utilization of glycerine.

10.3 Applications and valorization of glycerol

The traditional application of glycerol has been in the food, pharmaceutical and personal care sectors. It functions as a moistening agent and emollient, preserving and giving texture to the products. It can also be used as a lubricant in compressors and food processing machines. Nevertheless, all these applications consume small amounts of glycerol, which can be mainly considered to be an additive in many products. With the increasing supply of glycerol from biodiesel, it is mandatory to find new applications. The world's glycerol production is expected to reach 1.2 million tonnes in 2012 (Zhou *et al.*, 2008) and the forecast points to an increasing production owing to the start up of new biodiesel plants. The case in Brazil may illustrate this scenario. According to the Brazilian Chemical Industry Association (ABIQUIM), the traditional glycerol market consumes approximately 30 thousand tonnes per year. On the other hand, glycerol from biodiesel plants has reached 250 thousand tonnes in 2010, with the addition of 5 vol% of biodiesel to regular diesel. The Brazilian government expects to implement a mandatory 7 vol% addition of biodiesel in the coming years, which will put more pressure into the glycerol balance.

Apart from its traditional use in cosmetics, soaps and pharmaceuticals, glycerol from biodiesel production has found utilization in animal food, secondary petroleum recovery, heat generation, as well as feedstock for producing other chemicals. The use of glycerol in horse, poultry and cattle nutrition is expanding, although many countries impose rigid limits to the presence of methanol in animal diets. Glycerol is about 60% as sweet as sugar and provides, roughly, the same energy content per mass unit. The use of glycerol in formulation of fluids used in secondary oil recovery is an attractive option for regions in which oil production is decreasing (Quintella *et al.*, 2009).

Many wells show a decrease in oil production owing to the reduction of the pressure in the field. This makes the extraction of the oil more difficult and secondary techniques, such as injection of fluids through a neighbouring well, is widely used. An advantage of this application is that one can use the crude glycerol as it comes from the biodiesel production, with practically no purification. On the other hand, applicability is limited to biodiesel plants located near the oil wells, because the costs of transportation do not justify the economic viability of the process. The direct burning of glycerol in biodiesel plants has also being considered in many countries. The complete combustion of glycerol generates 4195 kcal kg^{-1}, but there are many inconvenient aspects. Incomplete burning may generate acrolein, which is highly toxic to humans. The salts present in the crude glycerol from biodiesel production may deteriorate the equipment, leading to corrosion and other problems. All these drawbacks make these possibilities economically less attractive than the chemical transformation of glycerol. They all have limited uses and may be applied in regional terms or within a short period of time.

Glycerol is a triol and can be converted into different functional groups and compounds. Oxidation may afford aldehydes, carboxylic acids and ketones, whereas functionalization of the hydroxy groups produces ethers, esters, acetals and ketals. In addition, glycerol hydrogenolysis or dehydration produces chemicals of great industrial importance. The main attraction for expanding the use of glycerol as feedstock for producing other chemicals comes from its increasing supply from biodiesel production plants, which has an impact on the price, making the glycerol molecule an accessible and cheap raw material.

10.3.1 Glycerol as raw material for the chemical industry

The chemical industry is still based on oil and gas. In 2009, sales reached approximately 2 trillion euros, indicative of its huge and powerful economic situation. Naphtha is the main feedstock of the chemical industry. It is initially transformed into light olefins, such as ethene and propene, as well as aromatics, like benzene, toluene and xylenes. These compounds are then transformed into polymers and other chemicals by complex chemical processes, before being used in everyday life in plastic components, dyes, textiles, paints and other materials. The shortage of oil and its increasing price in the near future will force the chemical industry to diversify the feedstock, giving more importance to renewable materials. Bioethanol produced from sugarcane is opening up this new era. Recently, the major Brazilian chemical company, Braskem, started up a plant to dehydrate ethanol to ethylene, which is subsequently polymerized to polyethylene (Braskem, 2010). Glycerol from biodiesel production may follow the same path, being a substitute for propylene-based chemicals.

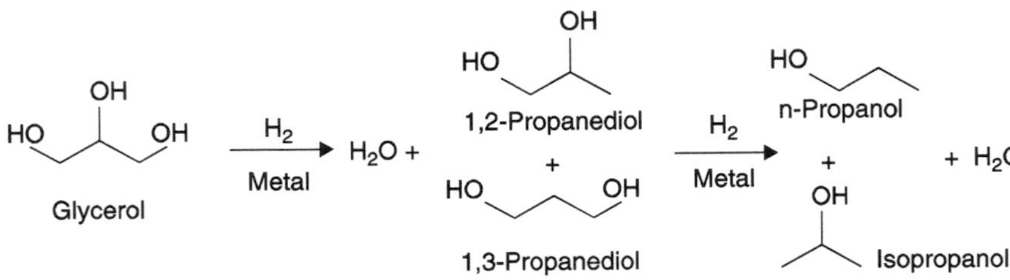

10.3 Metal-catalysed hydrogenolysis of glycerol to propanediols and propanols.

The hydrogenolysis of glycerol over supported metal catalyst can afford 1,2- and 1,3-propanediol (Chaminand *et al.*, 2004) (Fig. 10.3). The 1,2 isomer, also known as propylene glycol, has many uses, including as anti-freezing agent and in the production of polyurethane foams and other polymers. The present worldwide production of propylene glycol is about 1 million tonnes per year. The traditional process involves the hydrolysis of propylene oxide, which in turn is produced from propene. Major chemical companies, such as Dow, Hunstman and ADM, have developed technologies for the selective hydrogenolysis of glycerol into propylene glycol. The reaction can be carried out in batch or continuous flow conditions, in temperatures normally ranging from 180–250°C. Copper, palladium and ruthenium supported catalysts have usually been used (Dasari *et al.*, 2005), but other metals like iron, nickel and rhodium can be employed as well. The concomitant use of an acidic catalyst, such as sulphonic acid resins, improves the selectivity of the catalyst toward hydrogenolysis, allowing it to work at lower temperatures and reduced pressures in batch conditions (Miyazawa *et al.*, 2006).

More severe reaction conditions leads to deeper hydrogenolysis of the glycerol molecule, yielding n-propanol and isopropanol (Casale and Gomez, 1994). The reaction pathway is complex and may involve dehydration steps. This may explain why the concomitant use of acidic catalysts in the medium or as support, improves the selectivity. It has been shown that CO and CO_2 can also be formed by decomposition reactions over supported Pt catalysts (Wawrzetz *et al.*, 2010).

A process to hydrogenate glycerol selectively into propene was recently developed (Mota *et al.*, 2009) (Fig. 10.4). The selectivity for propene can reach up to 90% over a proprietary catalyst and specific continuous flow conditions.

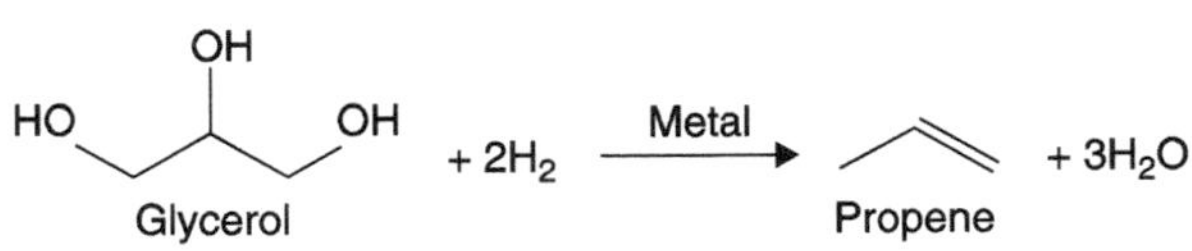

10.4 Selective hydrogenolysis of glycerol to propene.

Propene is an important raw material for producing many plastics, with worldwide production around 40 million tonnes per year. The search for a 'green' propene, produced from renewable materials, is a major goal of many chemical companies. The use of glycerol from biodiesel production may be an alternative for the 'green' propene.

In contrast to its 1,2 isomer, 1,3-propanediol has considerably fewer uses, with a global market of approximately 360 thousand tonnes per year. Its main use is in the reaction with terephthalic acid to produce a polyester fibre, commercially known as CORTERRA. The traditional route used by Shell involves the reaction of ethylene oxide with CO/H_2 at high pressures. Nevertheless, the product can be also obtained by glycerol hydrogenolysis. With the choice of proper reaction conditions and modifications of the metal catalyst, the ratio between 1,3 and 1,2-propanediol can go up to 5, but at moderate glycerol conversion (Gong *et al.*, 2009). The 1,3-propanediol can also be produced through biotechnological processes, using genetically modified bacteria. Nevertheless, the long reaction time limits the widespread use of this route.

The Belgian chemical company Solvay has developed a process to transform glycerol into epichloridrin. This chemical is mostly used in the production of epoxy resins, as well as water and paper treatment. The process involves the reaction of glycerol with two moles of HCl in the presence of Lewis acid catalysts, followed by controlled alkaline hydrolysis (Fig. 10.5). The company has built a pilot plant and has plans to construct an industrial unit in Asia.

Glycerol gasification to synthesis gas, a mixture of CO and H_2, has also been studied (Soares *et al.*, 2006). The reaction is endothermic by 83 kcal mol^{-1}, but can be carried out at temperatures around 350°C over Pt and Pd catalysts. Synthesis gas is used in many industrial processes, like methanol production and the Fischer-Tropsch synthesis of hydrocarbons.

Acid-catalysed glycerol dehydration can follow two pathways: dehydration of the primary hydroxy group affording hydroxy-acetone, also known as acetol, or dehydration of the secondary hydroxy group yielding 3-hydroxy-propanal. This latter compound can be subsequently dehydrated to form acrolein (Fig. 10.6). Oxidation of acrolein over Mo and V based catalysts produces acrylic acid (Kampe *et al.*, 2007), an important chemical used in the fabrication of superabsorbent polymers, paints and adhesives. The reaction runs in the liquid phase with the use of mineral acids, such as H_2SO_4. However, dehydration in the

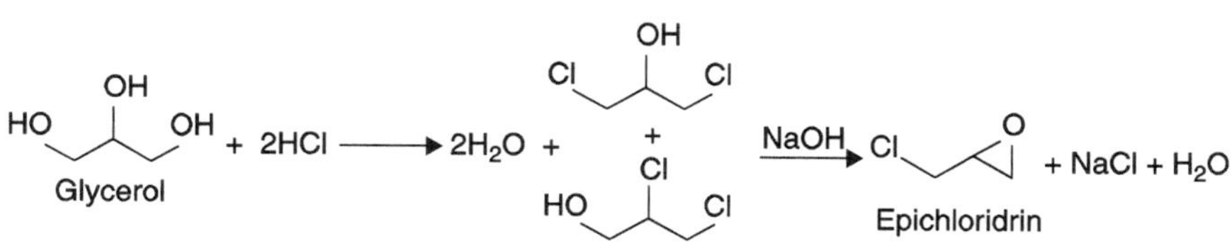

10.5 Production of epichloridrin from glycerol.

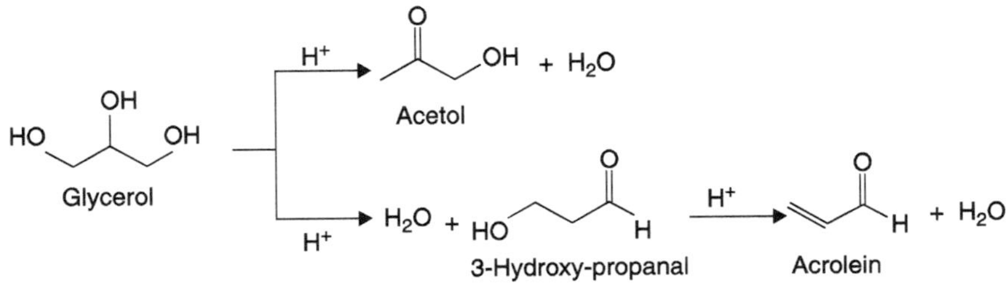

10.6 Acid-catalysed dehydration of glycerol.

gas phase, under continuous flow conditions and using heterogeneous catalysts, has received more attention lately. Different acidic catalysts, such as zeolites, metal oxides, clays and heteropolyacids have been tested (Chai *et al.*, 2007). The conversion and selectivity depends on the catalysts and conditions used. For instance, supported heteropolyacids can show 86% selectivity for acrolein for a glycerol conversion of 98% (Tsukuda *et al.*, 2007), but deactivation of the catalyst is still a major problem.

The one-step synthesis of acrylic acid from glycerol, combining acidic and oxidant properties in the same catalyst, is an interesting approach. The oxidative dehydration of glycerol has been studied over mixed oxide catalysts in the presence of air (Deleplanque *et al.*, 2010). Mixed molybdenum and vanadium oxides, as well as vanadium/tungsten oxides were active for this reaction. They show 100% glycerol conversion with selectivity to acrylic acid between 24 and 28%. Acetic acid, probably from oxidation of acetaldehyde formed upon cracking of the 3-hydroxy-propanal, was also observed as a by-product (Fig. 10.7).

Vanadium-impregnated zeolite Beta can also be used in the oxidative dehydration of glycerol to acrylic acid (Pestana, 2010). Only the catalysts prepared by wet impregnation showed significant selectivity. These catalysts were prepared

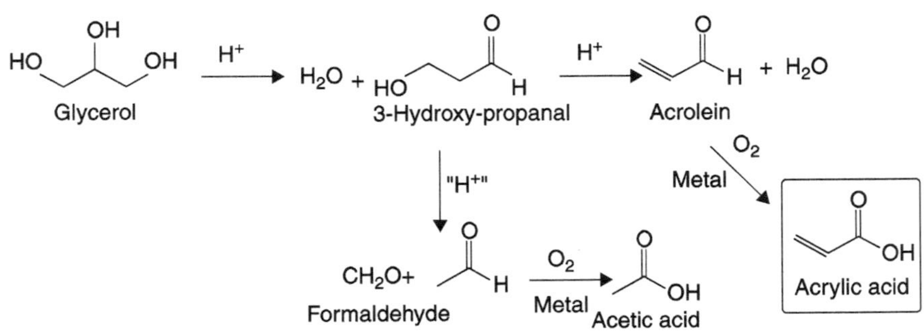

10.7 Oxidative dehydration of glycerol to acrylic acid.

by evaporation of a suspension of ammonium metavanadate solution with ammonium-exchanged zeolite Beta, followed by air calcination. Apparently, this procedure leads to a better dispersion of vanadium inside the zeolite pores, showing selectivity for acrylic acid of around 20% for 70% glycerol conversion.

Glycerol can also be used in the synthesis of speciality chemicals. Glycerol oxidation can produce several important compounds (Fig. 10.8). The selective oxidation of the secondary hydroxyl group leads to 1,3-dihydroxy-acetone (DHA), which is utilized in artificial tanning and has a global production of more than two thousand tonnes per year. DHA is normally produced from glycerol fermentation (Hekmat *et al.*, 2007), but can also be prepared by electrochemical methods (Ciriminna *et al.*, 2006).

Glyceraldehyde is an intermediate in the carbohydrate metabolism. A good method of preparation involves the oxidation of glycerol over Pt-supported catalysts. Selectivity of 55% to glyceraldehyde with a glycerol conversion of 90% can be observed with the use of a Pt/C catalyst (Garcia *et al.*, 1995).

Glyceric acid is selectively produced by the oxidation of glycerol over Au/C catalysts in the presence of oxygen (Carrettin *et al.*, 2004). Bimetallic gold-platinum or gold-palladium catalysts show higher turnover frequencies, but are also less selective, yielding C–C bond cleavage products, such as oxalic and glycolic acids (Bianchi *et al.*, 2005).

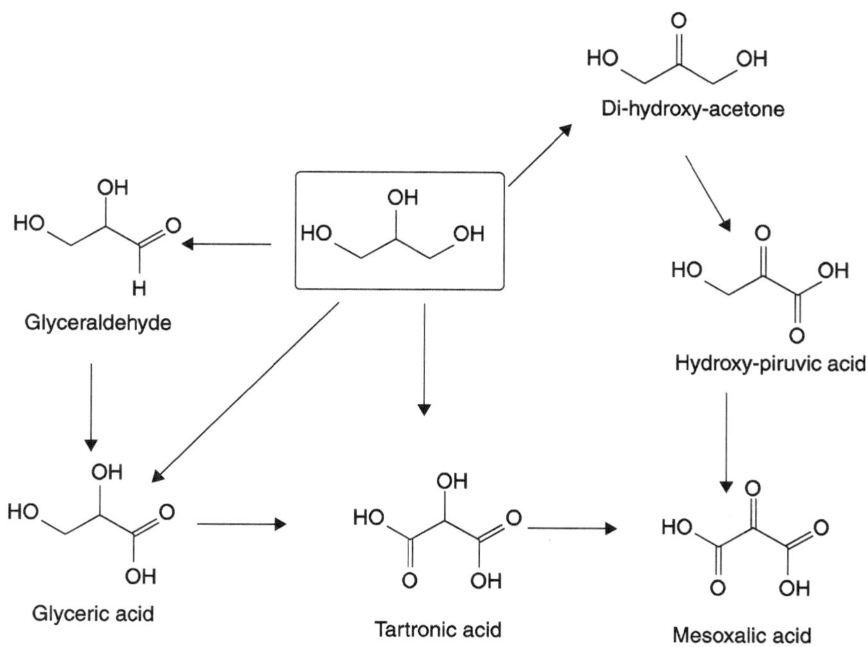

10.8 Products formed in the oxidation of glycerol.

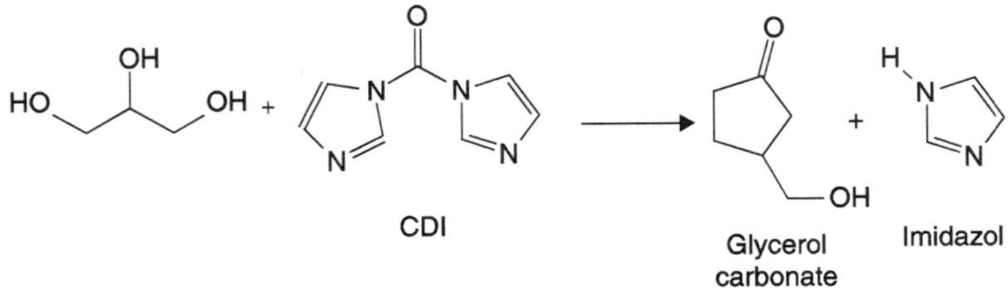

10.9 Production of glycerol carbonate in the reaction of glycerol with carbonyl-diimidazol (CDI).

Glycerol carbonate is a product with increasing applications. It can be used as solvent and monomer for the production of polycarbonates, polyesters and polyamides. There are many routes for its production, such as the reaction of glycerol with cyclic organic carbonates (Vieville *et al.*, 1998), direct carbonation of glycerol with CO_2 catalysed by organotin compounds (Aresta *et al.*, 2006) and by the reaction of glycerol with urea. This later procedure has become popular over the last few years, but involves long reaction times (6 to 8 hours) and high temperatures (160°C). A simple procedure to produce glycerol carbonate involves the reaction of glycerol with *N,N'*-carbonyl-diimidazole (CDI) (Mota *et al.*, 2007) (Fig. 10.9). The product can be obtained in quantitative yields in 15 min at room temperature. The use of crude glycerine, from biodiesel production, does not affect the yield, as the alkaline catalyst dissolved in the glycerine phase accelerates the reaction. The major drawback is the cost of CDI and its production route, which still employs phosgene ($COCl_2$), a highly toxic gas. Studies involving the *in situ* formation of CDI by reacting CO_2 and imidazole are still on course.

10.3.2 Glycerol as a fuel additive

Vehicular fuels are, traditionally, composed of hydrocarbons. Gasoline is a complex mixture of hydrocarbons, ranging from 5 to 11 carbon atoms, whereas diesel is heavier, going from 16 to about 22 carbon atoms. In spite of the major hydrocarbon composition, many fuels have additives. In the past, tetraethyl lead (TEL) was blended with the gasoline to improve the octane number (anti-knocking property). Today, owing to environmental restrictions, TEL is no longer added to the gasoline. In the same manner, cetane boosters are used to improve the cetane number of diesel oil.

Recently, there has been an increasing demand for oxygenated additives. They improve the burning process inside the engine, reducing the emission of carbon monoxide. In addition, many oxygenated additives are octane boosters for gasoline, or improve the lubricity and reduce particle emissions in diesel. The methyl-tert-butyl ether (MTBE) was the main oxygenated additive used worldwide in gasoline (Nadim *et al.*, 2001). However, because of environmental and health problems, its use has been phased out in recent years (Ahmed, 2001). Today, ethanol produced from sugarcane, corn or sugar beet, is the main oxygenated blend in gasoline. The worldwide production of MTBE reached 19 million tonnes in 2003, whereas ethanol still faces some concerns as an additive to gasoline in many countries. These facts open up opportunities for other oxygenated additives, especially from renewable sources.

Glycerol is insoluble in hydrocarbons and poorly volatile, not being suitable to blend directly in gasoline or diesel. On the other hand, glycerol derivatives, such as ethers, acetals/ketals and esters meet the criteria for blending in hydrocarbon fuels. They are less polar and more volatile than glycerol, being potential candidates for new fuel additives.

The acid-catalysed reaction of glycerol with isobutene affords tert-butyl-glyceryl ethers (Klepacova *et al.*, 2005) (Fig. 10.10). The use of sulphonic acid resins gives better results and the selectivity for mono-, di- and tri-ethers depends on the reaction conditions and a proper choice of the catalyst. The tri-ether is only significantly formed at molar ratios above 4. tert-Butanol can replace isobutene in the formation of the ethers, but conversions are normally lower because of water formation, which may weaken the acid strength of the catalyst (Klepacova *et al.*, 2003).

Glycerol can react with alcohols in the presence of acidic catalysts to produce ethers. Reaction with benzyl alcohol produces mono- and di-benzyl glyceryl ethers (da Silva *et al.,* 2009a), whereas reaction with ethanol yields the respective ethyl glyceryl ethers (Pariente *et al.*, 2009) (Fig. 10.11). A drawback in all these reactions is the formation of dibenzyl ether and diethyl ether, respectively, formed through the self-etherification of the alcohols. In addition, formation of water may affect the conversion.

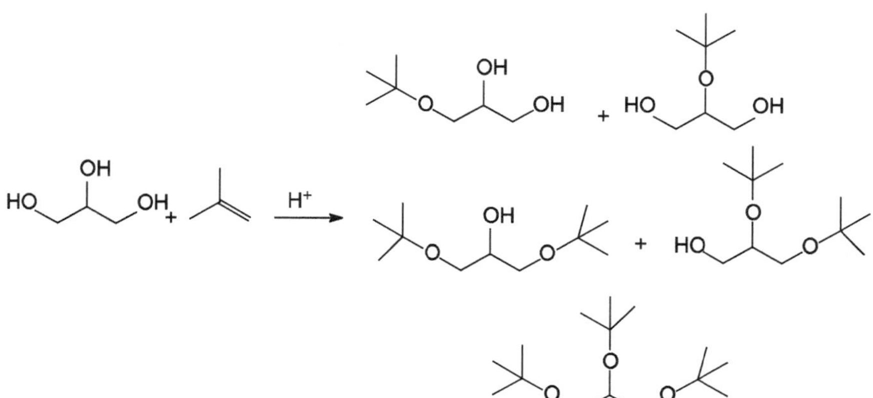

10.10 Etherification of glycerol with isobutene in the presence of acid catalysts.

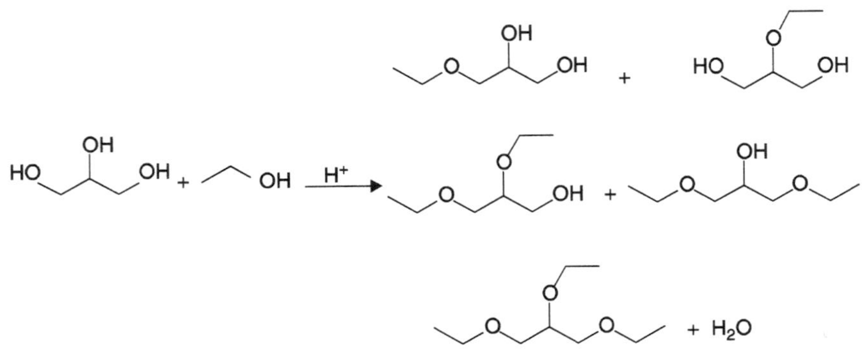

10.11 Reaction of glycerol with ethanol in the presence of acid catalysts.

The method of Williamson, involving alkoxides and alkylating agents, such as halides and sulphates, is a traditional route for the synthesis of ethers. Reaction of glycerol with methyl sulphate, in the presence of potassium or sodium hydroxide affords the tri-methyl glyceryl ether in 100% selectivity (Mota and Gonçalves, 2007). The product can be obtained in 30 min reaction time, even with the use of crude glycerine from biodiesel production. The presence of the alkaline catalyst in the glycerine phase favours the reaction.

The tert-butyl glyceryl ethers are considered to be octane boosters for gasoline (Wessendorf, 1995). The ethyl glyceryl ethers are potential additives for biodiesel, improving the cold flow properties (Pinto, 2009). Table 10.2 shows the results of cold flow properties of palm and soybean oil biodiesel with a mixture of mono- and di-ethyl glyceryl ethers. One can see that with 0.5 vol% there can be a reduction of up to 5°C in the pour point, indicating that these ethers can be used in blends with biodiesel.

Table 10.2 Results of cold flow properties of biodiesel blended with ethyl glyceryl ethers, measure according to the ASTM-D97 method

Sample	Cold point (°C)	Freezing point (°C)	Pour point (°C)
Palm oil B100	18	15	18
Palm oil B100 + 0.1 vol% ethers	15	12	15
Palm oil B100 + 0.5 vol% ethers	14	11	14
Palm oil B100 + 1.0 vol% ethers	14	11	14
Soybean oil B100	0	–5	–2
Soybean oil B100 + 0.1 vol% ethers	0	–5	–2
Soybean oil B100 + 0.5 vol% ethers	–2	–10	–7
Soybean oil B100 + 1.0 vol% ethers	–1	–6	–3

Another class of glycerol derivatives with potential for use as fuel additives are the acetals and ketals. They are produced through the acid-catalysed reaction of glycerol with aldehydes and ketones, respectively. Many glycerol acetals and ketals have industrial applications, such as surfactants (Piasecki *et al.*, 1997), flavouring agents (Climent *et al.*, 2002) and in drug formulations (Sari *et al.*, 2004). The reaction of acetone with glycerol produces one ketal, known as solketal, whereas reaction with formaldehyde solution affords two acetal isomers (da Silva *et al.*, 2009b) (Fig. 10.12).

Figures 10.13 and 10.14 show the glycerol conversion in the reaction with acetone and formaldehyde solution, respectively, for different acidic catalysts. One can see that the use of a heterogeneous catalyst is particularly indicated, especially in the reaction with formaldehyde solution, owing to the large amount of water in the medium. Zeolite HBeta with a high Si/Al ratio shows the best performance, with conversion in the range of 90% within 60 min reaction time. Silicon-rich zeolites are known to possess hydrophobic properties and are particularly relevant in reactions where water is present (Okuhara, 2002). On the other hand, zeolites HUSY and HZSM–5 give poor results. The first is rich in aluminium and, in this way, is hydrophilic, probably being affected by the water formed or present in the medium, whereas zeolite HZSM–5 is known for its shape-selective properties (Smith and Maessen, 2008) and probably cannot fit the transition state for acetal/ketal formation within the pores.

The glycerol/acetone ketal, solketal, was tested in mixtures with gasoline. Within 5 vol% addition, the octane number improved and gum formation was significantly reduced (Fig. 10.15), without affecting other important properties of the gasoline, such as the vapour pressure (Mota *et al.*, 2010). These data were obtained for gasoline with 25% ethanol, the regular gasoline used in Brazil, and without ethanol. It is interesting to note that glycerol/formaldehyde acetals showed the opposite trend. Gum is normally formed by complex polymerization reactions, mostly involving unsaturated hydrocarbons present in the gasoline. It is

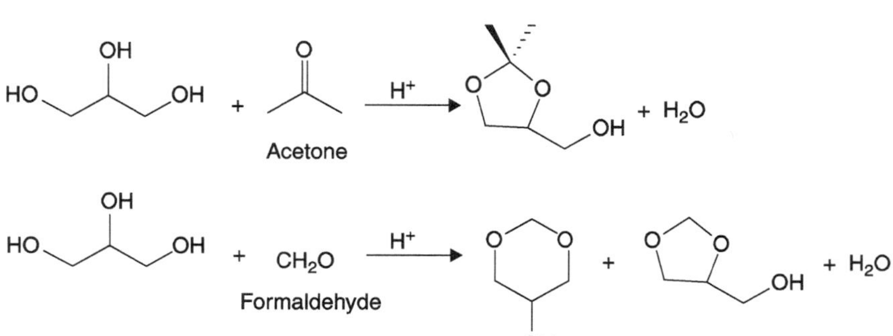

10.12 Reaction of glycerol with acetone and formaldehyde in the presence of acid catalysts.

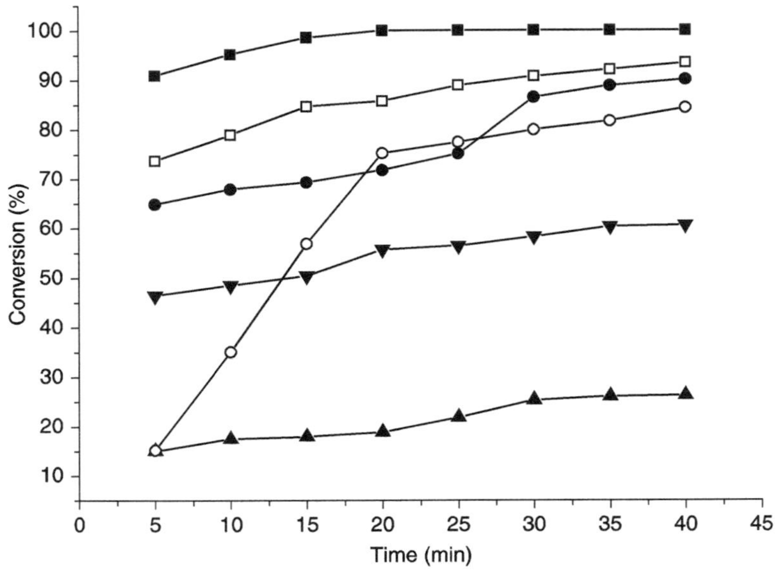

10.13 Kinetics of the glycerol reaction with acetone at 70°C over various acid catalysts. (■) Amberlyst-15, (□) zeolite Beta, (•) K-10, (O) *p*-toluene sulphonic acid, PTSA, (▼) HUSY, (▲) zeolite ZSM-5 (reprinted with permission from RSC).

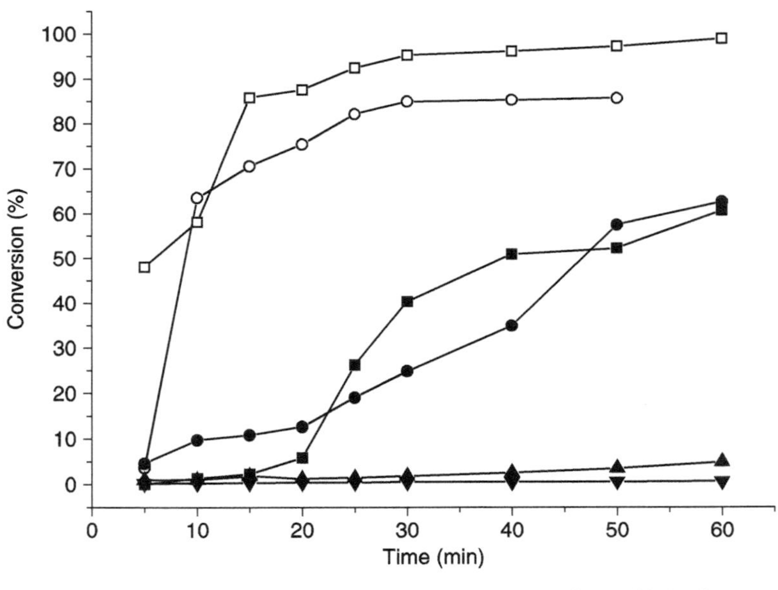

10.14 Kinetics of the glycerol reaction with aqueous formaldehyde at 70°C over various acid catalysts. (■) Amberlyst-15, (□) zeolite Beta, (•) argila K-10, (O) PTSA, (▼) HUSY, (▲) zeolite ZSM-5 (reprinted with permission from RSC).

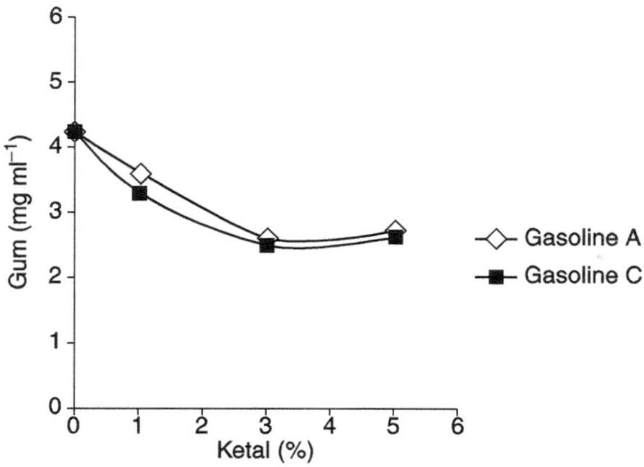

10.15 Gum formation in gasoline blended with solketal. Gasoline A refers to a gasoline without ethanol and gasoline C refers to a gasoline with 25 vol% ethanol.

particularly important to prevent gum formation in gasoline produced from catalytic cracking of gasoil, where the concentration of olefins is usually around 30%. This type of gasoline is common in many countries owing to the popularity and versatility of the catalytic cracking process. Although acetone is produced today from petrochemical feedstock, it can be produced from sugars, by fermentation procedures (Jones and Woods, 1986), making solketal a completely renewable oxygenated compound with potential for use in gasoline.

Acetals produced in the reaction between glycerol and n-alkylaldehyde have found application as additives for biodiesel, improving the cold flow properties (Silva *et al.*, 2010a). The best results were found with butyraldehyde. As the hydrocarbon chain increases, the effect on the pour point is less relevant. In addition, the glycerol conversion decreases with the increase of the hydrocarbon chain.

Glycerol esters are an important class of compounds. Mono and diacylglycerides are normally produced by partial hydrolysis of fats and oils and have use as surfactants. A useful procedure for selectively producing monoacylglycerides involves the reaction of glycerol with acetone to form solketal. In the sequence, the free hydroxy group of solketal can undergo transesterification with methyl esters, followed by acid hydrolysis (Fig. 10.16). Direct esterification of glycerol usually affords a mixture of mono, di and sometimes triglycerides. The selectivity for monoacylglycerides can be improved using a large excess of glycerol. For instance, the esterification of glycerol with oleic acid in the presence of sulphonic acid resin yields the monoacylglyceride in 90% selectivity for a conversion of 54%, using a 6-fold molar excess of the carboxylic acid (Pouilloux *et al.*, 1999).

The acetins or glycerol acetates are useful compounds. Monoacetin is a dense hygroscopic liquid used as a gelling agent and in the production of explosives.

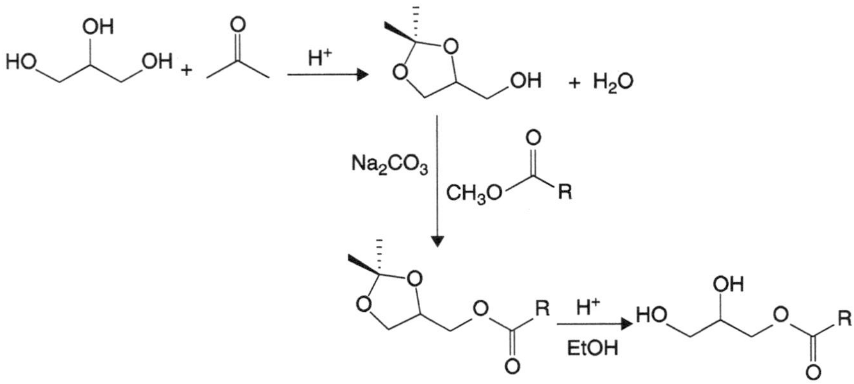

10.16 Three-step route to monoacylglycerides.

Diacetin has application as a lubricant and solvent, whereas triacetin is important in the tobacco industry. More recently, triacetin has been tested as a fuel additive, especially for biodiesel, improving the viscosity and the pour point (Melero *et al.*, 2007).

The most traditional method of preparation of the acetins is the direct esterification of glycerol with acetic acid in the presence of an acidic catalyst (Figure 10.17). The reaction affords a mixture of the acetins. To increase the selectivity for triacetin, a large excess of acetic acid should be used. The production of triacetin in high selectivity by the direct reaction with acetic acid is difficult,

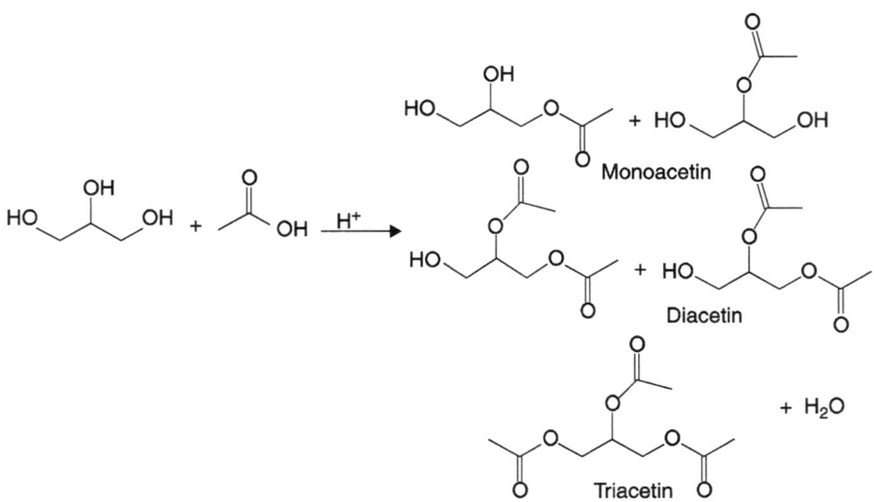

10.17 Acetylation of glycerol with acetic acid: production of acetins.

because the reaction is consecutive and the water formed may shift the equilibrium. Many heterogeneous catalysts have been tested, such as sulphonic acid resins, clays and zeolites (Gonçalves *et al.*, 2008), as well as heteropolyacids (Ferreira *et al.*, 2009) and mixed oxides.

The selectivity for triacetin can be significantly increased with the use of acetic anhydride. It can be used together with acetic acid (Liao *et al.*, 2009) or directly in the reaction with glycerol. In both cases, the selectivity for triacetin can be 100%. The reaction of glycerol with acetic anhydride can be carried out in the presence of heterogeneous acid catalysts (Silva *et al.*, 2010b). Zeolite HBeta and K–10 montmorillonite show the best results, with 100% selectivity for triacetin at 60°C and 20 min reaction time. Under the same conditions, the uncatalysed reaction affords a mixture of 10% monoacetin, 56% diacetin and 34% triacetin. It is interesting to note that zeolite HBeta presents the opposite behaviour when acetic acid is used as reagent. The product distribution favours formation of mono and diacetin in similar amounts in the uncatalysed reaction. These opposite results have been explained in terms of the formation of the acyl cation intermediate. Whereas the formation of the cation may be nucleophilically assisted by the zeolite framework when acetic anhydride is the reactant, it is formed without assistance when acetic acid is employed. This latter case may lead to a higher activation energy, slowing down the reaction.

Esterification of the free hydroxy group of glycerol ketals and acetals has also been reported in the literature (Garcia *et al.*, 2008). The reaction of solketal with acetic anhydride in the presence of triethylamine produces solketal acetate in 90% yield (Fig. 10.18). This product may be used to improve the viscosity of the biodiesel, without affecting the flash point. It may also reduce the formation of particulates in diesel oil.

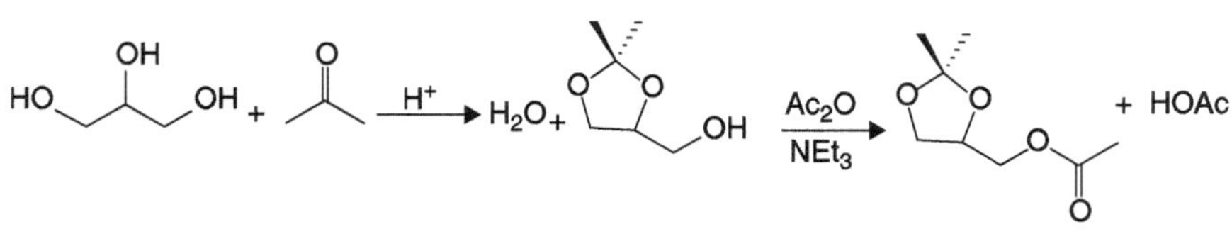

10.18 Production of solketal acetate.

10.4 Advantages and limitations of the use of crude glycerol from biodiesel production

Any chemical process becomes more attractive if the price of the raw materials is competitive. The main interest in developing new processes and products from glycerol resides in its large and cheap supply from biodiesel production. Thus, any direct use of the crude glycerol, as it comes from the biodiesel plants, will be economically more attractive. However, the glycerol phase contains many impurities, as mentioned in Table 10.1 and some of them may be deleterious to the catalysts used in further chemical transformations. The dissolved alkaline catalyst used for transesterification affects any acid material that could be employed in glycerol etherification, esterification and acetalization. Thus, neutralization is essential to assure high efficiency of the process. Nevertheless, the salts produced in neutralization procedures may also be a problem. In any gas phase process of glycerol conversion, the ashes will be deposited on the catalyst surface, altering its properties.

There are few studies on the use of crude glycerol in chemical processes, in part because many processes are still in development, it being more important to find better catalysts and reaction conditions, rather than working with impure glycerol, which would make the task still more difficult and complex. On the other hand, the purity of glycerol from biodiesel production may vary from case to case and generalizations should be made with care.

Recently, a study of the effect of impurities on the acid-catalysed reaction of glycerol with acetone was carried out (da Silva and Mota, 2011). It was found that the concomitant presence of water and salts, such as NaCl, significantly affects the conversion when sulphonic acid resin or zeolite HBeta were used as catalysts. The zeolite is more resistant to the presence of water, but salts neutralize the acid sites, leading to lower conversions. It is interesting to note that neither NaCl nor water alone, within 15 wt%, are so detrimental to the conversion. The biggest problem appears when both impurities are present in the medium, suggesting that ionic mobility plays a decisive role. The results with adulterated glycerol were similar to the results obtained with a crude glycerine sample of an industrial biodiesel plant, presenting roughly the same contaminants. These data indicate that purification of the glycerol, as well as modifications in the biodiesel transesterification processes are required in order to have a glycerol that can be used in further chemical transformations.

10.5 Future trends

In the next few years, glycerol will certainly occupy an important position as a renewable feedstock for the chemical industry. As the worldwide utilization of biodiesel increases, the surplus of this bio-product will become attractive and new applications will emerge. Notwithstanding, purity is still a major concern. Most biodiesel plants in the world still use homogeneous basic catalysts, which virtually end in the glycerol phase upon process work up. Neutralization of the catalyst with mineral acids leads to the dissolution of salts, such as NaCl, in the glycerol phase and may affect the performance of the heterogeneous catalysts used in further chemical transformations.

The development of better transesterification processes, especially using heterogeneous catalysts, may significantly improve the quality of the glycerol produced, with less impurities and virtually no ashes that can be deleterious to the catalysts used in further transformations. In addition, fast and cheap purification processes, using membranes and polymers should be consolidated (Carmona *et al.*, 2009). The main impurity affecting further utilization of glycerol is dissolved salts. If they can be removed, or at least significantly reduced in the glycerol phase, without the need of distillation, the cost and quality of the glycerol that is to be further used in chemical processes will be competitive.

There are many developments of glycerol-based processes applied to polymer production. The hydrogenolysis to propylene glycol is an established one, with industrial facilities being constructed all over the world. The process is operated in continuous flow mode and uses metal-supported catalysts. A variation of this reaction is the more severe hydrogenolysis to propene. This is a unique development, but it still requires major development before being operated on the industrial scale. It may open up new perspectives for the use of glycerol, because propene is one of the largest polymer building blocks nowadays.

The Epicerol process, developed by Solvay, offers an opportunity to produce epichloridrin from glycerol, whereas the French company, Arkema, is intensely studying a process to convert glycerol into acrylic acid. Today, acrylic acid is produced in a two-step process from propene. The possibility of developing a one-step process, through the use of bifunctional catalysts performing the oxidative dehydration of glycerol would be a major achievement. All these developments have the support of major chemical companies and can be considered technically feasible.

The synthesis of glycerol ethers, acetals/ketals and esters offers the possibility of utilizing glycerol in the fuel sector. Many studies show the potential of these glycerol derivatives as additives for gasoline, diesel and biodiesel, but the production process of many of these derivatives requires further optimization. An important concern is economic viability, because the fuel market holds large quantities and the profit gains reside on the amount sold, rather than on the aggregated value. Whereas ethyl glyceryl ethers, produced in the reaction of glycerol with ethanol are completely renewable compounds, solketal and triacetin are only partly renewable. Acetone and acetic acid are still produced from fossil sources, although they can both be produced from renewable ones. Yet, the main problem is the economics of the renewable procedures, which are still not competitive with respect to the traditional petrochemical route for these compounds.

An emerging area of study of glycerol conversion is the production of chiral molecules or synthons (construction blocks) useful for the synthesis of pharmaceuticals. The chiral derivatives can be either produced by separation methods or by direct synthesis, using chirality inductors or enzymes. Enantiomeric enriched solketal can be produced in three steps with the use of di-butyltin oxide and kinetic resolution (Mukaiyama *et al.*, 1984). Enzymes are also effective in producing glycerol derivatives with enantiomeric excess (e.e.). Chiral monoacylglicerides can be produced in the reaction of glycerol and anhydrides in

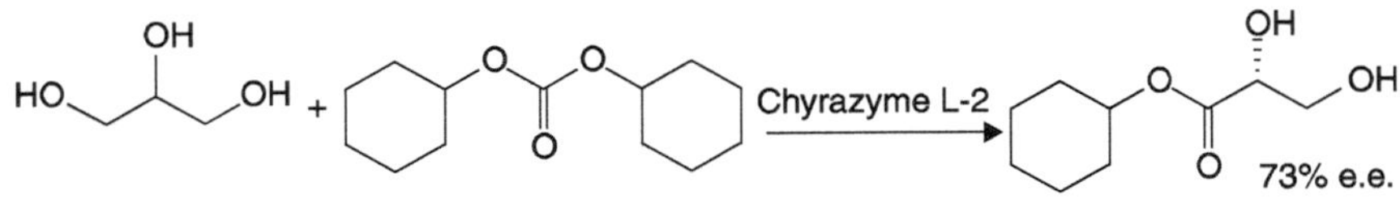

10.19 Production of 1-mono-acyl-glyceryl esters with enantiomeric excess.

the presence of lipase enzymes (Batovska *et al.*, 2004). The yields are poor to moderate. For instance, the reaction of glycerol with *p*-methylbenzoic anhydride produces the 1-mono-acyl-glyceryl ester in 23% yield with 63% e.e., whereas with cyclohexanoic anhydride the yield goes to 60%, with e.e. of 73% (Fig. 10.19).

In summary, glycerol will be one of the most important renewable feedstocks in the near future. Together with bioethanol, glycerol will be an important raw material for the production of plastics. Bioethanol is a good option for ethylene-based polymers and glycerol will be the best option for propylene-based polymers. Some developments are a reality nowadays, whereas others still need further development. As fuel additives, glycerol derivatives can occupy an increasing market share, especially in the biodiesel industry. Many additives in biodiesel can be made from glycerol, integrating the whole industrial chain and draining huge amounts of the glycerol produced in transesterification.

10.6 Sources of further information

In recent years, there have been many reviews and books regarding the chemical transformation of glycerol. The reader may find additional information in the following published material:

Book

Pagliaro M and Rossi M (2008), *The Future of Glycerol. New Usages for a Versatile Raw Material*, RSC, Cambridge, UK.

Reviews

Beatriz A, Araújo Y J K and de Lima D P (2011), 'Glycerol: A brief historic and application in stereoselective synthesis', *Quim Nova*, **34**, 306–19.
Behr A, Eilting J, Irawadi K, Leschinski J and Lindner F (2008), 'Improved utilisation of renewable resources: New important derivatives of glycerol', *Green Chem*, **10**, 13–30.
Huber G W, Iborra S and Corma A (2006), 'Synthesis of transportation fuels from biomass: chemistry, catalysts and engineering', *Chem Rev*, **106**, 4044–98.
Jérôme F, Pouilloux Y and Barrault J (2008), 'Rational design of solid catalysts for the selective use of glycerol as a natural organic building block', *ChemSusChem*, **1**, 586–613.

Mota C J A, da Silva C X A and Gonçalves V L (2009), 'Glycerochemistry: New products and processes from glycerine of biodiesel production', *Quim Nova*, **32**, 639–48.
Pagliaro M, Ciriminna R, Kimura H, Rossi M and Pina C D (2007), 'From glycerol to value-added products', *Angew Chem Int Ed*, **46**, 4434–40.
Zeng Y, Chen X and Shen Y (2008), 'Commodity chemicals derived from glycerol, an important biorefinery feedstock', *Chem Rev*, **108**, 5253–77.

10.7 References

Ahmed F E (2001), 'Toxicology and human health effects following exposure to oxygenated or reformulated gasoline', *Toxicology Lett*, **123**, 89–113.
Anastas P and Eghbali N (2009), 'Green chemistry: Principles and practice', *Chem Soc Rev*, **39**, 301–12.
Aresta M, Dibenedetto A, Nocito F and Patore C (2006), 'A study on the carboxylation of glycerol to glycerol carbonate with carbon dioxide: The role of the catalyst, solvent and reaction conditions', *J Molec Catal A*, **257**, 149–52.
Batovska D I, Tsubota S, Kato Y, Asanoa Y and Ubukatab M (2004), 'Lipase-mediated desymmetrization of glycerol with aromatic and aliphatic anhydrides', *Tetrahedron: Assymetry*, **15**, 3551–9.
Bianchi C L, Canton P, Dimitratos N, Porta F and Prati L (2005), 'Selective oxidation of glycerol with oxygen using mono and bimetallic catalysts based on Au, Pd and Pt metals', *Catal Today*, **102**, 203–12.
Braskem (2010), *Braskem Inaugurates Green Ethylene Plant in Triunfo (Rs), Becoming the Global Biopolymer Leader*, São Paulo. Available from: http://www.braskem.com.br/site/portal_braskem/en/sala_de_imprensa/sala_de_imprensa_detalhes_10338.aspx.
Carmona M, Lech A, de Lucas A., Perez A and Rodrigues J F (2009), 'Purification of glycerol/water solutions from biodiesel synthesis by ion exchange: sodium and chloride removal Part II', *J Chem Tech Biotech*, **84**, 1130–5.
Carrettin S, McMorn P, Johnston P, Griffin K, Kiely C J, *et al.* (2004), 'Oxidation of glycerol using supported gold catalysts', *Top Catal*, **27**, 131–136.
Casali B and Gomez A M, 1994. *Catalytic Method of Hydrogenating Glycerol*. US patent application 5276181. 1994-Jan–4.
Chai S H, Wang H P, Liang Y and Xu B Q (2007), 'Sustainable production of acrolein: investigation of solid acid–base catalysts for gas-phase dehydration of glycerol', *Green Chem*, **9**, 1130–6.
Chaminand J, Dajakovitch L, Gallezot P, Marion P, Pinel C and Rosierb C (2004), 'Glycerol hydrogenolysis on heterogeneous catalysts', *Green Chem*, **6**, 359–61.
Ciriminna R, Palmisano G, Della Pina C, Rossi M and Pagliaro M (2006), 'One-pot electrocatalytic oxidation of glycerol to DHA', *Tetrahedron Lett*, **47**, 6993–5.
Climent M J, Velty A and Corma A (2002), 'Design of a solid catalyst for the synthesis of a molecule with blossom orange scent', *Green Chem*, **4**, 565–9.
Dasari M A, Kiatsimkul P P, Sutterlin W R and Suppes J (2005), 'Low-pressure hydrogenolysis of glycerol to propylene glycol', *Appl Catal A*, **281**, 225–31.
Deleplanque J, Dubois J L, Devaux J F and Ueda W (2010), 'Production of acrolein and acrylic acid through dehydration and oxydehydration of glycerol with mixed oxide catalysts', *Catal Today*, **157**, 351–8.
Ferreira P, Fonseca I M, Ramos A M, Vital J and Castanheiro J E (2009), 'Esterification of glycerol with acetic acid over dodecamolybdophospohoric acid encaged in USY', *Catal Commun*, **10**, 481–4.

Garcia E, Laca M, Pérez E, Garrido A and Peinado J (2008), 'New class of acetal derived from glycerine as a biodiesel fuel component', *Energy Fuels*, **22**, 4274–80.
Garcia R, Bessom M and Gallezot P (1995), 'Chemoselective catalytic-oxidation of glycerol with air on platinum metals', *Appl Catal A*, **127**, 165–76.
Gonçalves V L C, Pinto B P, Silva J C and Mota C J A (2008), 'Acetylation of glycerol catalyzed by different solid acids', *Catal Today*, **673**, 133–5.
Gong L, Lu Y, Ding Y, Lin R, Li J, *et al.* (2009). 'Solvent effect on selective dehydroxylation of glycerol to 1,3-propanediol over a Pt/WO_3/ZrO_2 catalyst', *Chin J Catal*, **30**, 1189–91.
Hekmat D, Bauer R and Neff V (2007), 'Optimization of the microbial synthesis of dihydroxyacetone in a semi-continuous repeated-fed-batch process by in situ immobilization of *Gluconobacter oxydans*', *Proc Biochem* **42**, 71–6.
Jones D T and Woods D R (1986), 'Acetone–butanol fermentation revisited', *Microbiol Rev*, **50**, 484–524.
Kampe P, Giebelder L, Smuelis D, Kunert J, Drochner A, *et al.* (2007), 'Heterogeneously catalysed partial oxidation of acrolein to acrylic acid – structure, function and dynamics of the V-Mo-W mixed oxides', *Phys Chem Chem Phys*, **9**, 3577–89.
Klepacova K, Mravec D, Hajekova E and Bajus M (2003), 'Etherification of glycerol', *Petroleum and Coal*, **45**, 54–7.
Klepacova K, Mravec D and Bajus M. (2005), 'tert-Butylation of glycerol catalysed by ion-exchange resins', *Appl Catal A*, **294**, 141–7.
Liao X, Zhu Y, Wang S G and Li Y (2009), 'Producing triacetylglycerol with glycerol by two steps: esterification and acetylation', *Fuel Proc Tech*, **90**, 988–93.
Melero J A, van Grieken R, Morales G and Paniagua M (2007), 'Acidic mesoporous silica for the acetylation of glycerol: synthesis of bioadditives to petrol fuel', *Energy Fuel*, **21**, 1782–91.
Miyazawa T, Kusunoki Y, Kunimori K and Tomishige K (2006), 'Glycerol conversion in the aqueous solution under hydrogen over Ru/C plus an ion-exchange resin and its reaction mechanism', *J Catal*, **240**, 213–21.
Mota C J A and Gonçalves V L C (UFRJ) (2007). *Etherification Processes of Glycerine and Additives for Fuels*. Brazilian patent application 0700063–4. 2007-Jan–19.
Mota C J A, Gonçalves V L C and Rodrigues R C (UFRJ) (2007), *Process for the Production of Glycerine Carbonate*. Brazilian patent application 0706121–8. 2007-Jul–23.
Mota C J A, Gonçalves V L C, Gambetta R and Fadigas J C (Quattor Petrochemicals) (2009), *Preparation of Heterogeneous Catalysts used in Selective Hydrogenation of Glycerine to Propene and a Process for the Selective Hydrogenation of Glycerine to Propene*. WO patent application 155674 A2. 2009-Dec–30.
Mota C J A, Silva C X A, Rosenbach N, Costa J and Silva F (2010), 'Glycerine derivatives as fuel additives: the addition of glycerol/acetone ketal (solketal) in gasolines', *Energy Fuels*, **24**, 2733–6.
Mukaiyama T, Tanabe Y and Shimizu M (1984), 'Highly efficient asymmetric-synthesis of (s)–2,2-dimethyl–1,3-dioxolane–4-methanol from glycerol', *Chem Lett*, **3**, 401–4.
Nadim F, Zack P, Hoag G E and Liu S L (2001), 'United States experience with gasoline additives', *Energy Policy*, **29**, 1–5.
Okuhara T (2002), 'Water-tolerant solid acid catalysts', *Chem Rev*, **102**, 3641–3666.
Pariente S, Tanchoux N and Fajula F (2009), 'Etherification of glycerol with ethanol over solid acid catalysts', *Green Chem*, **11**, 1256–61.
Pestana C F M (2010), *Dehydration of Glycerol with Heterogeneous Acid Catalysts: Formation of Acrolein and Acrylic Acid.* Master's Thesis, Federal University of Rio de Janeiro, Rio de Janeiro.

Piasecki A, Sokolowski A, Burczyk B and Kotlewska U (1997), 'Chemical structure and surface activity. 30. Synthesis and surface properties of chemodegradable anionic surfactants: Sodium (2-n-alkyl–1,3-dioxan–5-yl) sulfates', *J Am Oil Chem Soc*, **74**, 33–7.

Pinto B P (2009), *Etherification of Glycerol with Alcohols Catalyzed by Solids Acids*. Master's Thesis, Federal University of Rio de Janeiro, Rio de Janeiro.

Pouilloux Y, Abro S, Vanhove C and Barrault J (1999), 'Reaction of glycerol with fatty acids in the presence of ion-exchange resins – Preparation of monoglycerides', *J Mol Catal A*, **149**, 243–54.

Quintella C M A L T M H, Barbosa L C F, Mocyzdlower P and Pegoraro R T (Petrobras) (2009). *Process of advanced petroleum recovery by the use of solutions of co-products of biodiesel production*. Brazilian application 0901604–0 A2. 2009-May–07.

Sari P, Razzak M and Tucker I G (2004), 'Isotropic systems of medium-chain mono- and diglycerides for solubilization of lipophilic and hydrophilic drugs', *Pharm Dev Technol*, **9**, 97–106.

da Silva C X A and Mota C J A (2011), 'The influence of impurities on the acid-catalyzed reaction of glycerol with acetone', *Biomass Bioenergy*, **35**, 3547–51.

da Silva C R B, Gonçalves V L C, Lachter E R and Mota C J A (2009a), 'Etherification of glycerol with benzyl alcohol catalyzed by solid acids', *J Braz Chem Soc*, **20**, 201–4.

da Silva C X A, Gonçalves V L C and Mota C J A (2009b), 'Water-tolerant zeolite catalyst for the acetalisation of glycerol', *Green Chem*, **11**, 38–41.

Silva P H R, Gonçalves V L C and Mota, C J A (2010a), 'Glycerol acetals as antifreezing additives for biodiesel', *Bioresour Technol*, **101**, 6225–9.

Silva L N; Gonçalves V L C and Mota C J A (2010b), 'Acetylation of glycerol with acetic anhydride', *Catal Commun*, **11**, 1036–9.

Smith B and Maessen T L M (2008), 'Towards a molecular understanding of shape selectivity', *Nature*, **451**, 671–8.

Soares R R, Simonetti D A and Dumesic J A (2006), 'Glycerol as a source for fuels and chemicals by low-temperature catalytic processing', *Angew Chem Int Ed*, **45**, 3982–5.

Tsukuda E, Sato S, Takahashi R and Sodesawa T (2007), 'Production of acrolein from glycerol over silica-supported heteropoly acids', *Catal Commun*, **8**, 1349–53.

Vieville C, Yoo J W, Pelet S and Mouloungui Z (1998), 'Synthesis of glycerol carbonate by direct carbonatation of glycerol in supercritical CO_2 in the presence of zeolites and ion exchange resins', *Catal Lett*, **56**, 245–7.

Wawrzetz A, Peng B, Hrabar A, Jentys, A, Lemonidou A A and Lercher J A (2010), 'Towards understanding the bifunctional hydrodeoxygenation and aqueous phase reforming of glycerol', *J Catal*, **269**, 411–20.

Wessendorf R (1995), 'Glycerinederivate als kraftstoffkomponenten', *Erdoel & Kohle, Erdgas, Petrochemie*, **48**, 138–43.

Zhou C, Beltramini J N, Fan Y X and Lu G Q (2008), 'Chemoselective catalytic conversion of glycerol as a biorenewable source to valuable commodity chemicals', *Chem Soc Rev*, **37**, 527–49.

Appendix

Supply chains, techno-economic assessment and market development for second generation biodiesel

D. M. YAZAN, J. CLANCY, and J. C. LOVETT,
University of Twente, The Netherlands

Abstract: This chapter provides a brief review of the environmental and economic assessment of second generation biodiesel supply chains (SGBSC) and the development and coordination of emerging feedstock markets. Common problems faced in SGBSC processes are briefly explained within four categories. An enterprise input–output model is proposed to evaluate total environmental benefits of second generation biomass use instead of first generation biomass in extended supply chains (SC). The bargaining power and willingness to cooperate among extended SC actors are measured by four extreme scenarios in order to understand how an emerging feedstock market is coordinated. Moreover, potential government incentives for chain actors are also proposed and discussed.

Key words: enterprise input–output, feedstock markets, market coordination, second generation biodiesel supply chain.

A1.1 Introduction

Supply chains represent the pathway that a product takes during the process of transformation from raw material to finished product. They can be complex and interlinked, with different supply chains (SC) being integrated and actors along the SC interacting in a range of different ways. In the framework of SC coordination, economic, operational and environmental drivers influence the willingness of the firms involved to cooperate, and affect the bargaining power of suppliers, consequently influencing benefit sharing among them. Therefore, firms must engage in collaboration as a strategy to overcome resource and technology limitations (Dautzenberg and Hanf, 2008). Second generation biodiesel supply chains (SGBSC) are particularly affected by collaborative strategies because of the current limited viability owing to the present relatively high cost of production. However, second generation biodiesel (SGB) has a high potential if these limitations can be overcome. A better managed SC within operational and organizational perspectives would increase the environmental, economic and social performance of the whole chain.

The conversion of biomass to liquid fuels that can be used to power internal combustion engines, particularly for road transport, is a logical way to reduce reliance on fossil fuels. The transition from oil to biomass, even if it is only partial, has a number of policy drivers (Poudyal and Lovett, 2010; Lovett *et al.*, 2011).

First, and this is probably one of the most important political dimensions, is the issue of energy security. There is increased competition for fossil fuel resources from emerging and renascent economies which, combined with financial speculation on commodities and political unrest in oil producing regions, has led to price volatility and increase. Biofuels offer an alternative source of fuel which can either be 'home-grown' or imported on the open market from a wide range of countries, as supply is not constrained by geology as in the case of oil. Second, home-grown biofuels can be the basis for creation of a new land-use economy. For example they were considered to provide potential for European harmonization through increasing land-use productivity when Bulgaria and Romania joined the EU in 2007 (Poudyal and Lovett, 2010). Bulgaria has since increased its biodiesel consumption to 38 911 tons in 2010 from 4260 and 6566 tons in the previous two years (EBRD, 2011). Third, in principle, biofuels are derived from renewable sources in that they harness photosynthesis to create biomass. They can thus be used to meet obligations for reduction in greenhouse gas (GHG) emissions under United Nations Framework Convention on Climate Change (UNFCCC) negotiations and help to mitigate global warming whilst enabling present generations to meet their needs, in other words, they are sustainable (Lovett *et al.*, 2011). Fourth, in line with the 'Porter Hypothesis' increased environmental regulation, such as that associated with international and national commitments to reduce GHG emissions, enhances competition and innovation (Porter and van der Linde, 1995) as witnessed by the rapid growth in renewable energy technologies (Johnstone *et al.*, 2010), including those for second generation biodiesel (Luque *et al.*, 2010).

The socio-economic, environmental and policy issues of biodiesel production are covered in Chapter 3 of this volume (De Lucia and Datta, 2011). Here we briefly review some of the numerous concerns that have been raised over the sustainability of biofuels in order to put the supply chains into perspective. Essentially, the core issue is that to supply the quantity of biofuel needed to provide a substantial substitution for fossil fuels, huge amounts of biomass are required. If first generation biofuels derived from food crops produced by industrial agriculture are used in the fuel supply chain, then GHG emission savings may not be as great as might be at first thought because fossil fuel inputs to crop fertilizers and GHG losses from soil and vegetation caused by opening up new agricultural land engender emissions from existing carbon stocks (Fargoine *et al.*, 2008; Royal Society, 2008; Searchinger *et al.*, 2008). Moreover, there are concerns about the environmental impact, like water use and pollution, and social impacts, like land rights conflicts, that it is feared a rapid expansion in growing biofuels could bring (Clancy, 2008).

Publicity has also been given to the idea that converting food crops into biofuel can cause food price rises, although analysis of food price rise increases shows that the underlying causes are complex, involving climate-related crop failure, increased demand from rapidly emerging economies and financial speculation on crop futures (Headey and Fan, 2008; Trostle, 2008; Wiggins *et al.*, 2009; Headey *et al.*, 2010; Gilbert, 2010). Second generation biofuels from non-food sources such as crop residues, forestry and food processing waste could overcome some of the sustainability issues in that they are essentially a by-product of existing (albeit perhaps unsustainable) industry or, in the case of forestry for biofuels, they retain the ecological integrity of the landscape (Dale *et al.*, 2011). In terms of the scale of production needed to meet demand, the forestry sector is well placed. For example, pulp and paper mills have to be large to be economic and are often located with access to marine docking facilities or other means of freight transport. If second generation biofuel manufacture was included in their operation then the industrial scale makes them suitable for production and movement of the quantity of biofuels required under policy and legal frameworks such as the European Renewable Energy Directive (EU, 2009).

Because of concerns about biofuels, sustainability criteria have been introduced into both mandatory policies, such as the EU Renewable Energy Directive and into voluntary industry guidelines such as the Roundtable on Sustainable Palm Oil. Compliance with these criteria requires analysis of supply chains and inputs and outputs of production. Debates on SGB usually focus on technical (e.g. raw material supply, conversion technology, investments), economic (e.g. fuel costs), organizational (e.g. cooperation among actors), and/or environmental (e.g. GHG emissions, efficient land use) factors influencing its acceptability (Festel, 2008). These factors are components of the whole SGBSC and an overall approach is required for their comprehensive evaluation.

SG biofuels are expected to be better than first generation biofuels in some supply chain components such as energy balance, GHG emission reductions and competition for land and food. Despite these advantages and the potential for large scale SGB production from crop residues and the forestry industry, the main reason SGB do not have a strong market position is that the necessary conversion technologies have not yet been proven on a commercial scale and their production costs (including SC-related costs) are estimated to be higher than those of first generation biodiesels (FGB) (Carriquiry *et al.*, 2010; James *et al.*, 2010; Sims *et al.*, 2010). Significant developments will be required to improve SGB technology pathways and the performance and reliability of the biomass conversion process (IEA, 2008) and a move towards SG biomass will require a shift in known business models which need government support in order to enhance SGB. This would reduce the perceived risk in SGB investments which are probably the biggest barrier to SGB production.

According to the International Energy Agency (IEA, 2008), misunderstandings about environmental/energy tradeoffs require a systematic evaluation of the impact of expanded SGB production. This indicates the importance of a complete economic and environmental analysis for emerging SCs. In the next section we provide general information about the processes of four categories of SGBSCs. In Section A1.3 we introduce an enterprise input–output (EIO) model to predict the environmental returns of extended SCs including SGB. The useful capability of the input–output (IO) approach that distinguishes it from other methods is the ability to dissect complicated systems of physical flows and effectively trace the path of flows (Bailey *et al.*, 2004). We use this advantage and model SGBSCs and the other SCs providing feedstock to SGBSCs and receiving by-products of SGBSCs by using an EIO approach to compute total material/energy/waste flows. How the environmental analysis results are integrated into economic analysis is explained in Section A1.4 with an explanation of how it relates to the four feedstock options discussed in Section A1.5.

Bioenergy systems require functioning and organized SCs and their development is based on the reciprocal economic benefits of suppliers and energy companies (McCormick and Kåberger, 2007). Reciprocal dependency between actors is an important factor: investing in biomass feedstock is only possible if there are companies also investing in related technology; and advancing these technologies is only possible if the suppliers provide enough biomass (Johansson and Goldemberg, 2002). Therefore, cooperation between potential chain actors is a central consideration in the coordination of SGBSCs.

Each actor in a SGBSC has their own conditions for involvement in the extended SC and finding ways of organizing actors for mutually beneficial cooperation may sometimes be problematic (e.g. in the case of competition for feedstock) while in some cases cooperation between the actors (e.g. centralized management of bioenergy chain) is inevitable. To understand how a local market of SG biomass can emerge in a specific location and how the associated SC can be coordinated, we make an analysis of actor decisions in Section A1.4. Supplier bargaining power and the willingness to cooperate by all actors are analysed in this section. Considering that, in many countries, incentives are a vital support for the viability of biofuels (see e.g. Wiesenthal *et al.*, 2009; Banse *et al.*, 2008), we also discuss potential incentive mechanisms in Section A1.6 including future trends for SGBSCs.

A1.2 Second generation biodiesel supply chains

Even though classification of SG biofuels is usually based on the technology adopted in their production, or characteristics of the final product, to classify SGBSCs we first consider the structural differences between the SCs. In this chapter we use the definition of SG biomass as 'biomass not used as food' (IEA, 2008). This covers a broad range of feedstocks and here we review SGBSCs from four contrasting categories: (1) ligno-cellulosic feedstock, (2) forestry by-products, (3) waste cooking oil and (4) jatropha-based SC.

A1.2.1 Ligno-cellulosic feedstock supply chain

Ligno-cellulosic biomass can be categorized as (1) agricultural residues and (2) forestry residues. Since forestry residues are mostly derived from wood-cutting processes, their SCs are placed in the second group. Ligno-cellulosic biomass is abundant and its estimated worldwide production is 10–50 billion dry tons including cereal straws, corn stoves, nut shells, sugarcane bagasse, and so on (Galbe and Zacchi, 2002). It is made up of cellulose, hemicellulose and lignin.

The ligno-cellulosic SC is initiated by harvesting ligno-cellulosic biomass from areas of cultivation. The removal of ligno-cellulosic residues from cultivated land is a costly process owing to high volume. Next, in the pre-treatment process, the moisture content of biomass is reduced to minimize weight and space to reduce operational and transportation costs. Feedstock moisture is also important from a technical perspective for the later thermo-chemical processing.

The next process is storage, which is inevitable because of seasonal biomass production. Storage is required to provide homogenous feedstock and to keep continuous supply and production in the conversion plant over the course of the year. To produce biodiesel from ligno-cellulosic feedstock, a crushing process is first applied to obtain a biocoke and a gas-containing tar in a low temperature gasifier. A tar-free synthesis gas is produced in a gasification reactor and then liquefied in a Fischer-Tropsch reaction to form carbohydrogen (Festel, 2008). This is a thermo-chemical conversion and does not depend on the cellulose breaking problem unlike the biochemical conversion where the physical condition (purity, water content, etc.) of the feedstock is an important factor (IEA, 2008). As in the SCs of FG biomass, spatial variables may have also a negative impact on SGBSC. A large dispersion of feedstock lands may render them costly in terms of collection and area accessibility increasing loading and transportation costs (Yazan *et al.*, 2011a).

A1.2.2 Forestry by-products supply chain

Forestry residues (e.g. sawdust, shavings, off-cuts) are mostly collected from wood processing plants and are particularly interesting for thermo-chemical processing owing to their low moisture content. Much of this type of feedstock can also be used for other options (IEA, 2008). Therefore, the potential presence of other SC actors competing for feedstock in the local area is also a possibly challenging barrier to the economic viability of forestry by-product SCs.

Compared to ligno-cellulosic feedstock SCs mentioned in the first category, forestry by-product SC are more developed primarily due to the longer experience of suppliers in the wood and paper sector. Bulky feedstock transportation can be considered to be the biggest challenge along the SC while the dispersion degree of feedstock is relatively less compared to ligno-cellulosic feedstock lands and so has a comparatively positive impact on collection costs.

A1.2.3 Waste cooking oil supply chain

Waste cooking oil feedstock has an advantage over pure vegetable oil owing to its lower cost. China has been using 40 000 to 60 000 tons waste cooking oils annually to make biodiesel (Worldwatch Institute, 2007). In most developed countries the return of waste cooking oil to the food chain, even if indirectly when used as animal feed, is banned (e.g. in EU since 2002) since dangerous compounds (Cvengroš and Cvengrošová, 2004) are formed during the frying process. Its discharge in drainage is also forbidden as the high potential for dispersion of the oil can contaminate a large volume of water and it can also congeal and block pipes. These legal restrictions indirectly support production of biodiesel from waste cooking oil by increasing availability and reducing the feedstock cost.

The collection cost of waste oil has a strong effect on feedstock cost, emphasizing the importance of collection vehicle routing, which must be optimized. Some biodiesel companies manage their own waste cooking oil SC while some purchase the feedstock from collection companies. In such cases, the presence of a third player along the SC may increase the feedstock cost due to a mark-up. The presence of third party collectors means that both vertical and horizontal integration within the SC are possible. Biodiesel companies tend to make vertical integration while collectors tend to adopt horizontal integration. Horizontal integration can keep the market competitive, pushing collectors to find solutions to reduce collection costs. On the other hand, vertical integration brings the advantage of reduced feedstock cost. Obviously, these all depend on the operational efficiency of whoever is performing the collection task. Good knowledge of feedstock sources is an advantage for reducing costs.

A1.2.4 Jatropha supply chain

The concerns expressed about using food crops as fuel feedstocks have generated interest in non-edible seeds such as *Jatropha curcas* and *Pongamia pinnata*. Jatropha has attracted the most attention particularly because, under certain conditions, it seems to have a promising oil content for energy production. However, high yields are only achieved in high quality soil with a high level of input. Irrigation and fertilization are unavoidable in order to obtain higher yields (Rajagopal and Zilberman, 2007). Although Jatropha is drought tolerant, planting of Jatropha for biofuel production in dry lands is controversial because of low yields and displacement of traditional agro-pastoral land use systems. After harvesting, the Jatropha fruit is dried. The residues of tree cutting are considered as co-product which can still be useful for energy production as a green manure.

After transportation, the Jatropha fruit oil is extracted, refined and then converted into biodiesel. Conversion of Jatropha oil into biodiesel is relatively easy compared to other cellulose-based SG biomass and opens up the opportunity for small-scale production. In the near future, small-scale plants in remote areas are expected to be constructed rather than centralized large-scale plants, which also gives an advantage of lower transportation costs by creating biodiesel supplies for local use (Prueksakorn and Gheewala, 2008).

A1.3 Enterprise input–output model for second generation biodiesel supply chain analysis

According to IEA (2008), misunderstandings about environmental/energy tradeoffs require a systematic evaluation of the impact of expanded SGB production. This indicates the importance of a complete economic and environmental analysis for emerging SCs and we will review EIO modelling to predict potential benefits and complete flows of SGBSCs.

In the examples considered here, since SG biomass is considered as waste or by-product of other SCs (e.g. food chain, wood processing chain); its use as a feedstock in SGB production has economic and environmental returns for these chains as well as to SGBSCs. There are also by-products (e.g. glycerine from a transesterification process) deriving from biodiesel SC processes and used in the processes of other SCs (e.g. cosmetics). Therefore, all the SCs involved comprise a unique joint supply chain (JSC). So, the potential returns of this alternative feedstock use must be evaluated for the whole JSC. This would enable us to foresee the potential total benefits of such JSCs and to understand how the SGBSCs and feedstock markets may emerge, develop and evolve. In Fig. A1.1 a

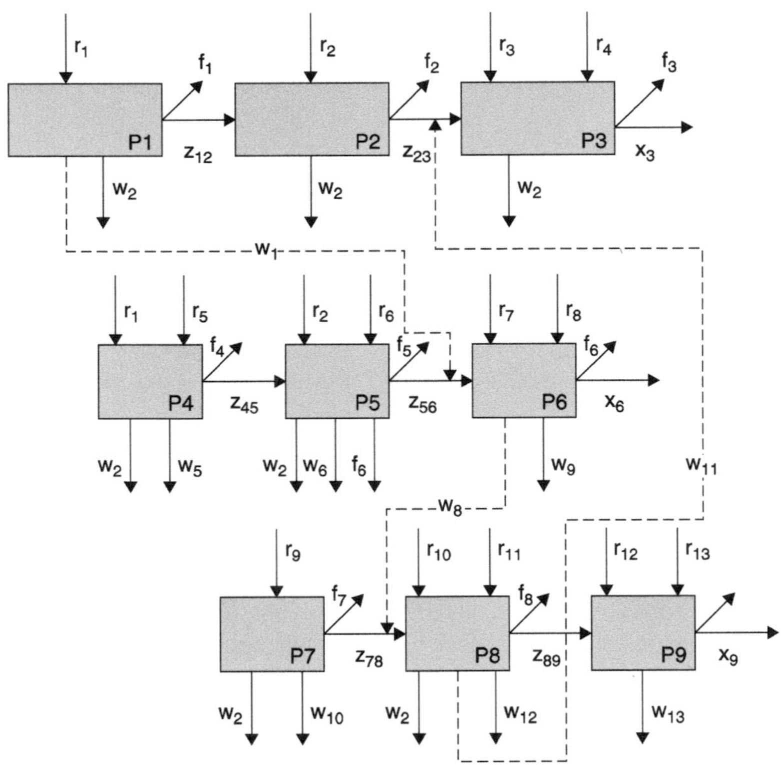

A1.1 A joint supply chain flowchart. See text for a full explanation.

JSC composed of three supply chains and nine processes is represented. Moreover there are 13 types of primary input (r) and 13 types (w) of waste.

The main output of a process can be supplied to another process as an intermediate input (e.g. P1 provides its main output to P2 as main input, z_{12}) or may just have to satisfy an external final demand (e.g. P3 provides its main output just to satisfy external final demand, f_3) or both cases may occur (e.g. P1 provides its main output for P2, z_{12} and for external final demand f_1). Three SCs have no relationship in terms of material flow before any waste recycling takes place. These chains create a JSC through recycling any waste in any chain and its use in another chain substituting for the main output of any process of that chain (e.g. w_1 of P1 is recycled and used instead of the main output of P5 or w_8 instead of the output of P7 or w_{11} instead of the main output of P2). This will cause a reduction in the production of the substituted main output inducing reductions in the production of the main outputs of its precedent processes (e.g. main output of P2 is reduced by waste substitution meaning that P1 also will reduce what it produces for P2). Furthermore, a reduction in a 'process' and 'precedent processes' outputs causes a decrease in waste quantities (and also other environmental impacts) emitted by them. Therefore, input–output modelling is an effective tool to predict what economic and environmental benefits are obtainable from a JSC and can be used for short and long term policy making.

A1.3.1 Input–output analysis

Input–output (IO) analysis has been widely applied to examine the production structure of a national economy. It is an accounting system which is characterized by a double-entry bookkeeping principle and goes back to the Nobel prize winner Leontief (1936). It can be viewed as the first full empirical implementation of a general equilibrium for a national economy. For an introduction, see Miller and Blair (2009) and for a broad selection of seminal papers see Kurz *et al.* (1998).

A1.3.2 Enterprise input–output models

In the 1990s, IO was also introduced as a tool to analyse a wide variety of questions within a production unit (Lin and Polenske, 1998), an enterprise (see e.g. Marangoni and Fezzi, 2002), a group of enterprises in an industrial district (see e.g. Albino *et al.*, 2003), or a SC (Grubbström and Tang, 2000). This type of model has become known as enterprise input–output (EIO) model.

Albino *et al.* (2007) propose an EIO model to measure the effect of different coordination policies of freight flows on the logistics and environmental performance of an industrial district. Yazan *et al.* (2011a) integrate the physical and monetary EIO modelling to measure economic and environmental performance of bioenergy production chains. Additionally, Yazan *et al.* (2011b) measure the impact of transportation in re-engineered SCs incorporating process specialization and disaggregation.

A1.3.3 The model

According to Mollenkopf *et al.* (2010) in order for managers to implement sustainable SC strategies successfully, practical tools and guidelines are needed that enable decision making among SC participants. They emphasize the importance of a measured approach through time and SC because decisions made at one phase may have important implications for efficiency and environmental performance in another echelon within the SC at different places and at different times.

The model is dynamic and able to assess the environmental performance of infinite number of SCs including infinite number of processes linked through reuse/recycling (the terms reuse and recycling are used to mean that the waste is regained in another supply chain) of infinite number of wastes and by-products (e.g. second generation biomass feedstock, by-products derived from biodiesel production). The term 'waste' mostly refers to second generation biomass feedstock, but it is used to make the model global and also valid for other reuse of other wastes in the chain. It is important that the processes of all involved chains and primary inputs and wastes are pre-defined and located within the flow chart.

In our model we consider a group of SCs in which n total processes can be distinguished and assume that each process has a single main product as its output. First, all the material/energy/waste flows of each chain are represented in the same model in order to visualize all flows in a unique IO table in the case of no waste recycling. Then a transition phase is introduced where the waste recycling is represented by a transformation (i.e. waste into main input) matrix. We assume that the recycled waste in time period t is reused instead of a main output in time period $t+1$ (e.g. substitution of SG biomass by FG biomass where SG biomass is considered as the waste of a SC and FG biomass is considered as the main output of a process of FGBSC). Using the transformation matrix, the flows are computed again for time period $t+1$. Finally, same procedure is applied in an iterative approach to compute maximum benefit in the longest possible period.

The matrix **Z** gives intermediate deliveries from a process to any other process and the demand from outside the SC for products is given by the vector **f**. The gross output of process i (= 1, . . ., n) is given by the element x_i of the vector **x**. It is obtained as the sum of the quantities of good i delivered to other processes within the chain and the quantity delivered as final demand, i.e. $x_i = \Sigma_j z_{ij} + f_i$.

We assume that each process may require intermediate inputs from the other processes, but not from itself so that the entries on the main diagonal of the matrix **Z** are zero. Further, we assume that the required amount of intermediate input i by process j is proportional to the production (or total output) in process j. That is, $z_{ij} =$

$a_{ij}x_j$ which holds for all i and j. The elements a_{ij} of the matrix **A** are the intermediate input coefficients and they are assumed to be fixed. Substituting $z_{ij} = a_{ij}x_j$ into $x_i = \Sigma_j z_{ij} + f_i$ yields $x_i = \Sigma_j a_{ij}x_j + f_i$, which can be written in matrix notation as:

$$\mathbf{x} = \mathbf{Ax} + \mathbf{f} \quad \text{[A1.1]}$$

Its solution is given by $\mathbf{x} = (\mathbf{I} - \mathbf{A})^{-1}\mathbf{f}$, where **I** indicates the $n \times n$ identity matrix. In empirical applications, information is usually available for **Z**, **f** and (thus) **x**. The intermediate input coefficients are then calculated as $a_{ij} = z_{ij}/x_j$, or $\mathbf{A} = \mathbf{Z}\hat{\mathbf{x}}^{-1}$ in matrix notation.

In addition, suppose that there are s types of primary input to purchase from outside the system and that m types of wastes are generated. Let the $s \times n$ matrix **R** be the matrix of primary input coefficients and let its element r_{kj} denote the use of primary input k (= 1, . . ., s) per unit of output of process j. Let the $m \times n$ matrix **W** be the matrix of waste coefficients with its element w_{ij} denoting the quantity of waste type l (= 1, . . ., m) generated by the production of one unit of output of process j. Then, the total requirements of the primary inputs, (**r**), is an ($sx1$) vector and the total generation of wastes, (**w**), is represented as an ($mx1$) vector.

As a starting-point take an arbitrary final demand vector **f** and the matrix of intermediate input coefficients **A**. The gross outputs, the matrix of intermediate deliveries, the generation of wastes and required primary inputs are (respectively) given by:

$$\mathbf{x} = (\mathbf{I} - \mathbf{A})^{-1}\mathbf{f} \quad \text{[A1.2]}$$

$$\mathbf{Z} = \mathbf{A}\hat{\mathbf{x}} \quad \text{[A1.3]}$$

$$\mathbf{w} = \mathbf{Wx} \quad \text{[A1.4]}$$

$$\mathbf{r} = \mathbf{Rx} \quad \text{[A1.5]}$$

Since technology is assumed to be fixed, the values of **A**, **R** and **W** remain constant, when also there are fluctuations on final demands in different periods, which is a usual case for most of SCs. For such cases, let the vector of new final demands be denoted by $\mathbf{f}^{new}$, assuming that the intermediate input coefficients have not changed. The new gross outputs would be given by the vector $\mathbf{x}^{new} = (\mathbf{I} - \mathbf{A})^{-1}\mathbf{f}^{new}$. The intermediate deliveries between the processes in the new situation would then be given by $\mathbf{Z}^{new} = \mathbf{A}\hat{\mathbf{x}}^{new}$. For the new final demands, we then would find the total requirements of primary inputs and the total generation of wastes as $\mathbf{r}^{new} = \mathbf{Rx}^{new}$ and $\mathbf{w}^{new} = \mathbf{Wx}^{new}$.

Now, we can describe how the model is adapted in the case where some of the wastes are recycled into main inputs. This applies in our case in which SG biomass (e.g. agricultural residues, cutting-process residues, bagasse) substitutes FG biomass resource (e.g. sunflower seed, canola). Assume that one unit of waste of type l (= 1, . . ., m) can be recycled into t_{lj} units of product j (= 1, . . ., n). The total amount of the different waste types (i.e. **w**) can thus be transformed into main products. This yields $x_j^{rec} = \Sigma^m_{l=1} t_{lj} w_l$ for the amount of product j obtained through recycling. In matrix notation, we have

$$\mathbf{x}^{rec} = \mathbf{T}'\mathbf{w} \qquad [A1.6]$$

where a prime is used to indicate transposition of matrices and vectors. We assume that the amount of recycled products is relatively small when compared to the original gross outputs (i.e. for each main product i we assume $x_i^{rec} < x_i$). For ease of exposition, we also assume that each type of waste or by-product can be recycled. In most real-life cases this will not apply, which implies that we would have to distinguish between recyclable wastes and non-recyclable wastes. In this way the model permits us to compute the impact of recycling of all wastes if a real life case applies. If there is no recycling of a type of waste, then its transformation coefficient equals zero.

Next, these products obtained from recycling will be used in the production processes (i.e. one main input can be used in more than one process). Part of the recycled output x_i^{rec} of product i will be used as an intermediate input in one or more of the other production processes and part will be used to satisfy final demand. Let the $n \times n$ matrix **H** and the column vector **h** indicate the distribution of the recycled outputs. Element h_{ij} gives the part of x_i^{rec} that is used in process j (= 1, . . ., n) and element h_i gives the part that goes to final demands. Note that all recycled outputs are used so that $\Sigma_j h_{ij} + h_i = 1$ holds for all i (= 1, . . ., n) obtained from recycling. Given the recycled output x_i^{rec}, the intermediate deliveries are given by $z_{ij}^{rec} = x_i^{rec} h_{ij}$ and the products for final demands are given by $f_i^{rec} = x_i^{rec} h_i$. In matrix notation we have:

$$\mathbf{Z}^{rec} = \hat{\mathbf{x}}^{rec}\mathbf{H} \qquad [A1.7]$$

$$\mathbf{f}^{rec} = \hat{\mathbf{x}}^{rec}\mathbf{h} \qquad [A1.8]$$

Whereas we have determined the gross outputs in Eq. A1.2 on the basis of the final demand vector **f**, it follows from Eq. A1.8 that it would have been sufficient for the SC to satisfy a final demand equal to $\mathbf{f} - \mathbf{f}^{rec}$, because part of the final demand is satisfied by recycled products. In the same fashion, the gross output was obtained from the intermediate input coefficients in **A**. Expression A1.7 indicates that part of the requirements are met by recycled products. If we define:

$$\mathbf{A}^{rec} = \mathbf{Z}^{rec}\hat{\mathbf{x}}^{-1} \qquad [A1.9]$$

then the input requirements would be given by $\mathbf{A} - \mathbf{A}^{rec}$.

After this point, the changes in final demand and intermediate flows coefficients can be computed by an iterative procedure. Moreover, some wastes may not be appropriate to substitute the whole of the traditional input. This may be due to any main quality issues or environmental regulations that impose a certain level of substitution. Therefore, an ε, which indicates the minimum value of traditional main input or final demand to be produced, should be introduced to stop the iteration phase.

That is, define:

$$\mathbf{A}^{(1)} = \mathbf{A} - \mathbf{A}^{rec} \qquad [A1.10]$$

$$\mathbf{f}^{(1)} = \mathbf{f} - \mathbf{f}^{rec} \qquad [A1.11]$$

after which the process is rerun. Starting with Eqs A1.2–A1.4, we have $\mathbf{x}^{(1)} = (\mathbf{I} - \mathbf{A}^{(1)})^{-1}\mathbf{f}^{(1)}$, $\mathbf{Z}^{(1)} = \mathbf{A}^{(1)}\hat{\mathbf{x}}^{(1)}$, $\mathbf{w}^{(1)} = \mathbf{W}\mathbf{x}^{(1)}$ for gross output, intermediate deliveries and wastes, respectively. Next, the recycling of wastes and the distribution of the recycled products gives $\mathbf{x}^{rec(1)} = \mathbf{T}'\mathbf{w}^{(1)}$, $\mathbf{Z}^{rec(1)} = \hat{\mathbf{x}}^{rec(1)}\mathbf{H}$ and $\mathbf{f}^{rec(1)} = \hat{\mathbf{x}}^{rec(1)}\mathbf{h}$, according to Eqs A1.6–A1.8, taking account of recycling yields, an adaptation of the intermediate input coefficients. From Eq. A1.9, we have $\mathbf{A}^{rec(1)} = \mathbf{Z}^{rec(1)}(\hat{\mathbf{x}}^{(1)})^{-1}$ and the adaptations are obtained from Eqs A1.10–A1.11 as $\mathbf{A}^{(2)} = \mathbf{A}^{(1)} - \mathbf{A}^{rec(1)}$ and $\mathbf{f}^{(2)} = \mathbf{f}^{(1)} - \mathbf{f}^{rec(1)}$. After this, the next iterative step takes place. The procedure continues until 'convergence' is reached. For example, one may take $\max_i \left| f_i^{(t)} - f_i^{(t-1)} \right| < \varepsilon$ or $\max_i \left| x_i^{(t)} - x_i^{(t-1)} \right| < \varepsilon$ as the stopping-rule.

Suppose that the iterative procedure stops at step t. Then we can provide a full description of the deliveries related to the production chain. The final demands were exogenously specified as $\mathbf{f}$, part of which ($\mathbf{f}^{(t)}$) is provided by the production chain while the other part ($\mathbf{f}^{rec(t)}$) is obtained from recycling wastes. The deliveries of product i to process j are given by the elements of $\mathbf{Z}^{(t)}$ in case the product stems from the chain and by the elements of $\mathbf{Z}^{rec(t)}$ if they stem from recycling. The total amount of product is given by the vector $\mathbf{x}^{(t)}$ if the products are made in the chain and by $\mathbf{x}^{rec(t)}$ if they come out of the recycling process. A comparison between the case with and the case without recycling can be obtained.

A final remark is with respect to the primary inputs. They are given by $\mathbf{r} = \mathbf{R}\mathbf{x}$ (Eq. A1.5) if recycling is not taken into account and by $\mathbf{r}^{(t)} = \mathbf{R}\mathbf{x}^{(t)}$ in case of recycling. Our model can be expanded straightforwardly to cover the case where some of the wastes can be recycled into primary inputs. Because primary inputs originate outside the production chain (i.e. they are imported into the chain), this expansion does not affect the model. It only implies that the amount of imported primary inputs will be decreased.

A1.3.4 Case example

Let's consider two SCs, namely rapeseed SC for fast-food production and sunflower SC for FG biodiesel production. Moreover, let the rapeseed oil used for frying be recycled to produce SG biodiesel substituting for the sunflowers used in FG biodiesel production. The processes of both SCs are displayed in Fig. A1.2.

In the rapeseed SC, we assume a cultivation area size of 10 000 ha with an average harvest rate of 2.47 t of rapeseed/ha and 45% of the rapeseed is extracted as oil. Fertilizer use is considered to be 180 kg/ha = 0.073 t/t harvested rapeseed (Bernesson, 2004). 75% of the oil used in frying remains as waste cooking oil.

In the sunflower SC, cultivation is assumed to be performed in a 20 000 ha with a harvest rate of 2.5 t of sunflower/ha. Oil content is 42% which indicates that to

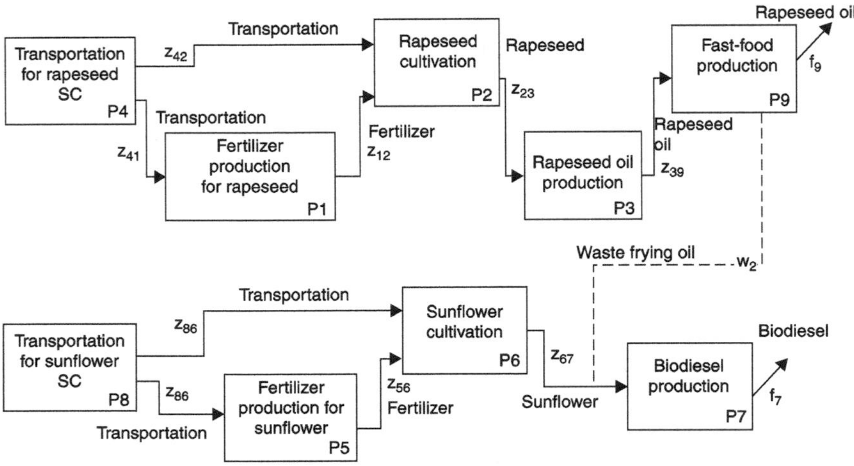

A1.2 Material flows in rapeseed and sunflower supply chains.

produce 1 t of biodiesel, 2.77 t of harvested sunflowers are required. Fertilizer use is 0.26 t/t harvested sunflowers (Yazan *et al.*, 2011a).

For both chains, industrial processes (i.e. fertilizer production, oil extraction, biodiesel production and fast-food preparation) are assumed to be 30 km away from cultivation areas and trucks performing transportation have a load capacity of 10 t.

In Table A1.1, all of the technical coefficients of intermediate flows, primary input use and waste emissions are displayed. Final demands are stochastically and endogenously defined (i.e. they just represent the values producible by two SCs based on the obtainable harvests).

Even though our model allows us to use an infinite number of primary inputs and wastes, for the sake of brevity, we will just consider four primary inputs (i.e. natural gas, butane, gas-oil, electricity) and two wastes (i.e. CO_2, waste cooking oil) in the SCs. Table A1.2 represents the balance input–output table where **Z** matrix, **f** and **x** vectors are represented as well as the total use of primary inputs and the total emission of wastes in the time period $t = 0$ (i.e. initial case where reuse of waste cooking oil does not take place). The displayed values are computed on a yearly basis.

Taking into account that from 1 t of sunflowers and 1 t of waste cooking oil, 0.37 t and 1 t of biodiesel are obtainable respectively, then the transformation coefficient is $t_{2,6}$=1/0,37=2,70 (i.e. waste type 2 – waste cooking oil – is reused instead of the main output of process 6 – sunflower). Then **T** matrix can be computed as:

$$\mathbf{T} = \begin{matrix} 0 & 0 & 0 & 0 & 0 & 0 & 0 & 0 & 0 \\ 0 & 0 & 0 & 0 & 0 & 2.70 & 0 & 0 & 0 \end{matrix}$$

Table A1.1 Coefficients matrix

Processes	p1	p2	p3	p4	p5	p6	p7	p8	p9	f
Main outputs										
p1 fertilizer for rapeseed (t)	0.00	0.07	0.00	0.00	0.00	0.00	0.00	0.00	0.00	0.00
p2 rapeseed (t)	0.00	0.00	2.22	0.00	0.00	0.00	0.00	0.00	0.00	0.00
p3 rapeseed oil (t)	0.00	0.00	0.00	0.00	0.00	0.00	0.00	0.00	0.19	0.00
p4 transportation rapeseed SC (km)	6.00	6.00	0.00	0.00	0.00	0.00	0.00	0.00	0.00	0.00
p5 fertilizer for sunflower (t)	0.00	0.00	0.00	0.00	0.00	0.26	0.00	0.00	0.00	0.00
p6 sunflower (t)	0.00	0.00	0.00	0.00	0.00	0.00	2.72	0.00	0.00	0.00
p7 biodiesel (t)	0.00	0.00	0.00	0.00	0.00	0.00	0.00	0.00	0.00	18 375.00
p8 transportation sunflower SC (km)	0.00	0.00	0.00	0.00	6.00	6.00	0.00	0.00	0.00	0.00
p9 french fries (t)	0.00	0.00	0.00	0.00	0.00	0.00	0.00	0.00	0.00	58 354.00
Primary inputs										
r1 electricity (MWh)	0.05	0.04	0.25	0.00	0.09	0.04	0.27	0.00	0.00	
r2 natural gas (m^3)	392.00	0.00	7.10	0.00	318.40	0.00	198.72	0.00	0.00	
r3 butane (kg)	69.00	0.00	0.00	0.00	56.00	0.00	0.00	0.00	0.00	
r4 gas-oil (litre)	0.00	0.75	0.00	0.18	0.00	0.74	0.00	0.18	0.00	
Wastes and by-products										
w1 CO_2 (t)	0.90	0.02	0.12	0.00	0.75	0.02	0.47	0.00	0.00	
w2 waste cooking oil (t)	0.00	0.00	0.00	0.00	0.00	0.00	0.00	0.00	0.14	

Table A1.2 Balance table for non-cooperation case

Processes	p1	p2	p3	p4	p5	p6	p7	p8	p9	f	x
Main outputs											
p1 fertilizer for rapeseed (t)	0	1800	0	0	0	0	0	0	0	0	1 800
p2 rapeseed (t)	0	0	24 700	0	0	0	0	0	0	0	24 700
p3 rapeseed oil (t)	0	0	0	0	0	0	0	0	11 115	0	11 115
p4 transportation rapeseed SC (km)	10 800	148 200	0	0	0	0	0	0	0	0	159 000
p5 fertilizer for sunflower (t)	0	0	0	0	0	12 950	0	0	0	0	12 950
p6 sunflower (t)	0	0	0	0	0	0	50 000	0	0	0	50 000
p7 biodiesel (t)	0	0	0	0	0	0	0	0	0	18 375	18 375
p8 transportation sunflower SC (km)	0	0	0	0	77 700	300 000	0	0	0	0	377 700
p9 french fries (t)	0	0	0	0	0	0	0	0	0	58 354	58 354
Primary inputs										r	
r1 electricity (MWh)	94	914	2 779	0	1 107	2 000	4 996	0	0	11 890	
r2 natural gas (m^3)	705 600	0	78 917	0	4 123 280	0	3651 373	0	0	8 559 169	
r3 butane (kg)	124 200	0	0	0	725 200	0	0	0	0	849 400	
r4 gasoil (litre)	0	186 000	0	28 131	0	37 200	0	66 824	0	150 755	
Wastes and by-products										w	
w1 CO_2 (t)	1615	458	1335	0	9674	990	8611	264	0	22 947	
w2 waste cooking oil (t)	0	0	0	0	0	0	0	0	8339	8 339	

Since there is no other process except biodiesel production to allocate reusable waste cooking oil, then **H** matrix and **h** vector are:

$$\mathbf{H} = \begin{matrix} 0 & 0 & 0 & 0 & 0 & 0 & 0 & 0 & 0 \\ 0 & 0 & 0 & 0 & 0 & 0 & 0 & 0 & 0 \\ 0 & 0 & 0 & 0 & 0 & 0 & 0 & 0 & 0 \\ 0 & 0 & 0 & 0 & 0 & 0 & 0 & 0 & 0 \\ 0 & 0 & 0 & 0 & 0 & 0 & 0 & 0 & 0 \\ 0 & 0 & 0 & 0 & 0 & 0 & 1 & 0 & 0 \\ 0 & 0 & 0 & 0 & 0 & 0 & 0 & 0 & 0 \\ 0 & 0 & 0 & 0 & 0 & 0 & 0 & 0 & 0 \\ 0 & 0 & 0 & 0 & 0 & 0 & 0 & 0 & 0 \end{matrix} \qquad \mathbf{h} = \begin{matrix} 0 \\ 0 \\ 0 \\ 0 \\ 0 \\ 0 \\ 0 \\ 0 \\ 0 \end{matrix}$$

When we adopt the upcoming equations for time period $t = 1$ and $t = 2$ then we obtain the following $\mathbf{x}^{(1)}$ and $\mathbf{x}^{(2)}$ vectors:

$$\mathbf{x}^{(1)} = \begin{matrix} 1800 \\ 24700 \\ 11115 \\ 159000 \\ 7119 \\ 27485 \\ 18375 \\ 207625 \\ 58354 \end{matrix} \qquad \mathbf{x}^{(2)} = \begin{matrix} 1800 \\ 24700 \\ 11115 \\ 159000 \\ 1287 \\ 4970 \\ 18375 \\ 37550 \\ 58354 \end{matrix}$$

If we define ε as 10% of total sunflower production in $t = 0$ (which is reasonable for a short term policy for feedstock substitution), then $\varepsilon = 50000*0.1 = 5000$. It is explicit from vector $x^{(2)}$ that $4970<5000$ and we can stop the iteration process at $t = 2$. This means that within two years of the implementation of a waste cooking reuse system, approximately 90% of the sunflower cultivation destined for biodiesel production can be saved. This implies that all fertilizer and transportation required for the production of the associated quantity of sunflower, as well as all associated energy use (i.e. electricity, natural gas, butane, gas-oil) and emitted waste (i.e. CO_2 emissions) will be saved, while the rapeseed oil destined for the food industry and the produced biodiesel for final use remain constant. Following Eqs A1.3–A1.5, (for $t = 2$) we can compute the total input use and waste emission. The difference between the values (for $t = 0$ and $t = 2$) equals the saved input and emitted waste at the end of time period $t = 2$:

$$\mathbf{Z}^{(\text{saved})} = \mathbf{Z} - \mathbf{Z}^{(2)} = \begin{matrix} 0 & 0 & 0 & 0 & 0 & 0 & 0 & 0 & 0 \\ 0 & 0 & 0 & 0 & 0 & 0 & 0 & 0 & 0 \\ 0 & 0 & 0 & 0 & 0 & 0 & 0 & 0 & 0 \\ 0 & 0 & 0 & 0 & 0 & 0 & 0 & 0 & 0 \\ 0 & 0 & 0 & 0 & 0 & 11663 & 0 & 0 & 0 \\ 0 & 0 & 0 & 0 & 0 & 0 & 45029 & 0 & 0 \\ 0 & 0 & 0 & 0 & 0 & 0 & 0 & 0 & 0 \\ 0 & 0 & 0 & 0 & 69975 & 270126 & 0 & 0 & 0 \\ 0 & 0 & 0 & 0 & 0 & 0 & 0 & 0 & 0 \end{matrix}$$

$$\mathbf{r}^{(\text{saved})} = \mathbf{r} - \mathbf{r}^{(2)} = \begin{matrix} 2798 \\ 3713365 \\ 653104 \\ 93682 \end{matrix} \qquad \mathbf{w}^{(\text{saved})} = \mathbf{w} - \mathbf{w}^{(2)} = \begin{matrix} 9731 \\ 0 \end{matrix}$$

The matrix $\mathbf{Z}^{(\text{saved})}$ shows the saved quantities of fertilizer (11 663 t), sunflowers (45 029 t) and transportation (69 975 km from fertilizer transport and 270 126 km from sunflower transport). In $\mathbf{r}^{(\text{saved})}$ vector, energy savings are illustrated (i.e. 2798 MWh of electricity, 3713 365 m^3 natural gas, 653 104 kg of butane and 93 682 l of gas-oil) while $\mathbf{w}^{(\text{saved})}$ represents CO_2 savings (9731 t).

If we consider the land use changes that are threatening sustainable agriculture in producing biodiesel, which sometimes results in a negative carbon balance and lower soil quality, by implementation of a system for recovering alternative feedstock, such lands can be gained back. This would increase the social and environmental sustainability of SGBSCs. From another perspective, even if there is no land use change, but (existing) cultivators supply their crops to the biodiesel industry, then with the help of the proposed model we can also predict how much harvested crop can be re-designated to the food industry (i.e. using $\mathbf{Z}^{(\text{saved})}$ matrix without primary input and waste emission savings). This would lead the food chain to a more socially sustainable position. Therefore, in both cases, sustainability of SGBSC as well as the food chain can be secured.

A1.4 Emerging second generation biodiesel feedstock markets and their coordination

Let's consider more than two actors who compete – for supply or for demand – in an emerging feedstock market. These actors can be waste suppliers (WS) or utility companies (UC), one supplying waste (e.g. feedstock) and other using the waste as an input of biodiesel production, respectively. We assume that the waste, if not used as feedstock on the demand side, causes discharge costs for WS. Therefore, if the waste is supplied to UC (i.e. biodiesel producer), it substitutes for traditional feedstock (i.e. FG biomass) and the discharge cost is saved. Use of alternative feedstock would result in savings from the purchasing cost of FG biomass for UC. UC also have an additional cost for production structure change for new adopted technology that processes SG biomass. Furthermore, we assume that all the associated costs related to waste supply are shared between the actors. This is because both the supply and demand sides accrue benefits from cooperation while added costs caused by waste supply are obstacles to the development of SGBSCs which can be (partly) sustained (if required) by government incentives. These are pre-treatment, transportation and storage costs and if their sum exceeds the aforementioned benefits, the market must be sustained by incentives until the costs are reduced and economic sustainability is obtained. Hence, the waste is regained in the economy and FG biomass can be supplied to the food market. In

Table A1.3 Potential extreme cases

Scenario	Number of waste suppliers (WS)	Number of utility companies (UC)	Produced waste – required waste
A	*M*	1	$w_{prod}^{tot} \leq w_{req}$
B	*M*	1	$w_{prod}^{tot} > w_{req}$
C	1	*M*	$w_{prod} \leq w_{req}^{tot}$
D	1	*M*	$w_{prod} > w_{req}^{tot}$

this way, the SGBSC becomes environmentally and socially sustainable, whereas economic sustainability depends on associated costs and (if necessary) government incentives.

Four different extreme scenarios can emerge according to the quantity of waste produced and the presence of competitors in the emerging waste markets as shown in Table A1.3.

In all cases we take into account a required minimum order quantity (MOQ) to be supplied to the UC to initiate cooperation. The bargaining power of one waste supplier (BPWS) is measured by its contribution to the reduction in purchasing cost of the UC (in the cases $w_{prod}^{tot} \leq w_{req}$ or $w_{prod} \leq w_{req}^{tot}$) where w_{prod}^{tot} is the total produced waste by *M* suppliers in the *M*–1 case, w_{req} is the total required waste to satisfy the demand of one utility company in the *M*–1 case, w_{prod} is the total produced waste by one supplier in the 1–*M* case and w_{req}^{tot} is the total required waste to satisfy the demand of *M* utility companies in the 1–*M* case; and by its loss of potential benefit in the case of no cooperation (inversely proportional, in the case $w_{prod}^{tot} > w_{req}$ or $w_{prod} > w_{req}^{tot}$).

Secondly we evaluate one WS's willingness to cooperate (WSWC) and one UC's unwillingness to cooperate (UCWC) by the total obtainable benefit of each actor in the case of cooperation (without considering the costs to be shared). How we compute the BPWS, WSCW and UCWC is described in Appendix A1.8. For each scenario some possible market/SC coordination policies are proposed.

In scenario A, since $w_{prod}^{tot} \leq w_{req}$, the WS producing the higher waste quantity has a higher BPWS (Fig. A1.3), because its contribution to the UC is higher, enabling it to reduce the purchasing cost of traditional material further. UCWC can be higher than, equal to, or less than WSWC depending on its potential benefit in the case of cooperation (Fig. A1.4). In such cases *negotiation* with small WSs can be a good coordination mechanism for the UC while *gaming* can be preferable with high quantity-supplying WSs. If the number of negotiated WSs increases, *standardization* can also be adopted to create a mature market reducing the uncertainties related to waste continuity and quality problems.

In scenario B, taking into account that there is an excess supply and many WSs compete among themselves, the higher the produced waste quantity by one WS the less will be its BP (Fig. A1.5). We assume that in the case of no cooperation other WSs can satisfy the demand of the UC (i.e. $w_{prod}^{tot} >> w_{req}$).

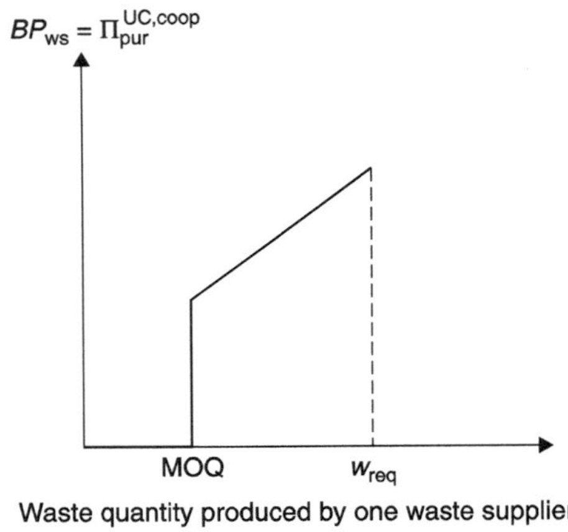

A1.3 Bargaining power (BP) of one waste supplier (WS) in *M*–1 case ($w_{prod}^{tot} \leq w_{req}$), where coop means cooperate and pur is purchasing cost.

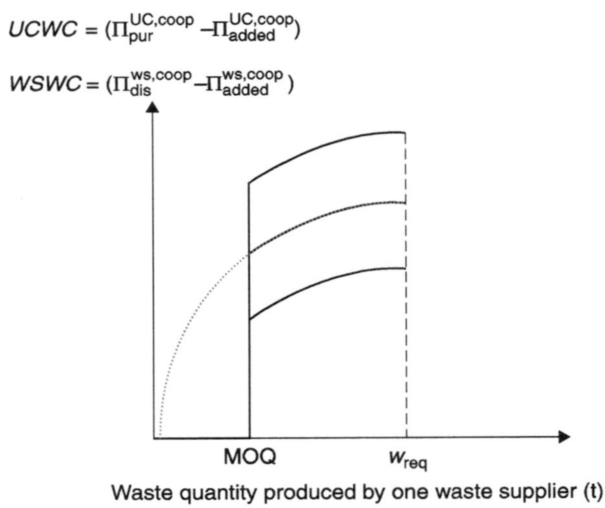

A1.4 WC of WS and UC in *M*–1 case.

Like scenario A, a similar case emerges for the WSWC and UCWC (Fig. A1.5). Having other alternatives for waste supply, if the UC prefers to cooperate with large-scaled WSs, then it can prefer *negotiation* or *standardization* as a coordination mechanism. Moreover, it can adopt *hierarchy* or *authority* with the advantage of being the unique demand side cooperating with smaller WSs. However this situation can cause higher coordination costs. Furthermore, to get more benefit from cooperation, UC can propose *bidding* to the WSs.

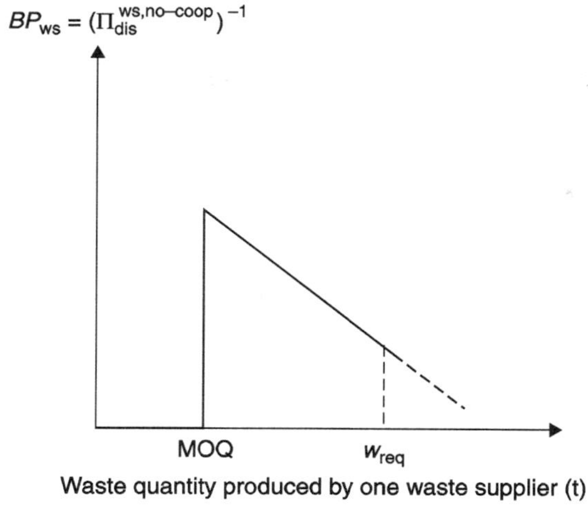

A1.5 BP of one WS in *M*–1 case ($w_{prod}^{tot} > w_{req}$), where no-coop means no cooperation and dis is discharge cost.

In scenario C, the higher the quantity of waste produced, the higher the BPWS (Fig. A1.6). The BP of a WS producing the same quantity of waste will be higher compared to scenario A because there is no competition with other WSs. WSWC and UCWC increase until the quantity of waste produced is equal to the quantity of waste demanded by one utility company (Fig. A1.8). In such a case WS can prefer *negotiation* with bigger UCs to reduce coordination costs or collect the bids (i.e. *bidding* as a coordination mechanism) from UCs. If the demand side includes many small UCs, then WS can try to use *authority* or create *hierarchy*.

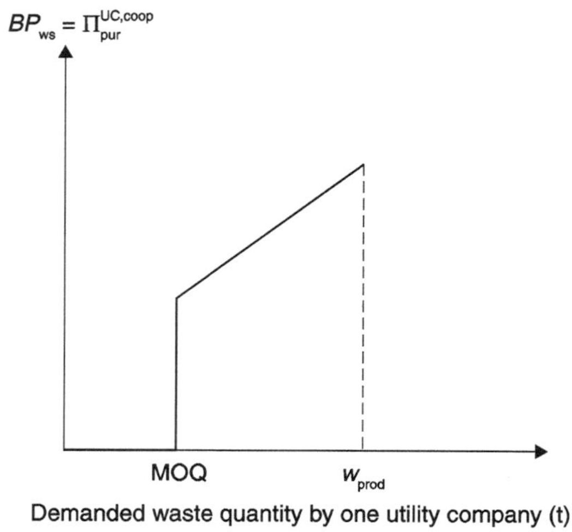

A1.6 BP of one WS in 1–*M* case ($w_{prod} \leq w_{req}^{tot}$).

In scenario D, since there is an excess supply, the BPWS goes down until the quantity demanded by one UC is satisfied (Fig. A1.7). WSWC will increase as well as UCWC while the quantity of waste produced increases (Fig. A1.8). Even though in mature markets higher amounts of available supply give advantage to the suppliers (e.g. high bargaining power, contractual power, quantity discount advantage), in this case the advantage converts into disadvantage because of the higher value of potential benefit loss. In such a case WS cannot use *hierarchy* if

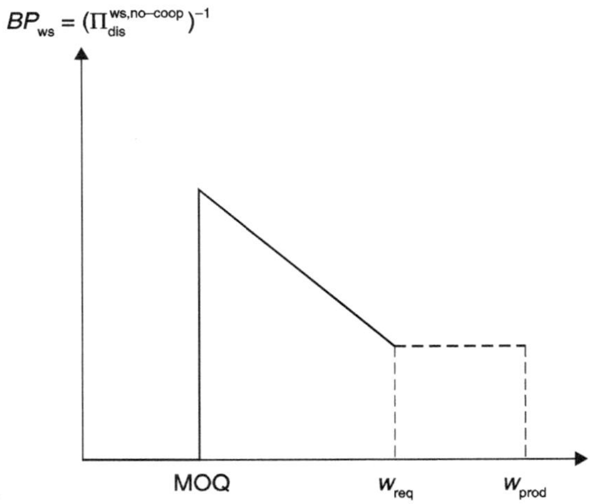

A1.7 BP of one WS in 1–*M* case ($w_{prod} > w_{req}^{tot}$).

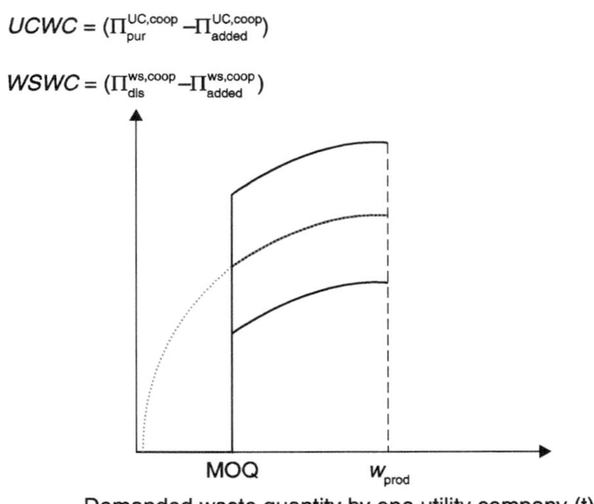

A1.8 WC of WS and UC in 1–*M* case.

the demand is not divided into small quantities by very small UCs. So, it can try to negotiate with the UCs (i.e. *negotiation* as coordination mechanism).

It is explicit that in our scenarios available feedstock quantity and presence of competitors have different impacts on the cooperation compared to usual forward SCs. In competitive waste markets, supplier selection strategies can vary according to the total available and requested waste in the local area. Large-scale WSs may have less BP owing to their higher loss of potential benefits in a non-cooperation case (i.e. 1–M and $w_{prod} > w_{req}^{tot}$ case) in contrast to mature product markets.

A1.5 Discussion

In this section we discuss how the proposed EIO model relates to the categories of SGBSCs. Since our model is applicable to infinite numbers of SCs which can be linked through infinite numbers of wastes, it can be adapted by (local) governments as well as involved SC actors to define FG biomass substitution policies for reducing their related environmental (e.g. increased fertilizer use, natural resource and fossil energy consumption along the SC) and social impacts (e.g. increased food prices caused by lack of supply).

The applicability of the EIO model is presented using the JSC flowchart in Fig. A1.9 where we display five forward supply chains composing a JSC. For the ease of presentation, used primary inputs and other emitted wastes (except reused ones) are not illustrated in the figure, neither are some other processes belonging to SCs such as irrigation, seeding, fertilization, storage, oil extraction and so on. The first SC is the chain of sunflower production for the fast food sector where sunflower stalk (i.e. ligno-cellulosic feedstock) is derived from cultivation and waste oil derives from a fast food production process. Similarly, the fifth SC is a chain of furniture production where off-cuts and shavings (i.e. forestry by-products) derive from tree-cutting and wood-processing, respectively. Second, third and fourth SCs are the chains of biodiesel production using three different FG biomass types, namely soybean, rapeseed and sunflower. We should mention that a Jatropha SC is not illustrated in the figure since it is a forward SC and the fruit of Jatropha is not categorized as waste differently from other sources.

If we consider that new lands are opened for FG biomass cultivation to produce biodiesel in the second to fourth SCs (i.e. no cultivation for food production), by the substitution of cultivated FG feedstock by the aforementioned wastes (i.e. SG biomass sources), transportation and fertilizer production associated with their cultivation are reduced proportionally. This induces savings from natural resources and the fossil energy (i.e. primary inputs) used for transportation and fertilization as well as in all of the non-illustrated processes (e.g. water savings). Moreover, other wastes (e.g. CO_2, NO_x emissions) caused by these processes are also avoided.

According to Fig. A1.9, the **T** matrix will be size of $4x21$ where we locate the substitution coefficients (e.g. one unit of waste oil – waste type 2 – substitutes t_{212}

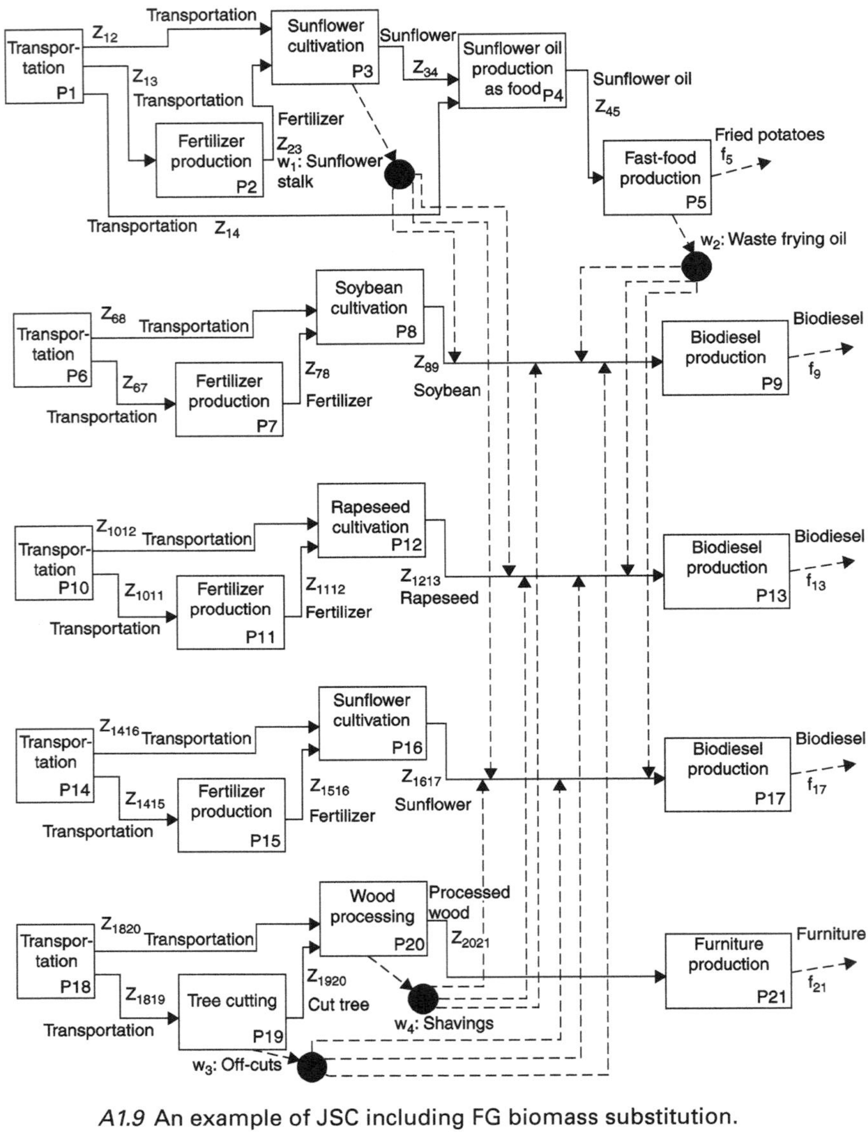

A1.9 An example of JSC including FG biomass substitution.

of rapeseed – main output type 12 – or one unit of sunflower stalk – waste type 1 – substitutes t_{18} of soybean – main output type 8). Once, the percentages of wastes to be distributed in various processes are decided, values of the **H** matrix become known. The **H** matrix can be used to predict which substitution has more benefits and its values are decided by chain actor(s). Therefore, net environmental returns can be calculated for the whole JSC. Moreover, using the substitution process we can also calculate how much food can be saved if the cultivation continues instead of being reduced by changing the last processes of the second to fourth SCs into the food production process. This would demonstrate how much each chain is able to contribute to the food sector, thereby increasing the social sustainability.

From a life cycle point of view, obviously, there will be added environmental impacts stemming from the substitution process (e.g. CO_2 emissions caused by collection and transportation of wastes and energy used in the pre-treatment and storage processes). Therefore, a SC of feedstock substitution can be designed and its related impact can be calculated using the same model by inserting 0 as coefficients in the **T** and **H** matrices.

Finally in relation to the sustainability of the SGBSC, the biodiesel produced can easily be turned into in its own SC (e.g. use of obtained biodiesel in collection or transportation processes), increasing the environmental sustainability of the chain itself (i.e. a self-sustainable supply chain, for details see Yazan *et al.*, 2011a).

A1.6 Conclusions and future trends

Development of SGB technology needs to be sustained by substitution of feedstock sources by SG biomass. Our proposed EIO model also allows us to measure how much FG feedstock is substitutable. At first sight, feedstock source change may not appear to be profitable to investors, but if the initial investment risks were reduced by government incentives, then its attractiveness could increase in the open market. In turn, financial viability would also induce research and development activities on SGB.

The major four SGB market actors are suppliers, processors, the oil industry and customers. Each of them requires certain conditions to enter the market. These decisions contain risks and a range of policies and incentives can be proposed by governments at every stage of the SC to cope with these risks. So, a risk comparison of agricultural technology and input use scenarios is required (Rajagopal *et al.*, 2007). In this chapter we applied a scenario analysis to understand how a SG biomass market emerges, where we measured the bargaining power of suppliers and the willingness of actors to cooperate. Our analysis shows explicitly that not only the potential benefits, but also the potential losses, drive their decisions.

Adoption of particular coordination mechanisms is also strongly correlated with the balance between waste supply and demand. Cooperation strategies are influenced by added costs such as recycling cost, transportation cost and cost of production structure change. How companies share these costs is a question of the contractual power of an actor and power asymmetry between the potential actors. Government incentives may help to increase competition in the market, both in supply and on the demand side, and might directly influence the bargaining power and willingness to cooperate of all actors. A dynamic and flexible subsidy system may give incentives to all SC actors proportional to their contribution to the whole chain performance. A pre-defined benefit-sharing contract can also be adopted according to the number of actors in the local market. This can reduce the uncertainty about profit but can also reduce the operational efficiency and effectiveness of each actor. Better information sharing and coordination with the actors of other SCs (e.g. food chain, paper industry) would increase the operational efficiency of the SC.

A1.7 References

Albino V, Dietzenbacher E and Kühtz S (2003), 'Analyzing material and energy flows in an industrial district using an enterprise input-output model', *Economic Systems Research*, **15**(4), 457–80.

Albino V, Kühtz S and Messeni Petruzzelli A (2007) 'Analysing logistics flows in industrial clusters using an enterprise input–output model', *Interdisciplinary Information Sciences*, **14**, 1–17.

Bailey R, Allen J K and Bras B (2001), 'Applying Ecological Input–Output Flow Analysis to Material Flows in Industrial Systems Part I: Tracing Flows', *Journal of Industrial Ecology*, **8**(1–2), 45–68.

Banse M, van Meijl H, Tabeau A and Woltjer G (2008), 'Will EU biofuel policies affect global agricultural markets?', *European Review of Agricultural Economics*, **35**, 117–41.

Bernesson S (2004). *Life Cycle Assessment of Rapeseed Oil, Rape Methyl Ester and Ethanol as Fuels – a comparison between large- and small-scale production*. Available from http://pub.epsilon.slu.se/3436/1/SLU_BT_R2004_01_LCA_ro_RME_ethanol_Sven_Bernesson.pdf [Accessed 28 September 2011].

Carriquiry M, Du X and Timilsina G R (2010), Second-generation Biofuels: economics and policies, *World Bank Policy Research Working Paper Series*. Available from http://www-wds.worldbank.org/servlet/WDSContentServer/WDSP/IB/2010/08/30/000158349_20100830090558/Rendered/PDF/WPS5406.pdf

Clancy J S (2008), 'Are biofuels pro-poor? Assessing the evidence', *The European Journal of Development Studies*, **20**(3), 416–31.

Cvengroš J and Cvengrošová Z (2004), 'Used frying oils and fats and their utilization in the production of methyl esters of higher fatty acids', *Biomass and Bioenergy*, **27**, 173–81.

Dale V H, Kline K L, Lynn L W, Perlack R D, Downing M and Graham R L (2011), 'Interactions among bioenergy feedstock choices, landscape dynamics and land use', *Ecological Applications*, **21**, 1039–54.

Dautzenberg K and Hanf J (2008), 'Biofuel chain development in Germany: organisation, opportunities and challenges', *Energy Policy*, **36**, 485–9.

De Lucia C and Datta B (2011), 'Socio-economic, environmental and policy perspectives of future generation of biodiesel production', in *Advances in Biodiesel Production: Processes and technologies*, Luque R and Melero, J A (eds). Woodhead Publishing, Cambridge, UK.

EU (2009), Directive 2009/28/EC of the European Parliament of 23 April 2009 on the Promotion of the use of Energy from Renewable Sources and Amending and Subsequently Repealing Directives 2001/77/EC and 2003/30/EC.

European Bank for Reconstruction and Development (EBRD) (2011), Bulgaria country profile. Available from http://ebrdrenewables.com/sites/renew/countries/Bulgaria/default.aspx [Accessed 1 September 2011].

Fargoine J, Hill J, Tilman D, Polasky S and Hawthorne P (2008), 'Land clearing and the biofuel carbon debt', *Science*, **319**, 1235–8.

Festel G W (2008), 'Biofuels: economic aspects', *Chemical Engineering and Technology*, **31**(5), 715–20.

Galbe M and Zacchi G (2002), 'A review of the production of ethanol from softwood', *Applied Microbiology Biotechnology*, **59**(6), 618–28.

Gilbert C L (2010), 'How to understand high food prices', *Journal of Agricultural Economics*, **61**(2), 398–425.

Grubbström R W, Tang O, (2000), 'An overview of input–output analysis applied to production-inventory systems', *Economic Systems Research*, **12**(1), 3–25.

Headey D and Fan S (2008), 'Anatomy of a crisis: the causes and consequences of surging food prices', *Agricultural Economics*, **39** (s1), 375–91.

Headey D, Malaiyandi S and Fan S (2010), 'Navigating the perfect storm: reflections on the food, energy and financial crises', *Agricultural Economics*, **41**(S1), 217–28.

International Energy Agency (IEA) (2008). *From First to Second Generation Biofuel Technologies: an overview of current industry and RD&D activities*. Available from: http://www.iea.org/papers/2008/2nd_Biofuel_Gen_Exec_Sum.pdf [Accessed 30 June 2011].

James L K, Swinton S M and Thelen K D (2010), 'Profitability analysis of cellulosic energy crops compared with corn', *Agronomy Journal*, **102**, 675–87.

Johansson T B and Goldemberg J (2002), 'Overview and a policy agenda', in *Energy for Sustainable Development: a Policy Agenda*, Johansson T B and Goldemberg J (eds), United Nations Development Programme, New York.

Johnstone N, Haščič I and Popp D (2010), 'Renewable energy policies and technological innovation: evidence based on patent counts', *Environmental and Resource Economics*, **45**, 133–55.

Kurz H D., Dietzenbacher E, Lager C. (1998), *Input–Output Analysis*, The International Library of Critical Writings in Economics 92, in three volumes, Edward Elgar, Aldershot.

Leontief W W, (1936), 'Quantitative input and output relations in the economic system of the United States', *General Linguistics and Literary Theory*, **105**.

Lin X and Polenske K R (1998), 'Input–output modelling of production processes for business management', *Structural Change and Economic Dynamics*, **9**, 205–26.

Lovett J C, Hards S, Clancy J and Snell C (2011), 'Multiple objectives in biofuels sustainability policy', *Energy and Environmental Science*, **4**, 261–8.

Luque R, Datta B, Lovett J C, Clancy J, Campelo J M and Romero A A (2010), 'Biodiesel as feasible petrol fuel-replacement: A multidisciplinary overview', *Energy and Environmental Science*, **3**, 1706–21.

Marangoni G and Fezzi G (2002), 'I–O for management control: the case of GlaxoSmithKline', *Economic Systems Research*, **14**(3), 245–56.

McCormick K and Kåberger T (2007), 'Key barriers for bioenergy in Europe: economic conditions, know-how and institutional capacity and supply chain coordination', *Biomass and Bioenergy*, **31**, 443–52.

Miller R E and Blair P D (2009), *Input–output Analysis: Foundations and Extensions*, Cambridge University Press, Cambridge.

Mollenkopf D, Stolze H, Tate W L and Ueltschy M (2010), 'Green, lean and global supply chains', *International Journal of Physical Distribution & Logistics Management*, **40**(1/2), 14–41.

Porter M E and van der Linde C (1995), 'Toward a new conception of the environment-competitiveness relationship', *The Journal of Economic Perspectives*, **9**, 97–118.

Poudyal M and Lovett J C (2010), 'Biodiesel as the potential alternative vehicle fuel: European policy and global environmental concern', in *Handbook of Environmental Management*, Lovett J C and Ockwell D G (eds), Edward Elgar, Cheltenham.

Prueksakorn K and Gheewala S H, (2008), 'Full chain analysis of biodiesel from jatropha curchas l. in Thailand', *Environmental Science and Technology*, **42**, 3288–3393.

Rajagopal D and Zilberman D (2007), *Review of Environmental, Economic and Policy Aspects of Biofuels*, The World Bank, Washington DC.
Rajagopal D, Sexton S E, Roland-Holst D and Zilberman D (2007), 'Challenge of biofuel: filling the tank without emptying the stomach', *Environmental Research Letters*, **2**.
Royal Society (2008), *Sustainable Biofuels: Prospects and Challenges*, Policy Document 01/08, Royal Society, London, UK.
Searchinger T, Heimlich R, Houghton R A, Dong F, Elobeid A, *et al.* (2008), 'Use of U.S. croplands for biofuels increases greenhouse gases through emissions from land-use change', *Science*, **319**, 1238–40.
Sims R E H, Mabee W, Saddler J N and Taylor M (2010), 'An overview of second generation biofuel technologies', *Bioresource Technology*, **101**, 1570–80.
Trostle R (2008), *Global Agricultural Supply and Demand: Factors Contributing to the Recent Increase in Food Commodity Prices*, United States Department of Agriculture, Washington.
Wiesenthal T, Leduc G, Christidis P, Schade B, Pelkmans L, *et al.* (2009), 'Biofuel support policies in Europe: lessons learnt for the long way ahead', *Renewable and Sustainable Energy Reviews*, **13**, 789–800.
Wiggins S, Keats S and Vigneri M (2009), *Impact of the Global Financial and Economic Situation on Agricultural Markets and Food Security*, ODI, London, UK.
Worldwatch Institute (2007), *Biofuels for Transport*, London, Earthscan.
Yazan D M, Garavelli A C, Messeni Petruzzelli A and Albino V (2011a), 'The effect of spatial variables on the economic and environmental performance of bioenergy production chains', *International Journal of Production Economics*, **131**, 224–233.
Yazan D M, Messeni Petruzzelli A and Albino V (2011b), 'Analyzing the environmental impact of transportation in re-engineered supply chains', *Transportation Research Part D: Transport and Environment*, **16**, 335–40.

A1.8 Appendix: computation of bargaining power and willingness to cooperate

$$BPWS = \Pi_{\text{pur}}^{\text{UC,coop}} = \Pi_{\text{pur}}^{l,i}$$

$$BPWS = (\Pi_{\text{dis}}^{\text{WS,no-coop}})^{-1} = (\Pi_{\text{dis}}^{l})^{-1}$$

$$WSWC = (\Pi_{\text{dis}}^{\text{WS,coop}} - \Pi_{\text{added}}^{\text{WS,coop}}) = (\Pi_{\text{dis}}^{l} - \Pi_{\text{added}}^{\text{WS,coop}})$$

$$UCWC = (\Pi_{\text{pur}}^{\text{UC,coop}} - \Pi_{\text{added}}^{\text{UC,coop}}) = (\Pi_{\text{pur}}^{l,i} - \Pi_{\text{added}}^{\text{UC,coop}})$$

where $\Pi_{\text{pur}}^{\text{UC,coop}}$ is the benefit of the UC from cooperation by reuse of waste l in process i, $\Pi_{\text{dis}}^{\text{WS,no-coop}}$ is the loss of the WS in case of no-cooperation caused by discharge of waste l, $\Pi_{\text{added}}^{\text{WS,coop}}$ is the total added costs for WS caused by cooperation excluding collection cost. (Some costs such as transportation and coordination costs are shared with the UC according to the benefit sharing contracts).

$\Pi_{\text{dis}}^{\text{WS,coop}}$ is the benefit of WS in case of cooperation by not discharging waste l, $\Pi_{\text{added}}^{\text{UC,coop}}$ is the total added costs for UC caused by cooperation excluding production structure change cost. (Some costs such as transportation and coordination costs are shared with the WS according to the benefit sharing contracts).

Index

Kolibri 360 GmbH
Ferdinand-Jühlke-Straße 7
99095 Erfurt, Deutschland
produktsicherheit@kolibri360.de